Lecture Notes in Mathematics 1518

H. Stichtenoth M. A. Tsfasman (Eds.)

Coding Theory and Algebraic Geometry

Proceedings of the International Workshop held in Luminy, France, June 17-21, 1991

Springer-Verlag
Berlin Heidelberg New York
London Paris Tokyo
Hong Kong Barcelona
Budapest

Editors

Henning Stichtenoth
Fachbereich 6 – Mathematik und Informatik
Universität GHS Essen
Universitätsstr. 3, W-4300 Essen 1, Fed. Rep. of Germany

Michael A. Tsfasman
Institute of Information Transmission (IPPI)
19, Ermolovoi st., Moscow, GSP – 4, 101447, Russia

Mathematics Subject Classification (1991): 14-06, 94-06, 11-06

ISBN 3-540-55651-6 Springer-Verlag Berlin Heidelberg New York
ISBN 0-387-55651-6 Springer-Verlag New York Berlin Heidelberg

Printed in Germany

Typesetting: Camera ready by author/editor
Printing and binding: Druckhaus Beltz, Hemsbach/Bergstr.
46/3140-543210 - Printed on acid-free paper

Foreword

The workshop "Algebraic Geometry and Coding Theory - 3" organized by the Institute of Information Transmission (Moscow), University of Essen, Équipe Arithmétique et Théorie de l'Information de C.N.R.S. (Marseille-Luminy), and Group d'Étude du Codage de Toulon took place in the Centre International de Rencontres Mathématiques, June 17-21, 1991.

The workshop was a continuation of AGCT-1 and AGCT-2 that took place in 1987 and 1989, respectively. It is to be followed by AGCT-4 in 1993, etc., each time held in C.I.R.M.

The list of participants follows.

It is our pleasure to thank the staff of C.I.R.M. for their hospitality, the participants for their interest, all supporting organizations for their financial support, and Springer-Verlag for the Proceedings.

Organizers,

H.Stichtenoth
M.Tsfasman
G.Lachaud
J.Wolfmann

AGCT 3 - List of Participants

Aubry, Yves (Marseille)
Blahut, Richard E. (Owego, N.Y.)
Boutot, Jean-François (Strasbourg)
Bruen, Aiden (London, Ontario)
Carral, Michel (Toulouse)
Chassé, Guy (Issy les Moulineaux)
Cherdieu, Jean-Pierre (Guadeloupe)
Cougnard, Jean (Besançon)
Deschamps, Mireille (Paris)
Driencourt, Yves (Marseille)
Duursma, Iwan M. (Eindhoven)
Ehrhard, Dirk (Düsseldorf)
Gillot, Valérie (Toulon)
Guillot, Ph. (Genevilliers)
Hansen, Johan P. (Aarhus)
Harari, Sami (Toulon)
Hassner, Martin (San Jose, Ca.)
Helleseth, Tor (Bergen)
Høholdt, Tom (Lyngby)
Katsman, Gregory (Moscou)
Kumar, Vijay (Los Angeles)
Kunyavskii, Boris E. (Saratov)
Lachaud, Gilles (Marseille)
Langevin, Philippe (Toulon)
Lax, Robert (Baton Rouge)
Le Brigand, Dominique (Paris)
Li, Winnie (Pennsylvania State)
Lopez, Bartolomé (Madrid)
Luengo, Ignacio (Madrid)
Michon, Jean-Francis (Paris)
Munuera, Carlos (Valladolid)
Nogin, Dimitri (Moscou)
Pedersen, Jens Peter (Lyngby)
Pellikaan, Ruud (Eindhoven)
Perret, Marc (Marseille)
Polemi, Despina (New York)
Rodier, François (Paris)
Rodriguez, M.-C. (Madrid)
Rolland, Robert (Marseille)
Rotillon, Denis (Toulouse)
Seguin, Gerald (Kingston, Ontario)
Serre, Jean Pierre (Paris)
Shahrouz, Henri (Cambridge, Ma.)
Shparlinski, Igor (Moscou)
Skorobogatov, Alexei (Moscou)
Smadja, René (Marseille)
Sole, Patrick (Valbonne)
Stichtenoth, Henning (Essen)
Stokes, Philip (Valbonne)
Thiongly, Augustin (Toulouse)
Tsfasman, Mikhail (Moscou)
Vladut, Serge (Moscou)
Voss, Conny (Essen)
Wolfmann, Jacques (Toulon)

Contents

Algebraic Geometry and Coding Theory
An Introduction

Henning Stichtenoth, Michael A. Tsfasman
H.St.: Fachbereich 6 - Mathematik, Univ.GHS Essen,
D-4300 Essen 1, Germany
M.Ts.: Institute of Information Transmission,
19 Ermolovoi st., Moscow GSP-4, U.S.S.R.

About ten years ago V.D.Goppa discovered an amazing connection between the theory of algebraic curves over a finite field $\mathbf{F}_q$ and the theory of error-correcting block q-ary codes. The idea is quite simple and generalizes the well known construction of Reed-Solomon codes. The latter use polynomials in one variable over $\mathbf{F}_q$ and Goppa generalized this idea using rational functions on an algebraic curve.

Here is the definition of an *algebraic geometric code* (or a *geometric Goppa code*). Let X be an absolutely irreducible smooth projective algebraic curve of genus g over $\mathbf{F}_q$. Consider an (ordered) set $\mathcal{P} = \{P_1, \ldots, P_n\}$ of distinct $\mathbf{F}_q$-rational points on X and an $\mathbf{F}_q$-divisor D on X. For simplicity let us assume that the support of D is disjoint from $\mathcal{P}$. The linear space $L(D)$ of rational functions on X associated to D yields the linear evaluation map

$$Ev_{\mathcal{P}} : L(D) \to \mathbf{F}_q^n$$
$$f \mapsto (f(P_1), \ldots, f(P_n))$$

The image of this map is the linear code $C = (X, \mathcal{P}, D)_L$ we study.

The parameters of such a code can be easily estimated. Indeed, let $\mathbf{P} = P_1 + \ldots + P_n$, then the dimension k is given by

$$k = \ell(D) - \ell(D - \mathbf{P})$$

and in particular if $0 < \deg D < n$ then

$$k = \ell(D) \geq \deg D - g + 1.$$

The minimum distance

$$d \geq n - \deg D$$

since the number of zeroes of a function cannot be greater than the number of its poles. We get the lower bound

$$k + d \geq n + 1 - g$$

which is by g worse than the simplest upper bound valid for any code

$$k + d \leq n + 1.$$

An equivalent description of these codes can be given in terms of algebraic function fields in one variable over $\mathbf{F}_q$. The curve X corresponds to the function field $F = \mathbf{F}_q(X)$, and $\mathbf{F}_q$-points on X correspond to places of F of degree one.

Originally, Goppa used the dual construction using differentials on X rather than functions, and the residue map.

Unfortunately, there are at least two different traditions of notation. The second one uses D for our $\mathbf{P}$ and G for our D , and the code is denoted $C_L(D, G)$.

The construction can be generalized in several directions. In particular one can use sheaves (or some other tricks) to avoid the condition $\mathcal{P} \cap \mathrm{Supp} D = \emptyset$. The generalization to the case of higher dimensional algebraic varieties looks very promising but so far the results are few.

There are several main streams of the development of the theory. Let us briefly discuss some of them.

Asymptotic problems. One of the fundamental problems of coding theory is to construct long codes with good parameters (rate and relative minimum distance). One of the starting points of the theory was the construction of long codes which are asymptotically better than the Gilbert-Varshamov bound. The other asymptotic question is which codes can be constructed in polynomial time.

Specific curves. There are many interesting examples of curves with many $\mathbf{F}_q$-points which lead to codes with good parameters. Sometimes such curves and codes have nice additional properties, such as large automorphism groups.

Spectra and duality. The study of weight distribution and of duality leads to interesting questions of algebraic geometry, such as the study of Weierstrass points and special divisors on a curve.

Decoding. Surprisingly enough the decoding problem can be set in purely algebraic geometric terms and again one needs information about special divisors.

Exponential sums. Another component of the picture is the theory of exponential sums closely related both to algebraic geometry and to coding theory.

Related areas. The theory of algebraic geometric codes has either analogues or applications in several other topics. Such are sphere packings and spherical codes, multiplication complexity in finite fields, graph theory, and so on. These applications also require subtle information about the geometry and arithmetic both of function fields and of number fields.

To conclude, the first ten years of development show that the connection between algebraic geometry and coding theory proves fruitful for both, giving new results and posing many exciting questions.

Several books and many papers on the subject are either published or in preparation. The papers are too numerous to list them here and we refer to the extensive bibliography in [Ts/Vl] and to references given in the papers of this volume. Here is the list of books.

[Go] **V.D.Goppa,** *Geometry and Codes.* Kluwer Acad. Publ., 1988

[Mo] **C.J. Moreno,** *Curves over Finite Fields.* Cambridge Univ. Press, 1991

[Sti] **H.Stichtenoth,** *Algebraic Function Fields and Codes.* Springer-Verlag (in preparation)

[Ts/Vl] **M.A.Tsfasman, S.G.Vladut,** *Algebraic Geometric Codes.* Kluwer Acad. Publ., 1991

[vG/vL] **J.H.van Lint, G.van der Geer,** *Linear Codes and Algebraic Curves.* Birkhäuser, 1988

Reed-Muller Codes Associated to Projective Algebraic Varieties

Yves AUBRY

Equipe CNRS "Arithmétique et Théorie de l'Information"

C.I.R.M. Luminy Case 916 - 13288 Marseille Cedex 9 - France.

Abstract

The classical generalized Reed-Muller codes introduced by Kasami, Lin and Peterson [5], and studied also by Delsarte, Goethals and Mac Williams [2], are defined over the affine space $\mathbf{A}^n(\mathbf{F}_q)$ over the finite field $\mathbf{F}_q$ with q elements. Moreover Lachaud [6], following Manin and Vladut [7], has considered projective Reed-Muller codes, i.e. defined over the projective space $\mathbf{P}^n(\mathbf{F}_q)$.
In this paper, the evaluation of the forms with coefficients in the finite field $\mathbf{F}_q$ is made on the points of a projective algebraic variety V over the projective space $\mathbf{P}^n(\mathbf{F}_q)$. Firstly, we consider the case where V is a quadric hypersurface, singular or not, Parabolic, Hyperbolic or Elliptic. Some results about the number of points in a (possibly degenerate) quadric and in the hyperplane sections are given, and also is given an upper bound of the number of points in the intersection of two quadrics.

In application of these results, we obtain Reed-Muller codes of order 1 associated to quadrics with three weights and we give their parameters, as well as Reed-Muller codes of order 2 with their parameters.

Secondly, we take V as a hypersurface, which is the union of hyperplanes containing a linear variety of codimension 2 (these hypersurfaces reach the Serre bound). If V is of degree h, we give parameters of Reed-Muller codes of order d < h, associated to V.

1. Construction of the Projective Reed-Muller codes

We denote by $\mathbf{P}^n(\mathbf{F}_q)$ the projective space of dimension n over the finite field $\mathbf{F}_q$ with q elements, q a power of a prime p. The number of (rational) points (over $\mathbf{F}_q$) of $\mathbf{P}^n(\mathbf{F}_q)$ is :

$$\pi_n = |\mathbf{P}^n(\mathbf{F}_q)| = q^n + q^{n-1} + \ldots + q + 1 = \frac{q^{n+1} - 1}{q - 1}.$$

Let W_i be the set of points with homogeneous coordinates $(x_0 : x_1 : ... : x_n) \in \mathbf{P}^n(\mathbf{F}_q)$ such that

$$x_0 = x_1 = ... = x_{i-1} = 0 \text{ and } x_i \neq 0.$$

The family $\{ W_i \}_{0 \leq i \leq n}$ is clearly a partition of $\mathbf{P}^n(\mathbf{F}_q)$.

Let $\mathbf{F}_q[X_0, X_1, ..., X_n]_d^0$ be the vector space of homogeneous polynomials of degree d with (n+1) variables and with coefficients in $\mathbf{F}_q$. Let **V** be a projective algebraic variety of $\mathbf{P}^n(\mathbf{F}_q)$ and let $|\mathbf{V}|$ denotes the number of theirs rational points over $\mathbf{F}_q$. Following G. Lachaud ([6]), we define the *projective Reed-Muller code* $\mathcal{R}(d,\mathbf{V})$ of order d associated to the variety **V** as the image of the linear map

$$c : \mathbf{F}_q[X_0, X_1, ..., X_n]_d^0 \rightarrow \mathbf{F}_q^{|\mathbf{V}|}$$

defined by $c(P) = (c_x(P))_{x \in \mathbf{V}}$, where

$$c_x(P) = \frac{P(x_0, ..., x_n)}{x_i^d} \quad \text{if } x = (x_0 : ... : x_n) \in W_i \; .$$

G. Lachaud has considered in [6] the case where $\mathbf{V} = \mathbf{P}^n(\mathbf{F}_q)$, with $d \leq q$.Moreover, A.B. Sorensen has considered in [12] the case where **V** is equal to $\mathbf{P}^n(\mathbf{F}_q)$ too, but with a weaker hypothesis on d.

Now we are going, firstly, to study the case where **V** is a quadric, degenerate or not, but before we have to establish results on quadrics and this is the subject of the following paragraph.

2. Results on quadrics

In what follows the characteristic of the field $\mathbf{F}_q$ is supposed to be arbitrary (the results hold in characteristic 2 as well as in characteristic different of 2).

2.1. The quadrics in $\mathbf{P}^n(\mathbf{F}_q)$.

In this paragraph, we recall some properties of quadrics in the projective space $\mathbf{P}^n(\mathbf{F}_q)$. J.F. Primrose has given in [8] the number of points in a nondegenerate quadric (see below the definition of the rank of a quadric), and D.K. Ray-Chaudhuri [9] gave more general results (which with, in a particular case, we recover those of Primrose's). We are going here to follow the notations of J.W.P. Hirschfeld in [4].

A quadric Q of $\mathbf{P}^n(\mathbf{F}_q)$ is the set of zeros in $\mathbf{P}^n(\mathbf{F}_q)$ of a quadratic form

$$F \in \mathbf{F}_q[X_0, X_1, ..., X_n]_2^0,$$

that is of an homogeneous polynomial of degree 2. We set $Q = Z_{P^n}(F)$ or simply Z(F) if no confusion is possible. The quadric Q is said to be *degenerate* if there exists a linear change of coordinates with which we can write the form F with a fewer number of variables. More precisely, if T is an invertible linear transformation defined over $\mathbf{P}^n(\mathbf{F}_q)$, denote by $F_T(X)$ the form F(TX). Let i(F) be the number of indeterminates appearing explicitly in F. The rank r(F) of F (and by abuse of language, of the quadric Q), is defined by :

$$r(F) = \min_T i(F_T)$$

where T ranges over all the invertible transformations defined over $\mathbf{F}_q$. A form F (and by abuse the quadric Q) is said to be *degenerate* if

$$r(F) < n + 1.$$

Otherwise, the form and the quadric are *nondegenerate.*
Let us remark that a quadric is degenerate if and only if it is singular (see [4]).

We recall after J.W.P. Hirschfeld (see [4]) that in $\mathbf{P}^n(\mathbf{F}_q)$, the number of different types of nondegenerate quadrics Q is 1 or 2 as n is even or odd, and they are respectively called *Parabolic* ($\mathcal{P}$), and *Hyperbolic* ($\mathcal{H}$) or *Elliptic* ($\mathcal{E}$).
The maximum dimension g(Q) of linear subspaces lying on the nondegenerate quadric Q is called the *projective index* of Q. The projective index has the following values (see [4]) :

$$g(\mathcal{P}) = \frac{n-2}{2}, \quad g(\mathcal{H}) = \frac{n-1}{2}, \quad g(\mathcal{E}) = \frac{n-3}{2}.$$

The character $\omega(Q)$ of a nondegenerate quadric Q of $\mathbf{P}^n(\mathbf{F}_q)$ is defined by :

$$\omega(Q) = 2g(Q) - n + 3.$$

Consequently, we have :

$$\omega(\mathcal{P}) = 1, \quad \omega(\mathcal{H}) = 2, \quad \omega(\mathcal{E}) = 0.$$

Then, we have the following proposition (for a proof see [4]) :

Proposition 1 : The number of points of a nondegenerate quadric Q of $\mathbf{P}^n(\mathbf{F}_q)$ is :

$$|Q| = \pi_{n-1} + (\omega(Q) - 1)\, q^{(n-1)/2}.$$

We want now to evaluate the number of points of a degenerate quadric Q = Z(F) of $\mathbf{P}^n(\mathbf{F}_q)$ of rank r (called a "cone" of rank r).
We have the following decomposition in disjoint union (an analogous decomposition is given by R.A. Games in [3]) :

$$Q = V_{n-r} \cup Q^*_{r-1}.$$

We have set

$$V_{n-r} = \{(0:0:\ldots:0:y_r:\ldots:y_n) \in \mathbf{P}^n(\mathbf{F}_q)\} \cong \mathbf{P}^{n-r}(\mathbf{F}_q),$$

if we suppose that the r variables appearing in the quadratic form F are $X_0, X_1, \ldots, X_{r-1}$. The set V_{n-r} is called the vertex of Q, and is the set of singular points of Q. We note also

$Q^*_{r-1} = \{(x_0:\ldots:x_{r-1}:y_r:\ldots:y_n) \in \mathbf{P}^n(\mathbf{F}_q) \mid F(x_0,\ldots,y_n) = 0$ and the x_i are not all zero$\}$.

Let Q_{r-1} be the nondegenerate quadric of $\mathbf{P}^{r-1}(\mathbf{F}_q)$ associated to Q, i.e. defined by

$$Q_{r-1} = Z_{P^{r-1}}(F_{r-1})$$

or more precisely,

$$Q_{r-1} = \{ (x_0 : \dots : x_{r-1}) \in \mathbf{P}^{r-1}(\mathbf{F}_q) \mid F_{r-1}(x_0, \dots, x_{r-1}) = 0 \},$$

where $F_{r-1}(X_0, \dots, X_{r-1}) = F(X_0, \dots, X_n)$. The (degenerate) quadric Q will abusively be said to be parabolic, hyperbolic or elliptic according to the type of its associated nondegenerate quadric Q_{r-1}. Its character $\omega(Q)$ is by definition the character $\omega(Q_{r-1})$ of Q_{r-1}.
Then, we have the following result which can be found in R.A. Games [3] :

Theorem 1 : The number of points of a quadric Q of $\mathbf{P}^n(\mathbf{F}_q)$ of rank r is :

$$|Q| = \pi_{n-1} + (\omega(Q) - 1)\, q^{(2n-r)/2}$$

and we have $\omega(Q) = 1$ if r is odd, and $\omega(Q) = 0$ or $\omega(Q) = 2$ if r is even.

In particular, a quadric of odd rank is necessarily parabolic, and a quadric of even rank is hyperbolic or elliptic.

Corollary : Let Q be a quadric of $\mathbf{P}^n(\mathbf{F}_q)$, with $n \geq 2$. We have :

$$\pi_{n-2} \leq |Q| \leq \pi_{n-1} + q^{n-1},$$

and the bounds are reached.

Observe that the lower bound is the Warning bound and that the upper bound reaches the following Serre bound, conjectured by Tsfasman, which says that (see [11]) if $F \in \mathbf{F}_q[X_0,\dots,X_n]_d^0$ is a nonzero form of degree $d \leq q$, with $n \geq 2$, then the number N of zeros of F in $\mathbf{F}_q^{\,n}$ is such that :

$$N \leq d\, q^{n-1} - (d-1)\, q^{n-2}.$$

2.2. Hyperplane sections of quadrics.

This paragraph deals with the number of points in the intersection of a quadric and a hyperplane. When the quadric is nondegenerate, the result is known (see for example [13]). R.A. Games has given the result when the quadric has the size of a hyperplane, provided the quadric itself is not a hyperplane (see [3]). Furthermore, I.M. Chakravarti in [1] has solved the case when the quadric is 1-degenerate, that is a quadric of rank n in $\mathbf{P}^n(\mathbf{F}_q)$.
We are going, here, to consider the general case, i.e. quadrics in $\mathbf{P}^n(\mathbf{F}_q)$ of any rank.

We begin by the known nondegenerate case. If Q is a nondegenerate quadric of $\mathbf{P}^n(\mathbf{F}_q)$ (i.e. of rank $r = n + 1$) and if H is a hyperplane of $\mathbf{P}^n(\mathbf{F}_q)$, with $n > 1$, then $Q \cap H$ can be seen as a quadric in a space of dimension $n - 1$. We know (see for example [8]) that the rank of $Q \cap H$ is $r - 1$ or $r - 2$. Then, either $Q \cap H$ is nondegenerate (in $\mathbf{P}^{n-1}(\mathbf{F}_q)$), or $Q \cap H$ is of rank $r - 2 = n - 1$ (whence degenerate in $\mathbf{P}^{n-1}(\mathbf{F}_q)$) ; one says in this last case that H is *tangent* to Q.

Now we have to know what is the value of $\omega(Q \cap H)$, i.e. what happens to the type of the quadric. If the hyperplane H is not tangent to Q, it is obvious that $Q \cap H$ becomes parabolic if Q is hyperbolic or elliptic (indeed $r(Q)$ is necessarily even, and if H is not tangent we have $r(Q \cap H) = r(Q) - 1$ hence odd, then $Q \cap H$ is parabolic) ; and $Q \cap H$ becomes hyperbolic or elliptic if Q is parabolic (same reason rest on the parity of the ranks).
Now if the hyperplane H is tangent to Q, we have the following proposition (see [13]) :

Proposition 2 : The quadric $Q \cap H$ is of the same type as the nondegenerate quadric Q if the hyperplane H is tangent to Q.

Then, we can give the result about the hyperplane sections of a quadric of any rank :

Theorem 2 : Let Q be a quadric of $\mathbf{P}^n(\mathbf{F}_q)$ of rank r whose decomposition is

$$Q = V_{n-r} \cup Q^*_{r-1}$$

and let H be a hyperplane of $\mathbf{P}^n(\mathbf{F}_q)$. Then :

a) If $H \supset V_{n-r}$ then

$$|Q \cap H| = \pi_{n-2} + (\omega(Q_{r-1} \cap H_*) - 1)\, q^{(2n-r-1)/2}$$

if H_* is not tangent to Q_{r-1}, and

$$|Q \cap H| = \pi_{n-2} + (\omega(Q) - 1)\, q^{(2n-r)/2}$$

if H_* is tangent to Q_{r-1}, where H_* is the hyperplane of $\mathbf{P}^{r-1}(\mathbf{F}_q)$ defined by

$$H_* = Z_{\mathbf{P}^{r-1}}(h)$$

where h is the linear form in $\mathbf{F}_q[X_0,\ldots,X_{r-1}]_1^0$ defining H ; moreover $\omega(Q_{r-1} \cap H_*)$ is equal to 1 if Q is hyperbolic or elliptic, and equal to 0 or 2 if Q is parabolic.

b) If $H \not\supset V_{n-r}$ then

$$|Q \cap H| = \pi_{n-2} + (\omega(Q) - 1)\, q^{(2n-r-2)/2}.$$

Proof : We suppose that the r variables appearing in the quadratic form F defining Q are $X_0, X_1, \ldots, X_{r-1}$.
If we set H_i the hyperplane whose equation is $X_i = 0$, we have

$$V_{n-r} = H_0 \cap H_1 \cap \ldots \cap H_{r-1}.$$

But

$$Q \cap H = (V_{n-r} \cup Q^*_{r-1}) \cap H = (V_{n-r} \cap H) \cup (Q^*_{r-1} \cap H),$$

Thus

$$|Q \cap H| = |V_{n-r} \cap H| + |Q^*_{r-1} \cap H| - |V_{n-r} \cap Q^*_{r-1} \cap H|;$$

but $V_{n-r} \cap Q^*_{r-1} = \varnothing$, thus :

$$|Q \cap H| = |V_{n-r} \cap H| + |Q^*_{r-1} \cap H|.$$

1°) Suppose that $H \supset V_{n-r}$.
Then, we have : $|V_{n-r} \cap H| = |V_{n-r}| = |P^{n-r}(F_q)| = \pi_{n-r}$.
Furthermore, the linear form h defining H is such that $h \in F_q[X_0,...,X_{r-1}]_1^0$. Indeed, if

$$h = \sum_{i=0}^{n} a_i X_i ,$$

we have for all $i \geq r$, $P_i = (0 : ... : 0 : 1 : 0 : ... : 0)$ where the 1 is at the ith - coordinate, $P_i \in V_{n-r}$ and $H \supset V_{n-r}$ thus $h(P_i) = 0$. But $h(P_i) = a_i$, thus $a_i = 0$ for all $i \geq r$. Hence,

$$|Q^*_{r-1} \cap H| = q^{n-r+1} |Q_{r-1} \cap H_*|.$$

The quadric $Q_{r-1} \cap H_*$ of $P^{r-2}(F_q)$ is degenerate or not, according as H_* is tangent or not to Q_{r-1}. Now :

— If H_* is not tangent to Q_{r-1} , then by proposition 1, (since $Q_{r-1} \cap H_*$ is nondegenerate in $P^{r-2}(F_q)$), we have :

$$|Q_{r-1} \cap H_*| = \pi_{r-3} + (\omega(Q_{r-1} \cap H_*) - 1)\, q^{(r-3)/2}.$$

Thus

$$|Q \cap H| = \pi_{n-r} + q^{n-r+1} |Q_{r-1} \cap H_*| = \pi_{n-2} + (\omega(Q_{r-1} \cap H_*) - 1)\, q^{(2n-r-1)/2} .$$

— If H_* is tangent to Q_{r-1} , then by theorem 1, we have :

$$|Q_{r-1} \cap H_*| = \pi_{r-3} + (\omega(Q_{r-1} \cap H_*) - 1)\, q^{(r-2)/2},$$

but by proposition 2 we know that $\omega(Q_{r-1} \cap H_*) = \omega(Q_{r-1})$,which is equal to $\omega(Q)$ by definition. Finally,

$$|Q \cap H| = \pi_{n-r} + q^{n-r+1} (\pi_{r-3} + (\omega(Q) - 1)\, q^{(r-2)/2})$$
$$= \pi_{n-2} + (\omega(Q) - 1)\, q^{(2n-r)/2} .$$

2°) Suppose now that H not contains V_{n-r}.
We have $V_{n-r} \cap H = H_0 \cap H_1 \cap ... \cap H_{r-1} \cap H$, thus

$$|V_{n-r} \cap H| = |P^{n-r-1}(F_q)| = \pi_{n-r-1} .$$

If $h = \sum_{i=0}^{n} a_i X_i$ is the linear form defining H, there exist necessarily one j, $r \leq j \leq n$, such that $a_j \neq 0$. Thus

$$Q^*_{r-1} \cap H = \{ (x_0 : ... : x_{r-1} : y_r : ... : y_{j-1} : t : y_{j+1} : ... : y_n) \in P^n(F_q)$$
$$\text{with } Q_{r-1}(x_0,...,x_{r-1}) = 0 \text{ and the } x_i \text{ are not all zero } \},$$

where t is such that

$$a_j t = - a_0 x_0 - ... - a_{r-1} x_{r-1} - a_r y_r - ... - a_{j-1} y_{j-1} - a_{j+1} y_{j+1} - ... - a_n y_n.$$

Thus

$$|Q^*_{r-1} \cap H| = q^{(n-r+1)-1} |Q_{r-1}|$$

with Q_{r-1} a nondegenerate quadric of $P^{r-1}(F_q)$, then :

$$|Q^*_{r-1} \cap H| = q^{n-r}(\pi_{r-2} + (\omega(Q_{r-1}) - 1)\, q^{(r-2)/2})$$

and finally :

$$|Q \cap H| = \pi_{n-r-1} + q^{n-r}(\pi_{r-2} + (\omega(Q_{r-1}) - 1)\, q^{(r-2)/2})$$
$$= \pi_{n-2} + (\omega(Q) - 1)\, q^{(2n-r-2)/2},$$

which concludes the proof.♦

2.3. Intersection of two quadrics in $\mathbf{P}^n(\mathbf{F}_q)$.

The subject matter of this paragraph is to estimate the number of points in the intersection of two quadrics in $\mathbf{P}^n(\mathbf{F}_q)$ with $n > 1$. We give an exact value of this number in a particular case, and an upper bound in the general case (Theorem 3), inspired by an another upper bound of W.M. Schmidt ([10] p.152). We need first a lemma :

Lemma : If Q_1 and Q_2 are two distinct quadrics in $\mathbf{P}^n(\mathbf{F}_q)$, then :

$$|Q_1 \cap Q_2| \le \pi_{n-1} + q^{n-2}.$$

Proof : By theorem 1, $|Q_1| = \pi_{n-1} + (\omega(Q_1) - 1)\, q^{(2n-r)/2}$ if r is the rank of Q_1. Thus :

— if $r \ge 4$, we have $\frac{2n-r}{2} \le n-2$ and then $|Q_1| \le \pi_{n-1} + q^{n-2}$, hence a fortiori

$$|Q_1 \cap Q_2| \le \pi_{n-1} + q^{n-2}.$$

— if $r = 3$ or $r = 1$ then Q_1 is parabolic and

$$|Q_1 \cap Q_2| \le |Q_1| = \pi_{n-1} < \pi_{n-1} + q^{n-2}.$$

— if $r = 2$: either Q_1 is elliptic, and then $|Q_1| = \pi_{n-1} - q^{n-1}$ and the result holds ; or Q_1 is hyperbolic, and then Q_1 is the union of two distinct hyperplanes. We can suppose that the quadric Q_2 is also hyperbolic of rank 2, otherwise the same reasoning which we have made to Q_1 must hold for Q_2.

We set $Q_1 = H_0 \cup H_1$ and $Q_2 = H_2 \cup H_t$, and without loss of generality, we can take for H_i the hyperplane $X_i = 0$. Since, by hypothesis, the quadrics Q_1 and Q_2 are distincts, two cases can appear :

1°) The four hyperplanes are distincts, i.e. t is different of 0 and 1. We obtain, simply in "counting" the points :

$$|Q_1 \cap Q_2| = \pi_{n-4} + 4\, q^{n-2} \le \pi_{n-1} + q^{n-2}$$

(the preceding inequality is equivalent to $(q-1)^2 \ge 0$).

2°) Q_1 and Q_2 have a common hyperplane, i.e. $t = 0$ or $t = 1$. Suppose that $t = 0$. Then, we have :

$$Q_1 \cap Q_2 = \{ (0 : x_1 : \dots : x_n) \in \mathbf{P}^n(\mathbf{F}_q) \} \cup \{ (1 : 0 : 0 : x_3 : \dots : x_n) \in \mathbf{P}^n(\mathbf{F}_q) \},$$

where the union is disjoint. Hence :

$|Q_1 \cap Q_2| = \pi_{n-1} + q^{n-2}$, and the upper bound of this lemma is reached in this case.♦

Theorem 3 : Let $F_1(X_0,...,X_n)$ and $F_2(X_0,...,X_n)$ be two non zero quadratic forms with coefficients in $\mathbf{F}_q$, and let Q_1 and Q_2 respectively the two associated quadrics of $\mathbf{P}^n(\mathbf{F}_q)$. Three cases can appear :

1°) the forms F_1 and F_2 are proportional (i.e there exists $\lambda \in \mathbf{F}_q^*$ such that $F_1 = \lambda F_2$) and then :

$$|Q_1 \cap Q_2| = |Q_1| = |Q_2|.$$

2°) F_1 and F_2 have a common factor of degree 1, and then :

$$|Q_1 \cap Q_2| = \pi_{n-1} + q^{n-2}.$$

3°) F_1 and F_2 have no common factor (no constant), and then :

$$|Q_1 \cap Q_2| \leq \pi_{n-2} + \frac{7\,q^{n-1}}{q-1} - \frac{6\,q^{n-2}}{q-1}$$

(for $q \geq 7$ this upper bound is indeed better than the lemma).

Proof : 1°) Trivial.

2°) We are necessarily in the case where Q_1 and Q_2 are the union of two hyperplanes with one in common ; it is proved in the lemma.

3°) Let F_1 and F_2 be two quadratic forms without nonconstant common factor.

The result is obvious if $q \leq 4$. Indeed, by the lemma, we have :

$$|Q_1 \cap Q_2| \leq \pi_{n-1} + q^{n-2}$$

and furthermore,

$\pi_{n-1} + q^{n-2} \leq \pi_{n-2} + \frac{7\,q^{n-1}}{q-1} - \frac{6\,q^{n-2}}{q-1}$ is equivalent to $q \leq 5$.

Suppose now that $q > 4$.

We set, for i equal 1 and 2 :

$$F'_i(X_0,...,X_n) = F_i(X_0\,,\, X_1+c_1X_0\,,\, X_2+c_2X_0\,,\, ...\,,\, X_n+c_nX_0)$$
$$= P_i(c_1,c_2,...,c_n)\, X_0^2 + ... \,.$$

The polynomials P_1 and P_2 are not the zero polynomial (otherwise F_1 and F_2 would be too), and are not also identically zero, since they have degree at most 2, and $q > 4$ implies that F_1 and F_2 have at most $2q^{n-1} < q^n$ zeros in $\mathbf{F}_q^{\,n}$ (because a polynomial of degree d in $\mathbf{F}_q[X_1,...,X_n]$ have at most dq^{n-1}zeros in $\mathbf{F}_q^{\,n}$, see for example [10]).

Moreover, the total number of zeros of P_1 added to those of P_2 is then at most

$$4\,q^{n-1}$$

which is $< q^n$ since $q > 4$.

Thus it is possible to choose $(c_1,...,c_n) \in \mathbf{F}_q^{\,n}$ such that

$$P_1(c_1,...,c_n) \neq 0 \text{ and } P_2(c_1,...,c_n) \neq 0.$$

Thus, after a nonsingular linear transformation and after divided by $P_1(c_1,...,c_n)$ and $P_2(c_1,...,c_n)$ respectively, we may suppose without loss of generality that :

$$F_1(X_0,...,X_n) = X_0^2 + X_0\, g_1(X_1,...,X_n) + g_2(X_1,...,X_n) \text{ and}$$
$$F_2(X_0,...,X_n) = X_0^2 + X_0\, h_1(X_1,...,X_n) + h_2(X_1,...,X_n)$$

where $g_1,h_1 \in \mathbf{F}_q[X_1,\dots,X_n]_1^0$ and $g_2,h_2 \in \mathbf{F}_q[X_1,\dots,X_n]_2^0$.

If we look at now the polynomials F_1 and F_2 as polynomials in X_0, their resultant is a homogeneous polynomial $R(X_1,\dots,X_n)$ of degree 4. By the well known properties of the resultant, we can say that for any common zero (in $\mathbf{F}_q^{\,n+1}$) $(x_0,x_1,\dots,x_n)$ of $F_1(X_0,\dots,X_n)$ and $F_2(X_0,\dots,X_n)$, we have $R(x_1,\dots,x_n) = 0$.
If we apply the Serre bound (see § 2.1) to the resultant R, we obtain that

the number of zeros in $\mathbf{F}_q^{\,n}$ of $R(X_1,\dots,X_n)$ is $\leq 4\,q^{n-1} - 3\,q^{n-2}$.

Moreover, for such n-uple , the number of possibilities for x_0 is at most 2, and the forms F_1 and F_2 are of degree 2, thus the total number of common zeros $(x_0,\dots,x_n)$ of F_1 and F_2 in $\mathbf{F}_q^{\,n+1}$ is $\leq 8\,q^{n-1} - 6\,q^{n-2}$.
And by the following usual equality :

$$N_A(F) = 1 + (q-1)\,N_P(F)$$

where $N_A(F)$ represent the number of zeros in $\mathbf{A}^{n+1}(\mathbf{F}_q) = \mathbf{F}_q^{\,n+1}$ of F and $N_P(F)$ the number of zeros in $\mathbf{P}^n(\mathbf{F}_q)$ of F, we deduce :

$$|\,Q_1 \cap Q_2\,| \leq \frac{8\,q^{n-1} - 6\,q^{n-2} - 1}{q-1}$$

$$= \pi_{n-2} + 6\,q^{n-2} + \frac{q^{n-1}}{q-1} = \pi_{n-2} + \frac{7\,q^{n-1}}{q-1} - \frac{6\,q^{n-2}}{q-1}. \blacklozenge$$

3. Projective Reed-Muller codes of order 1 associated to a quadric

Let Q be a quadric in $\mathbf{P}^n(\mathbf{F}_q)$ of rank r, decomposing in disjoint union of its vertex V_{n-r} and of Q^*_{r-1}, where Q_{r-1} is the nondegenerate associated quadric of $\mathbf{P}^{r-1}(\mathbf{F}_q)$. We will apply the results of § 2.2 to determine the parameters of the projective Reed-Muller codes of order 1 associated to Q. Since these parameters vary according to the type of the quadric Q, we have to distinguish three cases.

Theorem 4 (parabolic case) : Let Q be a parabolic quadric of $\mathbf{P}^n(\mathbf{F}_q)$ of rank $r \neq 1$. Then the projective Reed-Muller code of order 1 associated to Q is a code with three weights :

$$w_1 = q^{n-1} - q^{(2n-r-1)/2},\ w_2 = q^{n-1} + q^{(2n-r-1)/2},\ w_3 = q^{n-1}$$

with the following parameters :

$$\text{length} = \pi_{n-1},\ \text{dimension} = n+1,\ \text{distance} = q^{n-1} - q^{(2n-r-1)/2}.$$

Theorem 5 (hyperbolic case) : Let Q be an hyperbolic quadric of $\mathbf{P}^n(\mathbf{F}_q)$ of rank r. Then the projective Reed-Muller code of order 1 associated to Q is a code with three weights :

$$w_1 = q^{n-1} + q^{(2n-r)/2},\ w_2 = q^{n-1},\ w_3 = q^{n-1} + q^{(2n-r)/2} - q^{(2n-r-2)/2}$$

with the following parameters :

$$\text{length} = \pi_{n-1} + q^{(2n-r)/2},\ \text{dimension} = n+1,\ \text{distance} = q^{n-1}.$$

Theorem 6 (elliptic case) : Let Q be an elliptic quadric of $\mathbf{P}^n(\mathbf{F}_q)$ of rank $r > 2$. Then the projective Reed-Muller code of order 1 associated to Q is a code with three weights :

$$w_1 = q^{n-1} - q^{(2n-r)/2},\ w_2 = q^{n-1},\ w_3 = q^{n-1} - q^{(2n-r)/2} + q^{(2n-r-2)/2}$$

with the following parameters :

$$\text{length} = \pi_{n-1} - q^{(2n-r)/2},\ \text{dimension} = n+1,\ \text{distance} = q^{n-1} - q^{(2n-r)/2}.$$

Let us remark that we recover the results of J. Wolfmann as a particular case of these results (see [13]), indeed he had considered the case of nondegenerate quadrics : his results correspond to the case where the rank $r = n+1$. Note that, here, the case $H \not\supset V_{n-r}$ is excluded, and then we find only two weights for the hyperbolic and elliptic quadrics, but still three weights for the parabolic one. We recover also the results of I.M. Chakravarti (see [1]) : it corresponds to the case where the rank $r = n$.

Proof : The lengths of the respective codes are equal to the number of points of the respective quadrics : theorem 1 gives the result.
The map **c** defining the code (see § 1) is one to one, and thus the dimension of the code is equal to the dimension of $\mathbf{F}_q[X_0,...,X_n]_1^0$ over $\mathbf{F}_q$, i.e. $n + 1$: indeed, if H is a hyperplane of $\mathbf{P}^n(\mathbf{F}_q)$, (which amounts to taking a linear form of $\mathbf{F}_q[X_0,...,X_n]$), it is sufficient to apply the results of Theorem 2 to see that $|Q \cap H| < |Q|$, and to have also the different weights.♦

4. Projective Reed-Muller codes of order 2 associated to a quadric

The map $\mathbf{c} : \mathbf{F}_q[X_0,...,X_n]_2^0 \to \mathbf{F}_q^{|Q|}$ as introduced in § 1 defining the projective Reed-Muller code of order 2 associated to the quadric Q has for domain the vector space of quadratic forms over $\mathbf{F}_q$; this is why we gave previously some results on the intersection of two quadrics of $\mathbf{P}^n(\mathbf{F}_q)$.

Theorem 7 (parabolic case) : Let Q be a parabolic quadric in $\mathbf{P}^n(\mathbf{F}_q)$, $n \geq 2$. If $q \geq 8$ then the projective Reed-Muller code of order 2 associated to Q has the following parameters :

$$\text{length} = \pi_{n-1},\ \text{dimension} = \frac{n(n+3)}{2},\ \text{distance} \geq q^{n-1} - 6q^{n-2} - \frac{q^{n-1}}{q-1}.$$

Theorem 8 (elliptic case) : Let Q be an elliptic quadric in $\mathbf{P}^n(\mathbf{F}_q)$ of rank $r > 2$. If $q \geq 8$ then the projective Reed-Muller code of order 2 associated to Q has the following parameters :

$$\text{length} = \pi_{n-1} - q^{(2n-r)/2},\ \text{dimension} = \frac{n(n+3)}{2},$$

$$\text{distance} \geq q^{n-1} - q^{(2n-r)/2} - 6q^{n-2} - \frac{q^{n-1}}{q-1}.$$

We reserve the case where the quadric is hyperbolic of rank 2 for the theorem 10 (we have indeed more precise results).

Theorem 9 (hyperbolic case of rank $r \geq 4$) : Let Q be an hyperbolic quadric in $\mathbf{P}^n(\mathbf{F}_q)$ of rank $r \geq 4$. If $q \geq 8$ then the projective Reed-Muller code of order 2 associated to Q has the following parameters :

$$\text{length} = \pi_{n-1} + q^{(2n-r)/2},\ \text{dimension} = \frac{n(n+3)}{2},$$

$$\text{distance} \geq q^{n-1} + q^{(2n-r)/2} - 6\,q^{n-2} + \frac{q^{n-1}}{q-1}.$$

Let us remark that we can have, for the theorem 9, the same results with a weaker hypothesis on q when the rank of Q is equal to 4 or 6, namely $q > 5$.

Now we consider the case of *maximal* quadrics, that is hyperbolic quadrics of rank 2. By the corollary of theorem 1, the number of points of these quadrics reaches the maximum number of points of a quadric, and it is in this sense that we call them "maximal". We can remark that they are particular quadrics (they are the union of two distinct hyperplanes). The codes which are associated to them have a minimum distance precisely known. These codes will have a generalization in the next paragraph.

Theorem 10 (hyperbolic case of rank = 2) : Let Q be an hyperbolic quadric in $\mathbf{P}^n(\mathbf{F}_q)$ of rank 2. The projective Reed-Muller code of order 2 associated to Q has the following parameters :

$$\text{length} = \pi_{n-1} + q^{n-1},\ \text{dimension} = \frac{n(n+3)}{2},\ \text{distance} = q^{n-2}(q-1).$$

Proof : The length of the codes is the number of points of the quadric Q, and is given by Theorem 1.

Let $F' \in \mathbf{F}_q[X_0,\ldots,X_n]_2^0$ and $Q' = Z_{pn}(F')$, $Q = Z_{pn}(F)$.

Either F and F' are proportional, and then $Q = Q'$. Remark that there is $q-1$ such non zero forms F' ; thus there is at least q quadratic forms vanishing in Q, hence in the kernel of the map c defining these codes. We claim that there are no other forms in Ker(c), and thus the dimension of this codes is :

$$\dim(\mathrm{Im}\ c) = \dim \frac{\mathbf{F}_q[X_0,\ldots,X_n]_2^0}{\mathrm{Ker}(c)} = \frac{(n+1)(n+2)}{2} - \log_q(|\mathrm{Ker}(c)|)$$

$$= \frac{(n+1)(n+2)}{2} - 1 = \frac{n^2+3n}{2} = \frac{n(n+3)}{2}.$$

Indeed, suppose now that F and F' are not proportional, we have by Theorem 3 :

$$|Q \cap Q'| \leq \pi_{n-2} + \frac{7\,q^{n-1}}{q-1} - \frac{6\,q^{n-2}}{q-1}.$$

- if Q is parabolic (Th 7), we have

$$\pi_{n-2}+\frac{7\,q^{n-1}}{q-1}-\frac{6\,q^{n-2}}{q-1}<|Q|\Leftrightarrow q^2-8q+6>0\Leftrightarrow q\geq 8.$$

Moreover, F and F' cannot have a common factor of degree 1 since Q would be the union of two hyperplanes and thus would be hyperbolic.
The minimum distance follows from the same inequality of the Theorem 3.
- if Q is elliptic (Th 8), F and F' cannot also have a common factor of degree 1, and we have :
$\pi_{n-2}+\dfrac{7\,q^{n-1}}{q-1}-\dfrac{6\,q^{n-2}}{q-1}<|Q|=\pi_{n-1}-q^{(2n-r)/2}$ if and only if $q>8$ for $r=4$, and thus a fortiori for $r\geq 4$, i.e. since r is even, $r>2$.
- if Q is hyperbolic of rank ≥ 4 (Th 9), the same reasoning gives a fortiori the results (indeed the hypothesis $q\geq 8$ holds for more "smallest " quadrics).
- if Q is hyperbolic of rank = 2 (Th 10) :
* either F and F' have a common factor of degree 1, and by the Theorem 3 :

$$|Q\cap Q'|=\pi_{n-1}+q^{n-2}\text{ which is }<|Q|=\pi_{n-1}+q^{n-1}.$$

* or F and F' have not a common factor of degree 1, and by the lemma preceding Theorem 3 we have : $|Q\cap Q'|\leq\pi_{n-1}+q^{n-2}$ which is $<|Q|$.
The minimum distance in this case is :

$$|Q|-(\pi_{n-1}+q^{n-2})=q^{n-1}-q^{n-2}=q^{n-2}(q-1).\blacklozenge$$

5. Projective Reed-Muller codes associated to a maximal hypersurface

We consider here hypersurfaces of degree $h\leq q$ reaching the Serre bound, i.e. which are the union of h distinct hyperplanes containing a linear variety of codimension 2. The Serre bound enunciated in § 2.1 has the following projective version : if F is a non zero form of degree $h\leq q$ of $\mathbf{F}_q[X_0,\ldots,X_n]$, then

$$|Z_{\mathbf{P}^n}(F)|\leq\pi_{n-2}+h\,q^{n-1}.$$

The construction of such varieties (called maximal) is easy ; indeed we can take for example :

$$F=\prod_{1\leq i\leq h}(X_0-\lambda_iX_1)$$

where the λ_i are h distinct elements of $\mathbf{F}_q$. We are going to construct projective Reed-Muller codes associated to such varieties.

Theorem 11 : Let $V=Z_{\mathbf{P}^n}(F)$ be a variety of $\mathbf{P}^n(\mathbf{F}_q)$ which is the union of h distinct hyperplanes containing a linear variety of codimension 2, with $h\leq q$. Then the projective Reed-Muller code of order $d<h$ associated to V has the following parameters :

$$\text{length}=\pi_{n-2}+h\,q^{n-1},\ \text{dimension}=\binom{n+d}{d},\ \text{distance}=(h-d)\,q^{n-1}.$$

Let us remark that we find again the projective Reed-Muller codes of order 1 associated to a maximal quadric (in the particular case $h = 2$ and $d = 1$).

Proof : The length of the code is equal to the number of points of the variety V which is, by construction,

$$\pi_{n-2} + h\, q^{n-1}.$$

The map **c**: $\mathbf{F}_q[X_0,\ldots,X_n]_d^0 \rightarrow \mathbf{F}_q^{|V|}$ defining the code is obviously one to one since $d < h$. Thus the dimension of the code is equal to the dimension, over $\mathbf{F}_q$, of $\mathbf{F}_q[X_0,\ldots,X_n]_d^0$ i.e. $\binom{n+d}{d}$.

If $V = H_1 \cup \ldots \cup H_h$ then the subvariety V' of degree d of V defined by $V' = H_1 \cup \ldots \cup H_d$ where the d hyperplanes are taken among the h defining V, is such that :

$$|V'| = \pi_{n-2} + d\, q^{n-1}.$$

Thus the minimum distance of the code is equal to :

$$|V| - (\pi_{n-2} + d\, q^{n-1}) = h\, q^{n-1} - d\, q^{n-1} = (h-d)\, q^{n-1}.\blacklozenge$$

We can say more if we consider the particular case of the codes above of order 1. Indeed, it is easy to see that the hyperplane sections of such maximal varieties have three possible sizes, namely π_{n-1}, π_{n-2} or $\pi_{n-3} + h\, q^{n-2}$. Thus, the projective Reed-Muller code of order 1 associated to V (with $h > 1$) is a code with three weights :

$$w_1 = (h-1)\, q^{n-1}, \quad w_2 = h\, q^{n-1}, \quad w_3 = h\, q^{n-1} + (1-h)\, q^{n-2}$$

and with the following parameters :

$$\text{length} = \pi_{n-2} + h\, q^{n-1}, \ \text{dimension} = n+1, \ \text{distance} = (h-1)\, q^{n-1}.$$

References

[1] Chakravarti I.M., *Families of codes with few distinct weights from singular and non-singular hermitian varieties and quadrics in projective geometries and Hadamard difference sets and designs associated with two-weights codes*, Coding Theory and Design Theory - Part I : Coding Theory IMA vol. **20**.

[2] Delsarte P., Goethals J.M. and Mac Williams F.J., *On generalized Reed-Muller codes and their relatives*, Inform. and Control **16** (1970) 403-442.

[3] Games R.A. , *The Geometry of Quadrics and Correlations of sequences*, IEEE Transactions on Information Theory. Vol. IT-32, No. 3, May 1986, 423-426.

[4] Hirschfeld J.W.P., *Projective Geometries over Finite Fields*, Clarendon Press, Oxford, 1979.

[5] Kasami T., Lin S. and Peterson W.W., *New generalization of the Reed-Muller codes - Part I : Primitive codes*, IEEE Trans. Information Theory **IT-14** (1968), 189-199.

[6] Lachaud G., *The parameters of projective Reed-Muller codes*, Discrete Mathematics **81** (1990), 217-221.

[7] Manin Yu.I. and Vladut S.G., *Linear codes and modular curves*, Itogi Nauki i Tekhniki **25** (1984) 209-257 J. Soviet Math. **30** (1985) 2611-2643.

[8] Primrose E.J.F., *Quadrics in finite geometries*, Proc. Camb. Phil. Soc., **47** (1951), 299-304.

[9] Ray-Chaudhuri D.K., *Some results on quadrics in finite projective geometry based on Galois fields*, Can. J. Math., vol. **14**, (1962), 129-138.

[10] Schmidt W.M., *Equations over Finite Fields. An Elementary Approach*, Lecture Notes in Maths **536** (1975).

[11] Serre, J.-P., *Lettre à M. Tsfasman, 24 juillet 1989*, Journées Arithmétiques de Luminy, Astérisque, S.M.F., Paris, to appear.

[12] Sorensen A.B., *Projective Reed-Muller codes*, to appear.

[13] Wolfmann J., *Codes projectifs à deux ou trois poids associés aux hyperquadriques d'une géométrie finie*, Discrete Mathematics **13** (1975) 185-211, North-Holland.

Decoding Algebraic-Geometric Codes by solving a key equation

Dirk Ehrhard*

1 Introduction

The recent work about the problem of decoding Algebraic-Geometric Codes has led to an algorithm (e.g., see [2,3]). Another algorithm has been given by Porter, see [6,7,8,9,10], generalizing Berlekamp's decoding algorithm. The main step is to solve a so-called "key-equation". For this purpose, Porter gave a generalization of Euclid's algorithm for functions on curves. Unfortunately, therefore he had to impose some strong restrictions to the code and its underlying curve, such that the resulting algorithm works only for a very small class of Algebraic-Geometric Codes. Recently, the generalized Euclidian algorithm was investigated and corrected by Porter, Shen and Pellikaan ([11]) and Shen ([12]).

Here, we will show how to generalize Porters ideas to all Algebraic-Geometric Codes and moreover, how to solve the key equation by simple linear algebra operations. Two observations on Porter's methods have motivated our work:

1. The operations done by Porters algorithm at the so-called "resultant-matrix", may be considered as a Gaussian algorithm, applied to the transposed matrix.

2. The key equation may be viewed as linear.

The result is given in section 2: A decoding algorithm of complexity order $O(n^3)$, that corrects up to $\left\lfloor \frac{1}{2}(d^* - 1 - g) \right\rfloor$ errors, exactly as the well known algorithm does. In section 3 we describe how strongly both algorithms are connected. We conclude with section 4, giving a short overview over Porters algorithm and explaining, in what manner Porters work embeds in ours.

2 The decoding procedure

2.1 The code

We will use the notations of [1]: Let X be a curve of genus g (i.e. a non-singular, absolutely irreducible projective curve defined over the finite field $\mathbb{F}_q$), $P_1, \ldots, P_n$ rational

*The author is with the Mathematisches Institut IV der Heinrich-Heine-Universität, 4000 Düsseldorf 1, Germany. The contents of this paper are also part of the author's Ph. D. Thesis ([13])

points on X, and G a divisor which has support disjoint from the P_ν's. We will assume that $2g-2 < \deg G \leq n+g-1$ and define $D := P_1 + \ldots + P_n$. Then the code $C = C^*(D,G)$ is the image of the linear, injective map

$$\begin{array}{rlcl} \mathrm{Res}_D\colon & \Omega(G-D) & \longrightarrow & \mathbb{F}_q^n \\ & \eta & \longmapsto & (\mathrm{Res}_{P_1}\eta, \ldots, \mathrm{Res}_{P_n}\eta). \end{array}$$

This is a linear $[n,k,d]_q$ - code with $k \geq n-1-\deg G+g$ and $d \geq d^* = \deg G - (2g-2)$. For details see [1]. Note, that Res_D may be extended to $\Omega(X)$ in a canonical way.

2.2 A theorem for preparation

Let G' be a divisor with support disjoint from the P_ν's such that $G-G'$ is effective and $\dim L(G') = 0$. An easy consequence is: $\Omega(G-D) \subset \Omega(G'-D)$, which we will use to generalize [7, Theorem V.1, p. 16]:

Theorem 1 *There exists a vector space V: $\Omega(G-D) \subset V \subset \Omega(G'-D)$ such that $\mathrm{Res}_D|_V\colon V \to \mathbb{F}_q^n$ is an isomorphism*

Proof. Since $\mathrm{Res}_D|_{\Omega(G-D)}$ is injective, it suffices to prove that $\mathrm{Res}_D|_{\Omega(G'-D)}$ is surjective. We have

$$\ker\left(\mathrm{Res}_D|_{\Omega(G'-D)}\right) = \{\eta \in \Omega(G'-D) : \mathrm{Res}_{P_\nu}\eta = 0,\ \nu = 1\ldots n\} = \Omega(G').$$

Therefore

$$\begin{aligned} \mathrm{rank}\left(\mathrm{Res}_D|_{\Omega(G'-D)}\right) &= \dim\Omega(G'-D) - \dim\Omega(G') \\ &= g-1-\deg(G'-D) - (g-1-\deg G') \\ &= \deg D = n, \end{aligned}$$

using the Riemann-Roch-Theorem and the fact that $0 \leq \dim L(G'-D) \leq \dim L(G') = 0$.

The problem of decoding is: *For an arbitrary given $y \in \mathbb{F}_q^n$, find $c \in C$ with minimal Hamming distance* $\mathrm{wt}(y-c)$. According to Theorem 1, Res_D gives a correspondence of vector spaces:

$$\begin{array}{ccccc} \Omega(G-D) & \subset & V & \subset & \Omega(G'-D) \\ \downarrow{\scriptstyle \mathrm{Res}_D} & & \downarrow{\scriptstyle \mathrm{Res}_D} & & \\ C & & \subset\ \mathbb{F}_q^n & & \end{array}$$

If $\mathbb{F}_q^n \to V$, $w \mapsto \eta_w$ denotes the inverse map of $\mathrm{Res}_D|_V$, then we can describe the decoding problem in terms of differentials: *For a given η_y, find $\eta_c \in \Omega(G-D)$, such that $\eta_e := \eta_y - \eta_c$ has minimal number of Poles in $P_1, \ldots, P_n$, i.e. such that η_e may be written as a fraction of functions with low degree.* We will precise that in the next section.

2.3 The key equation

Let $y = c+e \in \mathbb{F}_q^n$ with $c \in C$ and F an arbitrary divisor on X. Let D_e denote the unique Divisor with $0 \leq D_e \leq D$ and $P_\nu \in \operatorname{supp} D_e \Leftrightarrow e_\nu \neq 0$, that is, $D_e = ((\eta_e)_\infty)|_{\operatorname{supp} D}$.

Definition 1 *By a solution of the key equation we will denote any tripel* $(B, \omega, \alpha) \in (L(F) \setminus \{0\}) \times \Omega(G' - F) \times \Omega(G - D - F)$, *satisfying* $B\eta_y = \alpha + \omega$.

Proposition 1 *If* $\deg F + \mathrm{wt}(e) < d^*$, *then any solution* (B, ω, α) *of the key equation satisfies* $\eta_c = \frac{\alpha}{B}$ *and* $\eta_e = \frac{\omega}{B}$.

Proof. Let (B, ω, α) be a solution of the key equation. From $\omega = B\eta_y - \alpha$ and $\eta_y = \eta_c + \eta_e$ one concludes $\omega - B\eta_e = B\eta_c - \alpha$. If the two sides of this equation do not vanish, we may estimate their divisors:

$$(\omega - B\eta_e) \geq \min((\omega), (B) + (\eta_e)) \geq \min(G' - F, G' - F - D_e) = G' - F - D_e$$

$$(B\eta_c - \alpha) \geq \min((B) + (\eta_c), (\alpha)) \geq \min(-F + G - D, G - D - F) = G - D - F$$

Since D and G' have disjoint supports and $G \leq G'$, we get

$$(\omega - B\eta_e) \geq \max(G' - F - D_e, G - D - F) = G - D_e - F,$$

but $\deg(G - D_e - F) = \deg G - \mathrm{wt}(e) - deg F > \deg G - d^* = 2g - 2$, what contradicts the assumption that $\omega - B\eta_e$ is a non-vanishing differential. Therefore $\omega - B\eta_e = B\eta_c - \alpha = 0$, what proves the statement.

Proposition 2 *If* $\deg F \geq \mathrm{wt}(e) + g$, *then there exists a solution of the key equation.*

Proof. By the Riemann-Roch-Theorem, $\dim L(F - D_e) \geq 1 + \deg F - \deg D_e - g \geq 1 + \mathrm{wt}(e) + g - \mathrm{wt}(e) - g = 1$, which guarantees the existence of $B \in L(F - D_e) \setminus \{0\} \subset L(F) \setminus \{0\}$. Now we have $B\eta_y = B\eta_c + B\eta_e$, $(B\eta_c) = (B) + (\eta_c) \geq -F + G - D$, and $(B\eta_y - B\eta_c) = (B\eta_e) = (B) + (\eta_e) \geq -F + D_e + G' - D_e = G' - F$, therefore $(B, B\eta_c, B\eta_e)$ is a solution of the key equation.

Corollary 1 *Let* F *be a divisor of degree* $\left\lfloor \frac{d^*+g-1}{2} \right\rfloor$. *For any* $y \in \mathbb{F}_q^n$ *such that there is* $c \in C$ *with* $\mathrm{wt}(e := y - c) \leq \left\lfloor \frac{d^*-g-1}{2} \right\rfloor$, *there exists a solution of the key equation. On the other hand, for any solution* (B, α, ω) *the equality* $\eta_e = \frac{\omega}{B}$ *holds.*

Proof. One easily checks, that we have $\deg F + \mathrm{wt}(e) < d^*$ and $\deg F \geq \mathrm{wt}(e) + g$. Now apply Propositions 1 and 2.

The proofs of the Propositions show, that already the assumptions $\dim \Omega(G - D_e - F) = 0$ resp. $\dim L(F - D_e) > 0$ suffice to guarantee existence and uniqueness of solutions. This coincides exactly with the assumptions needed by the well-known algorithm (see [3]).

2.4 Solving the key equation

Assume $t_0 := \left\lfloor \frac{d^*-g-1}{2} \right\rfloor$ to be non-negative (i.e. $d^* > g$) and F to be a divisor of degree $\left\lfloor \frac{d^*+g-1}{2} \right\rfloor = t_0 + g$. For an arbitrary $y \in \mathbb{F}_q^n$ consider the linear map

$$\begin{array}{rcccl} \delta_y & : & L(F) & \to & \Omega(G' - D - F), \\ & & B & \mapsto & B \cdot \eta_y \end{array}$$

Notice, that $\Omega(G - D - F) \cap \Omega(G' - F) = \Omega(G - F) = \{0\}$, since $\deg(G - F) = \deg G - \deg F > \deg G - d^* = 2g - 2$. Hence there exists a vector space W such that

$$\Omega(G' - D - F) = \Omega(G - D - F) \oplus \Omega(G' - F) \oplus W. \qquad (1)$$

Let π_W, $\pi_{\Omega(G'-F)}$ denote the natural projections onto W resp. $\Omega(G' - F)$.

If there is any codeword c with $\mathrm{wt}(y - c) \leq t_0$, then, by the Corollary, there will exist $B \in L(F) \setminus \{0\}$, $\alpha \in \Omega(G - D - F)$ and $\omega \in \Omega(G' - F)$ such that $\delta_y(B) = B\eta_y = \alpha + \omega$. Every such triple will suffice $\eta_e = \frac{\omega}{B}$. In this case, the following algorithm will compute the error vector e:

1. Compute the matrix describing δ_y.
2. Determine $B \in \ker(\pi_W \circ \delta_y) \setminus \{0\}$.
3. Compute $\omega := \pi_{\Omega(G'-F)}(\delta_y(B))$.
4. Compute $e := \mathrm{Res}_D \frac{\omega}{B}$.

If too many errors have occured, then either $\ker(\pi_W \circ \delta_y) = \{0\}$ or the computed e won't suffice the conditions $\mathrm{wt}(e) \leq t_0$ and $y - e \in C$.

2.5 Realisation and complexity

Let $(e_\nu)_{\nu=1\ldots n}$ denote the canonical basis of $\mathbb{F}_q^n$. If the matrices describing δ_{e_ν} are computed before the algorithm starts and once forever, the matrix of $\delta_y = \sum_{\nu=1}^n y_\nu \delta_{e_\nu}$ may be computed with complexity order $O(n^3)$ in run time. Then parts 2 and 3 of our algorithm may be done by simple linear algebra operations; the first one with complexity order $O(n^3)$ and the second with order $O(n^2)$. We will now describe shortly how to realize the remaining third part in $O(n^2)$ steps:

Assume ω and B have been determined; now we want to compute $\mathrm{Res}_P \frac{\omega}{B}$ for any $P \in \mathrm{supp}\, D$. Let t be a local coordinate on X around P, $j := -v_P(F)$ and $\omega = \sum_{i=j}^{\infty} \omega_i t^i \, dt$ and $B = \sum_{i=j}^{\infty} B_i t^i$ the local power series around P. Now,

$$\mathrm{Res}_P \frac{\omega}{B} = \frac{\alpha_{i_0-1}}{B_{i_0}}$$

where $i_0 = \min\{i : B_i \neq 0\} = v_P(B)$ and $\alpha_{j-1} = 0$. Hence the computation can be done as follows:

- Compute $B_j, B_{j+1}, \ldots$ until $B_{i_0} \neq 0$ occurs.
- Compute α_{i_0-1}
- Set $\mathrm{Res}_P \frac{\omega}{B} := \frac{\alpha_{i_0-1}}{B_{i_0}}$.

Since ω and B are given as linear combinations of certain basis differentials (resp. functions), any coefficient of their power series may be computed as an corresponding linear combination if the power series at every point of $\operatorname{supp} D$ of any basis differential (resp. function) is known a priori. The complexity to compute one coefficient that way is $O(n)$. Altogether, there have to be computed

$$\sum_{\nu=1}^{n}(1+1+v_{P_\nu}(B)+v_{P_\nu}(F)) \leq 2n+\deg((B)+F)|_{\operatorname{supp} D} \leq 2n+\deg F$$

coefficients, hence one can get along with complexity order $O(n^2)$ in the decoding algorithm's fourth part. Furthermore, an analogous computation shows that not more than $\deg F$ coefficients of each power series around each point of the basis differentials (resp. functions) must be known a priori.

3 Essentially that's nothing new

To demonstrate how our method of decoding is connected to that of [3] resp. [4], we will show that the main steps of each of the decoding procedures are equivalent. We first consider the map

$$\Phi: \Omega(G'-D-F) \to L(G-F)^\vee = \{\text{linear } \phi: L(G-F) \to \mathbb{F}_q\},$$

where $\Phi(\eta)(f) = \sum_{\nu=1}^{n} \mathrm{Res}_{P_\nu}(f \cdot \eta)$.

Proposition 3 *If the supports of F and D are disjoint, then $\Phi|_W$ is an isomorphism.*[1]

Proof. Since equation (1) holds, it suffices to show that

1. Φ is surjective,
2. $\Omega(G'-F) \oplus \Omega(G-D-F) \subset \ker \Phi$ and
3. $\dim W = \dim L(G-F)^\vee$.

1.: The map $L(G-F) \to \mathbb{F}_q^n, f \mapsto (f(P_1), \ldots, f(P_n))$ has kernel $L(G-D-F) = \{0\}$, hence is injective, therefore it suffices to prove that $\tilde{\Phi}: \Omega(G'-D-F) \to (\mathbb{F}_q^n)^\vee$, $\tilde{\Phi}(\eta)(v) = \sum \mathrm{Res}_{P_\nu} v_\nu$, is surjective. Let $(e_\nu^\vee)_{\nu=1..n}$ denote the canonical basis of $(\mathbb{F}_q^n)^\vee$. For any ν, $\Omega(G'-F-P_\nu) \setminus \Omega(G'-F)$ is not empty as a consequence of the Riemann-Roch-Theorem, and the image by $\tilde{\Phi}$ of any such differential is $e_\nu^\vee$. Hence $\tilde{\Phi}$ is surjective, hence Φ, too.

[1]Proposition 3 together with decomposition (1) show the exactness of

$$0 \to \Omega(G-D-F) \oplus \Omega(G'-F) \xrightarrow{\iota} \Omega(G'-D-F) \xrightarrow{\Phi} L(G-F)^\vee \to 0,$$

that may be also derived in a canonical way of a short exact sequence of sheaves on the curve X. For details see [13]

2.: If $\eta \in \Omega(G' - F)$ then for any ν, $\mathrm{Res}_{P_\nu}\eta = 0$, hence $\Phi(\eta) = 0$. Now take $\eta \in \Omega(G - D - F)$ and $f \in L(G - F)$. Then $f\eta \in \Omega(-D)$ and, by the residue theorem is: $\sum \mathrm{Res}_{P_\nu}(f\eta) = 0$.

3.:

$$\begin{aligned}\dim W &= \dim\Omega(G' - D - F) - \dim\Omega(G' - F) - \dim\Omega(G - D - F)\\ &= g - 1 - \deg(G' - D - F) - (g - 1 - \deg(G' - F)) -\\ &\qquad -(g - 1 - \deg(G - D - F))\\ &= 1 - g + \deg(G - F)\\ &= \dim L(G - F)\\ &= \dim L(G - F)^\vee,\end{aligned}$$

by the Riemann-Roch-Theorem, using the fact that $\deg(G - F) > 2g - 2$ and $\deg(G - D - F) < 0$.

In our decoding procedure, the fundamental step is to find an element of $\ker(\pi_W \circ \delta_y) \setminus \{0\}$ which is, by Proposition 3, equivalent to finding something contained in $\ker(\Phi|_W \circ \pi_W \circ \delta_y) \setminus \{0\}$. But

$$\begin{aligned}(\Phi|_W \circ \pi_W \circ \delta_y(B))(f) &= \Phi(\delta_y(B))(f) = \Phi(\eta_y \cdot B)(f)\\ &= \sum \mathrm{Res}_{P_\nu}(f \cdot \eta_y \cdot B)\\ &= \sum f(P_\nu) \cdot B(P_\nu) \cdot \mathrm{Res}_{P_\nu}(\eta_y)\\ &= \sum f(P_\nu) \cdot B(P_\nu) \cdot y_\nu,\end{aligned}$$

for any $B \in L(F)$ and any $f \in L(G - F)$. Hence $\Phi|_W \circ \pi_W \circ \delta_y$ is exactly the *error locating map* E_y in [4]. Finding a non-trivial element of its kernel is the main step in the known algorithm, too.

4 Porter's decoding algorithm

We will give a very short overview over Porter's algorithm and explain why we consider the presented algorithm as its generalisation. Porter makes some restrictions to the codes he treats. We will formulate them using the notations of Section 2:

1. $G' = -P_\infty$, where P_∞ is a rational point on X, distinct from $\operatorname{supp} D$.
2. G is lineary equivalent to $(\deg G) \cdot P_\infty$, $G + P_\infty$ is effective and $\deg G \geq 0$.
3. $(2g - 2)P_\infty$ is canonical.

Especially assumption 3 won't be realizable for most curves, but fortunately this restriction may be dismissed easily.

First remember the isomorphism $\mathrm{Res}_D \colon V \to \mathbb{F}_q^n$ of Section 2, and let $\varepsilon_1, \ldots, \varepsilon_n$ a basis of V such that $\mathrm{Res}_D(\varepsilon_1), \ldots, \mathrm{Res}_D(\varepsilon_n)$ is the canonical basis of $\mathbb{F}_q^n$. Furthermore, let γ be a rational function[2] with $(\gamma) = G - (\deg G) \cdot P_\infty$, ω_0 a differential with $(\omega_0) = (2g-2)P_\infty$, and $m := -\nu_{P_\infty}(\gamma)$. For any "received" vector $y = c + e \in \mathbb{F}_q^n$, Porter defines a rational function called "syndrom" by

$$S := \sum_{\nu=1}^{n} \left(1 - \frac{\gamma}{\gamma(P_\nu)}\right) y_\nu \frac{\varepsilon_\nu}{\omega_0}.$$

Note that $S \in R := \bigcup_{t \geq 0} L(tP_\infty)$, $S\omega_0 \equiv \eta_y \pmod{\gamma}$, i.e. $S\omega_0 = \eta_y + \gamma f$ for some $f \in R$, and $-\nu_{P_\infty}(S) \leq m + 2g - 1$.

In order to decode Porter looks for solutions of the "polynomial congruence" $A - BS \equiv 0 \pmod{\gamma}$ subject to a certain "condition at P_∞". More precisely, one may describe this by[3]

If there is an integer $t < d^* - wt(e)$ and A, B, $C \in R$ satisfying $-\nu_{P_\infty}(A) \leq t + 2g - 1$ and $-\nu_{P_\infty}(B) \leq t$ such that

$$A - BS = C\gamma, \tag{2}$$

then $\eta_e = \frac{A}{B}\omega_0$.

In the following, we give a short description of Porter's method of finding such A, B, C. Consider the linear map

$$\begin{array}{cccc} \delta : & L(tP_\infty) \oplus L((t+2g-1)P_\infty) & \longrightarrow & L(m+t+2g-1)P_\infty) \\ & (B, C) & \longmapsto & BS + C\gamma. \end{array}$$

The solutions of (2) clearly will satisfy $-\nu_{P_\infty}(C) \leq t + 2g - 1$, therefore they coincide exactly with the tripels

$$\delta(B,C), B, C \quad \text{where} \quad \delta(B,C) \in L(t + 2g - 1).$$

If one chooses elements $\phi_i \in R$ such that

$$L(\tau P_\infty) = \text{span}\left\{\phi_1, \ldots, \phi_{\dim L(\tau P_\infty)}\right\} \quad \text{for any } \tau \in \mathbb{N},$$

then the matrix M, describing δ with respect to those bases, is the transpose of Porter's resultant matrix. Clearly, searching for a solution of (2), may be done by looking for a non-trivial linear combinition of the columns of M with zeroes in the m lower positions. One way to do that is Porter's "row reduction process" at M^T, but clearly there are several other methods.

We conclude by explaining, how Porter's method and the one described in section 2, are connected. Multiplying equation (2) by ω_0 and substituting $A\omega_0$ by ω, $S\omega_0$ by σ and $C\omega_0$ by α, one do not need neither the differential ω_0 nor assumption 3 for the equivalent statement

---

[2]In genus 0 case, γ is the Goppa-Polynomial, if P_∞ denotes the infinite point of the projective line

[3]This are sligthly weaker assumptions than those in the original, where only solutions of minimal degree ($= -\nu_{P_\infty}$) are taken into account; for more detailed information see [13]

If there is an integer $t < d^* - wt(e)$ *and* $B \in L(tP_\infty)$, $\omega \in \Omega(-(t+1)P_\infty)$ *such that* $\omega - B\sigma = \alpha\gamma$ *for some* $\alpha \in \Omega(X)$, *then* $\eta_e = \frac{\omega}{B}$.

The underlying idea of this is to write $\sigma \equiv \frac{\omega}{B} \pmod{\gamma}$ where B has few zeros, i.e. few poles. More directly and without the need of S^4 one may, provided a given η_y, search for ω and B such that $\eta_y \equiv \frac{\omega}{B} \pmod{\Omega(G-D)}$ and B has few poles (e.g. $(B)+F \geq 0$ for some "small" divisor F). This is the main idea of the method of decoding described in Section 2.

References

[1] J.H. van Lint, G. van der Geer, *Introduction to Coding Theory and Algebraic Geometry.* DMV Seminar, Band 12. Birkhäuser Verlag 1988.

[2] J. Justesen, K.J. Larsen, A. Havemose, H.E. Jensen, T. Høholt, *Construction and decoding of a class of algebraic geometry codes.* IEEE-IT 35(4)(1989), pp. 811-821.

[3] A.N. Skorobogatov, S.G. Vlăduţ, *On the decoding of algebraic-geometric codes.* IEEE-IT 36(5)(1990),pp. 1051-1060.

[4] R. Pellikaan, *On a decoding Algorithm for Codes on maximal curves.* IEEE-IT 35(6)(1989), pp. 1228-1232.

[5] S.G. Vlăduţ, *On the decoding of algebraic-geometric codes for* $q \geq 16$. IEEE-IT 36(6)(1990), pp. 1961-1963.

[6] S.C. Porter, *Decoding Codes arising from Goppa's construction on algebraic curves.* Thesis, Yale University, 1988.

[7] S.C. Porter, *Decoding Geometric Goppa Codes.* Preprint.

[8] S.C. Porter, *Euclid's algorithm, resultants and rational function representation on algebraic curves with a single point at infinity.* Preprint.

[9] S.C. Porter, *Dense representation of affine coordinate rings of curves with one point at infinity.* Proceedings of ISSAC-89.

[10] S.C. Porter, *An efficient data structure for rational function on algebraic curves.* Preprint.

[11] S.C. Porter, B.Z. Shen, R. Pellikaan, *Decoding geometric Goppa codes using an extra place.* Preprint Eindhoven University, September 1991

[12] B.Z. Shen, *Subresultant sequence on a Weierstrass algebra and its application to decoding algebraic-geometric codes.* preprint Eindhoven University, May 1991

[13] D. Ehrhard, *Über das Dekodieren Algebraisch-Geometrischer Codes.* Thesis, Düsseldorf University, 1991.

[4] As a consequence, one doesn't need assumption 2

On the Different of Abelian Extensions of Global Fields

G. Frey, M. Perret, H. Stichtenoth

0. Introduction

Let q be a power of some prime number p, and let $\mathbf{F}_q$ be the field with q elements. Coding theorists are interested in explicitly described function fields over $\mathbf{F}_q$ having a large number of $\mathbf{F}_q$-rational places (or, equivalently, irreducible complete smooth algebraic curves over $\mathbf{F}_q$ with many $\mathbf{F}_q$-rational points). For small values of the genus, such function fields are often abelian extensions of the rational function field $\mathbf{F}_q(z)$. For instance, this is the case for Hermitian curves, some Fermat curves, and some Artin-Schreier extensions of $\mathbf{F}_q(z)$. Moreover, one way to exhibit families of function fields $E/\mathbf{F}_q$ of genus growing to infinity and having *good asymptotic behaviour* (i.e., the ratio (number of rational places/genus) has a limit > 0), is to construct a tower of function fields $E_0 \subseteq E_1 \subseteq E_2 \ldots$ over $\mathbf{F}_q$, each step E_{i+1}/E_i being Galois with an abelian Galois group. In other words, *solvable* extensions may have a good asymptotic behaviour, cf. [3].

One aim of our paper is to show that *abelian* extensions E_i/F (where F is some fixed function field over $\mathbf{F}_q$, and $\mathbf{F}_q$ is assumed to be the full constant field of F and all $E_i, i \geq 1$) are *asymptotically bad* (i.e., the ratio (number of rational places/genus) tends to 0 as the genus of $E_i/\mathbf{F}_q$ goes to infinity).

It should be pointed out that our method uses only elementary results from Hilbert's ramification theory, cf. [2,4], and the finiteness of the residue class fields. In the case of global fields, one may also use class field theory in order to obtain some results of this paper.

1. Hilbert's Ramification Theory for Locally Abelian Extensions

In this section, we consider the following situation. K is some field, $o \subseteq K$ a discrete valuation ring and $\wp \subseteq o$ the maximal ideal of o. Let L/K be a *finite abelian field extension* with Galois group G (i.e. L/K is Galois, and its Galois group G is abelian). Let $\mathcal{O} \subseteq L$ be a discrete valuation ring of L with $o \subseteq \mathcal{O}$ and maximal ideal $\mathcal{P}$, hence $\wp = \mathcal{P} \cap o$. Throughout section 1, we suppose that *$\mathcal{O}$ is the only discrete valuation ring of L containing o*. Let $k := o/\wp$ and $l := \mathcal{O}/\mathcal{P}$ denote the residue class fields of o resp. $\mathcal{O}$. Then l/k is a finite field extension, and we shall always assume *that l/k is separable.* We choose some $\mathcal{P}$-prime element $\pi \in \mathcal{P}$ (i.e., $\mathcal{P}$ is the principal ideal generated by π), and consider the groups

$$G_0 := \{\sigma \in G \mid \sigma x \equiv x \mod \mathcal{P} \quad \text{for all} \quad x \in \mathcal{O}\}$$

and, for $i \geq 1$,

$$G_i := \{\sigma \in G_0 \mid \sigma\pi \equiv \pi \mod \mathcal{P}^{i+1}\}.$$

It is well-known that the definition of G_i is independent of the choice of π, and $G \supseteq G_0 \supseteq G_1 \supseteq \ldots \supseteq G_n = \{1\}$ for sufficiently large $n \geq 1$, see [2,4]. The factor groups $\mathcal{P}^i/\mathcal{P}^{i+1}$ (for $i \geq 0$) are considered as vector spaces over l via

$$(x+\mathcal{P})\cdot(a+\mathcal{P}^{i+1}) := xa+\mathcal{P}^{i+1} \qquad (x \in \mathcal{O}, a \in \mathcal{P}^i),$$

and G acts on $\mathcal{P}^i/\mathcal{P}^{i+1}$ by

$$\tau(a+\mathcal{P}^{i+1}) := \tau(a)+\mathcal{P}^{i+1}$$

(in order to see that this action is well-defined observe that $\mathcal{O}$ is the only extension of o in L, hence $\tau(\mathcal{P}) = \mathcal{P}$ for all $\tau \in G$). We set

$$X_i := \{a+\mathcal{P}^{i+1} \in \mathcal{P}^i/\mathcal{P}^{i+1} \mid \tau(a+\mathcal{P}^{i+1}) = a+\mathcal{P}^{i+1} \quad \text{for all} \quad \tau \in G\}.$$

Clearly, X_i is a k-subspace of $\mathcal{P}^i/\mathcal{P}^{i+1}$.

Proposition 1: The dimension of X_i as a vector space over k is at most one.

Proof: By Hilbert's ramification theory l/k is a normal field extension. Due to our assumption l/k being separable we obtain that l/k is Galois. Moreover, any automorphism τ_0 in the Galois group of l/k is induced by some $\tau \in G$, i.e. $\tau_0(x+\mathcal{P}) = \tau(x)+\mathcal{P}$ for any $x+\mathcal{P} \in \mathcal{O}/\mathcal{P} = l$, see [2]. In order to prove the proposition we can assume that $X_i \neq \{0\}$. We choose $a+\mathcal{P}^{i+1} \in X_i$ with $a \in \mathcal{P}^i\backslash\mathcal{P}^{i+1}$. Since $\mathcal{P}^i/\mathcal{P}^{i+1}$ is a one-dimensional vector space over l (this is obvious) we have for all $a_1+\mathcal{P}^{i+1} \in X_i$: $a_1+\mathcal{P}^{i+1} = (c+\mathcal{P})\cdot(a+\mathcal{P}^{i+1})$ for some $c \in \mathcal{O}$. For any $\tau \in G$, the following holds:

$$\begin{aligned}(c+\mathcal{P})\cdot(a+\mathcal{P}^{i+1}) &= a_1+\mathcal{P}^{i+1} = \tau(a_1+\mathcal{P}^{i+1})\\ &= \tau(c+\mathcal{P})\cdot\tau(a+\mathcal{P}^{i+1}) = \tau(c+\mathcal{P})\cdot(a+\mathcal{P}^{i+1}),\end{aligned}$$

consequently $\tau(c+\mathcal{P}) = c+\mathcal{P}$ for any $\tau \in G$. Thus $c+\mathcal{P}$ is invariant under any automorphism of l/k, i.e. $c+\mathcal{P} \in k$. This proves Proposition 1. ∎

It is well known that the mappings

$$\psi : \begin{cases} G_0/G_1 & \longrightarrow \quad l^* \\ \sigma & \longrightarrow \quad \frac{\sigma\pi}{\pi} \mod \mathcal{P} \end{cases}$$

resp. for $i \geq 1$

$$\varphi_i : \begin{cases} G_i/G_{i+1} & \longrightarrow \quad \mathcal{P}^i/\mathcal{P}^{i+1} \\ \sigma & \longrightarrow \quad \frac{\sigma\pi}{\pi} - 1 \mod \mathcal{P}^{i+1} \end{cases}$$

are embeddings of G_0/G_1 into the multiplicative group of l (resp. of G_i/G_{i+1} into the additive group $\mathcal{P}^i/\mathcal{P}^{i+1}$), and the definition of ψ and φ_i is independent of the choice of the prime element π. For our purposes, the following refinement is essential:

Proposition 2: Under the hypotheses of this section, the image of ψ is contained in k^*, and the image of φ_i is contained in X_i for any $i \geq 1$.

Proof: (a) Let $\sigma \in G$. We have to show that $\tau_0(\psi(\sigma)) = \psi(\sigma)$ for all τ_0 in the Galois group of l/k. As before, τ_0 is induced by some $\tau \in G$, and we obtain

$$\tau_0(\psi(\sigma)) = \tau_0\left(\frac{\sigma\pi}{\pi} + \mathcal{P}\right) = \frac{\tau(\sigma\pi)}{\tau(\pi)} + \mathcal{P} = \frac{\sigma(\tau(\pi))}{\tau(\pi)} + \mathcal{P} = \psi(\sigma)$$

(we have used that G is abelian and $\tau(\pi)$ is a $\mathcal{P}$-prime element as well).
(b) An analogous argument proves that $\varphi_i(\sigma) \in X_i$ for any $\sigma \in G_i$. ∎

We recall some facts from ramification theory. Let $f = f(\mathcal{P} \mid \wp) = [l : k]$ denote the *residue class degree* and $e := e(\mathcal{P} \mid \wp)$ the *ramification index* of $\mathcal{P}$ over $\wp$, i.e. $\wp\mathcal{O} = \mathcal{P}^e$. Since $\mathcal{O}$ is the only extension of o in L and l/k is separable, we have $e \cdot f = [L : K]$. Let $s :=$ char(k) be the characteristic of the residue class field and $g_i :=$ ord G_i for any $i \geq 0$. Then G_1 is the unique s-Sylow subgroup of G_0, and $g_0 = e$. The extension $\mathcal{P} \mid \wp$ is said to be *tame* if $g_1 = 1$ (hence $(s, e) = 1$), otherwise $\mathcal{P} \mid \wp$ is *wildly ramified.*

Let $W \subseteq k^*$ be the group of roots of unity in k. If W is finite, we set $w := \#W$.

Corollary 3: In addition to the hypotheses of this section, suppose that k contains only finitely many roots of unity. If $\mathcal{P} \mid \wp$ is tame then $e \leq w$.

Proof: Consider the map $\psi : G_0 \to l^*$ as before. Since $G_1 = 1, \psi$ is a monomorphism. By Proposition 2, the image of ψ is contained in W. ∎

In order to obtain a similar estimate for the ramification index also in the case of wild ramification, we introduce the following notion: an integer $i \geq 0$ is called a *jump* (for $\mathcal{P} \mid \wp$) if $G_i \neq G_{i+1}$.

Corollary 4: In addition to the hypotheses of this section, assume that k is a finite field. Then $e \leq (\#k)^r$ where r denotes the number of jumps.

Proof: $e = g_0 = (g_0/g_1) \cdot (g_1/g_2) \cdot \ldots \cdot (g_n/g_{n+1})$ where n is chosen such that $g_{n+1} = 1$. Proposition 1 and 2 yield $g_i/g_{i+1} \leq \#k$ for any $i \geq 0$. The corollary follows immediately. ∎

Hilbert's formula [2,4] states that the *different exponent* $d := d(\mathcal{P} \mid \wp)$ is given by

$$d = \sum_{i \geq 0} (g_i - 1).$$

This formula can be restated as follows. We consider the set $\{\nu_1, \ldots, \nu_r\}$ of jumps (where $0 \leq \nu_1 < \nu_2 < \ldots < \nu_r$ and r is the number of jumps) and set

$$t_1 := \nu_1 + 1; \quad t_i := \nu_i - \nu_{i-1} \quad \text{for} \quad i = 2, \ldots, r.$$

Then

$$d = \sum_{i=1}^{r} t_i(g_{\nu_i} - 1).$$

Since G is abelian, the Hasse-Arf theorem [2] applies. It yields

$$t_i \cdot g_{\nu_i} \equiv 0 \mod e$$

for $i = 1, \ldots, r$. Combining this with Hilbert's formula we obtain

Proposition 5: Under the hypotheses of this section, the different exponent d satisfies the estimate

$$d \geq \frac{1}{2} re$$

where r denotes the number of jumps.

Proof:

$$d = \sum_{i=1}^{r} t_i g_{\nu_i}(1 - g_{\nu_i}^{-1})$$

$$\geq \frac{1}{2} \cdot \sum_{i=1}^{r} t_i g_{\nu_i} \geq \frac{1}{2} re$$

by the Hasse-Arf theorem. ∎

2. The Different of Abelian Extensions of Global Fields

In this section, F denotes a *global field.* This means that either F is a *number field,* or F is an *algebraic function field of one variable* over a finite field $\mathbf{F}_q$ (we assume that $\mathbf{F}_q$ is the full constant field of F). A *place* of F is the maximal ideal of a discrete valuation ring of F. If $\wp$ is a place of F, its corresponding valuation ring will be denoted by $o_\wp$. The residue class field $o_\wp/\wp$ is a finite field, and in the function field case we have $\mathbf{F}_q \subseteq o_\wp$. The *degree* of $\wp$ is defined by

$$\deg \wp := \log \#(o_\wp/\wp)$$

(in the number field case, log is taken with respect to the basis $e = 2,718\ldots$; if F is a function field over $\mathbf{F}_q$, we take $\log = \log_q$ - the logarithm with respect to the basis q). The definition of the degree is extended to *divisors* of F (a divisor is a formal finite sum of places) by linearity.

Let E/F be an *abelian extension* of F and $\mathrm{Gal}(E/F)$ be its Galois group. If $\wp$ is a place of F, there are $g = g(\wp)$ places $\mathcal{P}_1, \ldots, \mathcal{P}_g$ of E lying over $\wp$ (i.e. $\wp \subseteq \mathcal{P}_\nu$). All of them have the same ramification index $e(\wp) := e(\mathcal{P}_\nu \mid \wp)$ and the same residue class degree $f(\wp) := f(\mathcal{P}_\nu \mid \wp)$, and we have $e(\wp) \cdot f(\wp) \cdot g(\wp) = [E : F]$. For an extension $\mathcal{P} = \mathcal{P}_\nu$ of $\wp$ in E/F, we consider the *decomposition group*

$$G(\wp) := G(\mathcal{P} \mid \wp) := \{\sigma \in \mathrm{Gal}(E/F) \mid \sigma\mathcal{P} = \mathcal{P}\}$$

(this is independent of the choice of the extension $\mathcal{P}$ since E/F is abelian), and the *decomposition field* $Z = Z(\wp)$, i.e. $F \subseteq Z \subseteq E$ and $G(\wp) = \mathrm{Gal}(E/Z)$.

There exists the unique *maximal unramified subextension* $F \subseteq M \subseteq E$. This means that all places of F are unramified in M/F, and M is a maximal subfield of E with this property. Let $S := S(E/F)$ be the set of places of F which are ramified in E/F (it is well-known that S is finite).

Lemma 6: With the notations as above, we have

$$\sum_{\wp \in S} \log e(\wp) \geq \log[E : F] - \log[M : F].$$

Proof: For any $\wp \in S$, let $G_0(\wp) \subseteq G$ be the *inertia group* of $\wp$, see [2,4]. Its order is $e(\wp)$, and if $U \subseteq G$ is the subgroup of G generated by all $G_0(\wp)$ (with $\wp \in S$), then M is the fixed field of U. Therefore $\mathrm{ord}\, U = [E : M] = [E : F]/[M : F]$. Since G is abelian,

$$\mathrm{ord} \quad U \leq \prod_{\wp \in S} \mathrm{ord} \quad G(\wp) = \prod_{\wp \in S} e(\wp).$$

Taking logarithms yields the assertion of the lemma. ∎

Let $\mathcal{D}(E/F)$ be the *different* of E/F. The main result of this section is the following:

Theorem 7: Suppose that E/F is an abelian extension of global fields and $F \subseteq M \subseteq E$ is the maximal unramified subextension. In the function field case we assume, in addition, that E and F have the same constant field $\mathbb{F}_q$. Then the degree of the different $\mathcal{D}(E/F)$ satisfies

$$\deg \mathcal{D}(E/F) \geq \frac{1}{2}[E : F] \cdot (\log[E : F] - \log[M : F]).$$

Proof: For $\wp \in S$, let $d(\wp)$ be the different exponent of a place $\mathcal{P}$ of E lying over $\wp$, and $r(\wp)$ be the number of jumps, cf. section 1 (observe that we can apply the results of section 1 if F is replaced by the decomposition field $Z(\wp)$). We obtain

$$\begin{aligned}
\deg \mathcal{D}(E/F) &= \sum_{\wp \in S} \sum_{\mathcal{P}|\wp} d(\wp) \cdot \deg \mathcal{P} \\
&= \sum_{\wp \in S} g(\wp) \cdot d(\wp) \cdot f(\wp) \cdot \deg \wp \\
&\geq \frac{1}{2} \sum_{\wp \in S} g(\wp) \cdot r(\wp) \cdot e(\wp) \cdot f(\wp) \cdot \deg \wp \qquad \text{(by Proposition 5)} \\
&= \frac{1}{2} [E:F] \cdot \sum_{\wp \in S} r(\wp) \cdot \deg \wp \\
&\geq \frac{1}{2} [E:F] \cdot \sum_{\wp \in S} \log e(\wp) \qquad \text{(by Corollary 4)} \\
&\geq \frac{1}{2} [E:F] \cdot (\log[E:F] - \log[M:F]) \qquad \text{(by Lemma 6).}
\end{aligned}$$

■

3. Abelian Extensions of Function Fields Are Asymptotically Bad

We want to prove a slightly more general result than we announced in the introduction. For an algebraic function field $E/\mathbf{F}_q$ (with $\mathbf{F}_q$ as its full constant field) we set

$$\begin{aligned}
g(E) &= \text{genus of } \quad E \\
N(E) &= \text{number of rational places of } \quad E/\mathbf{F}_q.
\end{aligned}$$

If E/F is a Galois extension with Galois group G, we let G' be the *commutator subgroup* of G. The fixed field $E^{ab} \supseteq F$ of G' is the *maximal abelian extension* of F contained in E. In particular, if G is abelian, $G' = \{1\}$ and $E^{ab} = E$.

Theorem 8: Let $F/\mathbf{F}_q$ be an algebraic function field and $(E_\nu)_{\nu \geq 1}$ be a sequence of extension fields of F with the following properties:

(i) $\mathbf{F}_q$ is the full constant field of F and all E_ν.
(ii) E_ν/F is Galois with Galois group G_ν.
(iii) ord $(G_\nu/G'_\nu) \longrightarrow \infty$ as $\nu \longrightarrow \infty$.

Then the quotient $N(E_\nu)/g(E_\nu)$ tends to zero as $\nu \longrightarrow \infty$.

Proof: There is a constant h (the class number of F) such that any abelian unramified extension M/F with the same constant field $\mathbf{F}_q$ is of degree $[M:F] \leq h$, cf. [1]. We consider the maximal abelian extension $F_\nu \subseteq E_\nu$ of F contained in E_ν. By (iii), the degree $n_\nu := [F_\nu : F] \longrightarrow \infty$ as $\nu \longrightarrow \infty$, and the degree d_ν of the different $\mathcal{D}(F_\nu/F)$ satisfies the estimate

$$d_\nu \geq \frac{1}{2} n_\nu (\log n_\nu - \log h)$$

by Theorem 7. The Hurwitz genus formula yields

$$\begin{aligned} g(F_\nu) &\geq n_\nu (g(F) - 1) + \frac{1}{2} d_\nu \\ &\geq n_\nu (g(F) - 1 + \frac{1}{4}(\log n_\nu - \log h)). \end{aligned}$$

On the other hand, we have the trivial estimate $N(F_\nu) \leq n_\nu \cdot N(F)$, hence

$$\frac{N(F_\nu)}{g(F_\nu)} \leq \frac{N(F)}{g(F) - 1 + \frac{1}{4}(\log n_\nu - \log h)} \longrightarrow 0$$

for $\nu \longrightarrow \infty$. Eventually, since $N(E_\nu) \leq [E_\nu : F_\nu] \cdot N(F_\nu)$ and $g(E_\nu) \geq [E_\nu : F_\nu](g(F_\nu) - 1) \geq \frac{1}{2}[E_\nu : F_\nu] \cdot g(F_\nu)$ (observe that $g(F_\nu) \longrightarrow \infty$ for $\nu \longrightarrow \infty$), we obtain $N(E_\nu)/g(E_\nu) \longrightarrow 0$. ∎

References

[1] *Artin, E. and Tate, J.:* Class field theory. New York - Amsterdam 1967

[2] *Serre, J.P.:* Corps locaux. Paris 1962

[3] *Serre, J.P.:* Sur le nombre des points rationnels d'une courbe algébrique sur un corps fini. C.R. Acad.Sc. Paris, t. 296 (1983), 397-402

[4] *Zariski, O. and Samuel P.:* Commutative Algebra, Vol. I. Princeton 1958

Gerhard Frey
Institut für Experimentelle Mathematik
Universität GHS Essen
Ellernstr. 29, D-4300 Essen 12
Germany

Marc Perret
Equipe Arithmétique et Théorie de l'Information
CIRM, Luminy Case 916
F-13288 Marseille Cedex 9
France

Henning Stichtenoth
Fachbereich 6, Universität GHS Essen
D-4300 Essen 1
Germany

Goppa Codes and Weierstrass Gaps

Arnaldo Garcia
IMPA
Estrada Dona Castorina 110
22.460 Rio de Janeiro
Brasil

R.F. Lax
Department of Mathematics
LSU
Baton Rouge, LA 70803
USA

We generalize an example of Goppa to show how the gap sequence at a point may often be used to define Goppa codes that have minimum distance greater than the usual lower bound.

1. Let X denote a nonsingular, geometrically irreducible, projective curve of genus g defined over the finite field $\mathbf{F}_q$ with q elements. Assume that X has $\mathbf{F}_q$-rational points. Let D be a divisor on X defined over $\mathbf{F}_q$ (i.e., D is invariant under $\mathrm{Gal}(\overline{\mathbf{F}}_q/\mathbf{F}_q)$). Then $L(D)$ will denote the $\mathbf{F}_q$-vector space of all rational functions f on X, defined over $\mathbf{F}_q$, with divisor $(f) \geq -D$, together with the zero function, and $\Omega(D)$ will denote the $\mathbf{F}_q$-vector space of all rational differentials η on X, defined over $\mathbf{F}_q$, with divisor $(\eta) \geq D$, together with the zero differential. Put $l(D) = \dim_{\mathbf{F}_q} L(D)$ and $i(D) = \dim_{\mathbf{F}_q} \Omega(D)$. The Riemann-Roch Theorem states that

$$l(D) = \deg D + 1 - g + i(D).$$

Also, we will write D_0 (resp. D_∞) for the divisor of zeros (resp. divisor of poles) of D. Hence we have

$$D_0 \geq 0, D_\infty \geq 0, (\mathrm{Supp}\ D_0) \cap (\mathrm{Supp}\ D_\infty) = \emptyset, \text{ and } D = D_0 - D_\infty.$$

V. D. Goppa [3,4] realized that one could use divisor theory on a curve to define nice codes. A q-ary linear code of length n and dimension k is a vector subspace of dimension k of $\mathbf{F}_q^n$. The minimum distance of a code is the minimum number of places in which two distinct codewords differ. For a linear code, the minimum distance is also the minimum weight of a nonzero codeword, where the weight of a codeword is the number of nonzero places in that codeword. A linear code of length n, dimension k and minimum distance d is called an $[n, k, d]$-code. Let C denote an $[n, k, d]$-code over $\mathbf{F}_q$. A generator matrix of C is a $k \times n$ matrix whose rows form a basis for the code. A parity check matrix for C is an $(n-k) \times n$ matrix B of rank $n-k$ such that $AB^t = 0$ for some generator matrix A of C. Two codes C and C^* are called dual if a generator matrix of C^* is a parity check matrix of C. The code C has minimum distance d if and only if in any parity check matrix for C every $d-1$ columns are linearly independent and some d columns are linearly dependent.

The first author was supported by the Alexander von Humboldt-Stiftung while visiting Universität GHS Essen within the GMD-CNPq exchange program. The second author was partially supported by a grant from the Louisiana Education Quality Support Fund through the Board of Regents.

Let G be a divisor on X defined over $\mathbf{F}_q$ and let $D = P_1 + P_2 + \cdots + P_n$ be another divisor on X where $P_1, \ldots, P_n$ are distinct $\mathbf{F}_q$-rational points and none of the P_i is in the support of G. The geometric Goppa code $C(D, G)$ (cf. [6]) is the image of the linear map $\alpha : L(G) \to \mathbf{F}_q^n$ defined by

$$f \longmapsto (f(P_1), f(P_2), \ldots, f(P_n)).$$

The geometric Goppa code $C^*(D, G)$ (cf. [6]) is the image of the linear map $\alpha^* : \Omega(G - D) \to \mathbf{F}_q^n$ defined by

$$\eta \longmapsto (\mathrm{res}_{P_1}(\eta), \mathrm{res}_{P_2}(\eta), \ldots, \mathrm{res}_{P_n}(\eta)).$$

The codes $C(D, G)$ and $C^*(D, G)$ are dual codes ([6, II.3.3]). Alternatively, if one fixes a nonzero rational differential ω with (canonical) divisor K, then, the image of α^* is the same as the image of the map $\beta^* : L(K + D - G) \to \mathbf{F}_q$ defined by

$$f \longmapsto (\mathrm{res}_{P_1}(f\omega), \mathrm{res}_{P_2}(f\omega), \ldots, \mathrm{res}_{P_n}(f\omega)).$$

We will be mainly interested in the code $C^*(D, G)$. This code has length n and dimension

$$k = \dim\, L(K + D - G) - \dim\, L(K - G).$$

In particular, if G has degree greater than $2g - 2$, then $k = \dim\, L(K + D - G)$. If $k > 0$, then the minimum distance d of $C^*(D, G)$ satisfies the well-known bound

$$d \geq \deg(G) - 2g + 2.$$

2. Now let P denote a (closed) $\mathbf{F}_q$-rational point on X and let B be a divisor defined over $\mathbf{F}_q$. We call a natural number γ a B-gap at P if there is no rational function f on X such that

$$((f) + B)_\infty = \gamma P.$$

We say that $\gamma - 1$ is an order at P for the divisor $K - B$ (where K denotes a canonical divisor on X) if we have

$$K - B \sim (\gamma - 1)P + E,$$

where $E \geq 0$, $P \notin \mathrm{Supp}(E)$, and $\sim$ denotes linear equivalence. It follows from the Riemann-Roch Theorem that γ is a B-gap at P if and only if $\gamma - 1$ is an order at P for the divisor $K - B$. Note that with this terminology, the usual Weierstrass gaps at P are the 0-gaps at P.

Let γ_j and γ_k be B-gaps at P. Put

$$G = (\gamma_j + \gamma_k - 1)P + 2B. \tag{1}$$

Our divisor D used to define the code $C^*(D, G)$ will be of the form

$$D = P_1 + P_2 + \cdots + P_n,$$

where the P_i are n distinct $\mathbf{F}_q$-rational points, each not belonging to the support of G.

(2.1) THEOREM. Assume that the dimension of $C^*(D,G)$ is positive (where G is of the form specified in (1)). Then the minimum distance of $C^*(D,G)$ is at least $\deg(G) - 2g + 3$.

PROOF. Put $w = \deg G - (2g-2)$. If there exists a codeword of weight w, then there exists $\eta \in \Omega(G-D)$ with exactly w simple poles $P_1, P_2, \ldots, P_w$ in $\mathrm{Supp}(D)$. We then have

$$(\eta)_0 \geq G_0 \text{ and } (\eta)_\infty \leq G_\infty + P_1 + P_2 + \cdots + P_w.$$

Hence, $2g-2 = \deg(\eta) \geq \deg G_0 - \deg G_\infty - w = 2g-2$. It follows that $(\eta)_0 = G_0$ and $(\eta)_\infty = G_\infty + P_1 + P_2 + \cdots + P_w$. Thus there exists a canonical divisor K of the form

$$K = G - (P_1 + P_2 + \cdots + P_w). \tag{2}$$

Since γ_k is a B-gap at P, we have

$$K - B \sim (\gamma_k - 1)P + E, \tag{3}$$

where $E \geq 0$ and $P \notin \mathrm{Supp}(E)$. From (1), (2), and (3), we have

$$E + (P_1 + P_2 + \cdots + P_w) - (B + \gamma_j P) \sim 0.$$

Thus there exists a rational function f such that $((f) + B)_\infty = \gamma_j P$, contradicting the fact that γ_j is a B-gap at P. ∎

(2.2) REMARK. If $B = 0$, then $\deg G = \gamma_j + \gamma_k - 1$ and it is natural to ask which integers can be written as the sum of two Weierstrass gaps. In this direction, G. Oliveira [7] showed that if X is nonhyperelliptic, then each r, for $2 \leq r \leq 2g$, is the sum of two gaps, with the exception of $r = 2g-1$ in the case that $\gamma_g = 2g-1$.

Let $\varphi : X \to \mathbf{P}^N$ be a morphism with nondegenerate image. As in [10], we may view $\varphi : X \to \mathbf{P}^N$ as a parametrized curve in $\mathbf{P}^N$ and the points of X as its branches. To any point $P \in X$, one associates a sequence of natural numbers

$$\epsilon_0(P) < \epsilon_1(P) < \cdots < \epsilon_N(P),$$

which are the possible intersection multiplicities of the curve $\varphi(X)$ with the hyperplanes in $\mathbf{P}^N$ at the branch centered at P. We denote by $L_P^{(j-1)}X$ (cf. [10]) the osculating space of dimension $j-1$ at P (associated to the morphism φ); thus,

$$L_P^{(j-1)}X = \bigcap \{H : H \text{ is a hyperplane with } I(P; \varphi(X) \cdot H) \geq \epsilon_j(P)\},$$

where $I(P; \varphi(X) \cdot H)$ denotes intersection number.

Assume from now on that the linear system $|K - B|$ is base-point-free. If $\gamma_1(P) < \gamma_2(P) < \cdots < \gamma_{N+1}(P)$ are the B-gaps at P, then for the morphism φ associated to the linear system $|K - B|$, we have

$$\epsilon_j(P) = \gamma_{j+1}(P) - 1.$$

(2.3) THEOREM. With notation as in Theorem (2.1), assume moreover that $\gamma_{j+1} = \gamma_j + 1$ and that the osculating space $L_P^{(j-1)}X$ (with respect to the morphism defined by $|K - B|$) does not meet the curve $\varphi(X)$ at another $\mathbf{F}_q$-rational point. If $C^*(D, G)$ has positive dimension, then its minimum distance is at least $\deg G - 2g + 4$.

PROOF. By Theorem (2.1), the minimum distance of $C^*(D,G)$ is at least $\deg G - 2g + 3$. Put $w = \deg G - 2g + 3$ and suppose that there is a codeword of weight w. Proceeding as in the proof of Theorem 1, we see that there is a canonical divisor K of the form

$$K = G + Q - (P_1 + P_2 + \cdots + P_w), \tag{4}$$

where Q is an $\mathbf{F}_q$-rational point on X. Using equations (1), (3), and (4), we get

$$B \sim -\gamma_j P - Q + (P_1 + P_2 + \cdots + P_w) + E.$$

Since γ_j and γ_j+1 are both B-gaps at P, we see that $Q \notin \{P, P_1, P_2, \ldots, P_w\} \cup \operatorname{Supp}(E)$. Hence,

$$L(B + \gamma_j P) \neq L(B + \gamma_j P + Q).$$

By Riemann-Roch, we then have

$$L(K - B - \gamma_j P) = L(K - B - \gamma_j P - Q).$$

This equality means that for any divisor A linearly equivalent to $K - B$, if $A \geq \gamma_j P$, then $A \geq \gamma_j P + Q$. Now, $\epsilon_j(P) = \gamma_{j+1} - 1 = \gamma_j$, so if $A \geq \epsilon_j(P)P$, then $A \geq \epsilon_j(P)P + Q$. It follows from this that Q is in $L_P^{(j-1)}X$, contradicting our hypothesis. ∎

(2.4) REMARK. Suppose $B = 0$; i.e., $\varphi : X \to \mathbf{P}^{g-1}$ is the canonical morphism. If at an $\mathbf{F}_q$-rational point P, we have $\gamma_g = 2g - 1$, then $L_P^{(g-2)}X$, the osculating hyperplane at P, meets X only at P and hence $L_P^{(j-1)}X$, for $j = 1, 2, \ldots, g-1$, meets X only at P, so the condition on the osculating space in Theorem (2.3) will be satisfied. The next result gives another useful criterion for verifying this osculating space condition.

(2.5) PROPOSITION. Suppose that $B = 0$ and that X is not hyperelliptic. For $j \in \{1, 2, \ldots, g-2\}$, we have

$$\gamma_{j+1} = 2j \Longrightarrow (L_P^{(j-1)}X) \cap X = \{P\}.$$

PROOF. First we remark that $\gamma_{j+1} \leq 2j$ for $j = 1, 2, \ldots, g-2$. For suppose $\gamma_{j+1} > 2j$. Then there are at most j gaps less than or equal to $2j$. Hence the dimension of $L(2jP)$ would be at least $2j + 1 - j = j + 1$, contradicting Clifford's Theorem.

Now, the intersection divisor of X and $L_P^{(j-1)}X$ is of the form $A = (\gamma_{j+1} - 1)P + E$, where $E \geq 0$ and $P \notin \operatorname{Supp}(E)$. By the geometric version of the Riemann-Roch Theorem [1, p.12], we have

$$\dim L(A) = \deg A - \dim L_P^{(j-1)}X = \deg A - (j-1). \tag{5}$$

From Clifford's Theorem, we have

$$\dim L(A) < (\deg A)/2 + 1. \tag{6}$$

From (5) and (6), we get $\deg A \leq 2j-1$, which implies that $\deg E$ is at most $2j-\gamma_{j+1}$. So, if $\gamma_{j+1}=2j$, then $E=0$ and the Proposition follows. ∎

We note that Proposition (2.5) was proven by G. Oliveira [7] in the case $j=g-2$.

(2.6) EXAMPLE. We give an example to show that the converse of Proposition (2.5) is false, and that one can sometimes deduce the fact that $L_P^{(j-1)} \cap X = \{P\}$ by testing the condition $\gamma_{l+1}=2l$ for some $l>j$.

Let X denote the nonsingular curve with function field $\mathbf{F}_{64}(x,y)$ where $y^8+y=x^3$. If $P=(1,\alpha)$, with $\alpha^8+\alpha=1$, is an $\mathbf{F}_{64}$-rational point, then by Garcia-Viana [2] the gap sequence at P is 1, 2, 3, 4, 5, 10, 11. Taking $l=5$, we see that $\gamma_{l+1}=10=2l$; hence, $L_P^{(4)}X$ meets X only at P. It follows that $L_P^{(3)}X$ must also meet X only at P, but notice that for $j=4$, we have $\gamma_{j+1}=5\neq 2j=8$.

In the case where the next t consecutive integers after γ_j are also B-gaps, one can prove results similar to Theorem (2.3) by assuming more conditions on the osculating spaces.

(2.7) THEOREM. Suppose that $\gamma_{j+t}=\gamma_j+t$ for $t=1,2$. Suppose that $L_P^{(j)}X$ does not meet X in another $\mathbf{F}_q$-rational point. Also suppose that $L_P^{(j-1)}X$ does not meet any of the following lines:

(i) A line joining two other $\mathbf{F}_q$-rational points.
(ii) A line joining two $\mathbf{F}_{q^2}$-rational points that are conjugate over $\mathbf{F}_q$.
(iii) A tangent line at another $\mathbf{F}_q$-rational point.

If $C^*(D,G)$ has positive dimension, then the minimum distance of $C^*(D,G)$ is at least $\deg G-2g+5$.

PROOF. Put $w=\deg G-2g+4$. Since $L_P^{(j)}X$ does not meet X at another $\mathbf{F}_q$-rational point, the same is true for $L_P^{(j-1)}X$. Thus by Theorem (2.3), the code $C^*(D,G)$ has minimum distance at least w. Suppose that there is a codeword of weight w. Then there is a canonical divisor K of the form

$$K=G+A-(P_1+P_2+\cdots+P_w), \tag{7}$$

for some positive divisor $A=Q_1+Q_2$ of degree 2 defined over $\mathbf{F}_q$. From (1), (3), and (7), we have

$$B\sim -\gamma_jP-(Q_1+Q_2)+(P_1+P_2+\cdots+P_w)+E, \tag{8}$$

where $E\geq 0$ and $P\notin \operatorname{Supp}(E)$.

Now, either Q_1 and Q_2 are both $\mathbf{F}_q$-rational points or they are $\mathbf{F}_{q^2}$-rational points that are conjugate over $\mathbf{F}_q$. Since γ_j+2 is a B-gap at P, we cannot have $Q_1=Q_2=P$. By applying the proof of Theorem (2.3) to the divisor $G+P$ (note that γ_{j+1} and $\gamma_{j+1}+1$ are B-gaps), we see that neither Q_1 nor Q_2 can equal P, since then the other point would lie in $L_P^{(j)}X$. Also, we have that neither Q_1 nor Q_2 is in $\{P_1,P_2,\ldots,P_w\}\cup \operatorname{Supp}(E)$. It then follows from (8) that we have

$$L(B+\gamma_jP+Q_1)\neq L(B+\gamma_jP+Q_1+Q_2),$$

where these are now vector spaces of rational functions defined over $\mathbf{F}_{q^2}$. It follows from the Riemann-Roch Theorem that

$$L(K-B-\gamma_j P-Q_1)=L(K-B-\gamma_j P-Q_1-Q_2).$$

Thus for any divisor $H\sim(K-B)$, we have

$$H\geq\gamma_j P+Q_1=\epsilon_j(P)P+Q_1\Longrightarrow H\geq\gamma_j P+Q_1+Q_2=\epsilon_j(P)P+Q_1+Q_2.$$

Therefore, the osculating space $L_P^{(j-1)}X$ meets the line joining Q_1 and Q_2, contradicting our hypothesis. ∎

The conditions on $L_P^{(j-1)}X$ in Theorem (2.7) can be expressed as: $L_P^{(j-1)}X$ misses all the lines determined by degree two $\mathbf{F}_q$-rational divisors A, where $P\notin\operatorname{Supp}(A)$. By using induction, one can generalize Theorem (2.7) in the case of t consecutive gaps after γ_j to obtain the following result.

(2.8) THEOREM. Suppose that $\gamma_{j+t}=\gamma_j+t$ for some natural number t. Suppose that for each $s\in\{0,1,\ldots,t-1\}$ the osculating space $L_P^{(j-1+s)}X$ misses the linear space spanned by each degree $t-s$ $\mathbf{F}_q$-rational divisor with support disjoint from $\{P\}$. If $C^*(D,G)$ has positive dimension, then its minimum distance is at least $\deg G-(2g-2)+(t+1)$.

3. We present some examples illustrating the theorems in the previous section.

(3.1) EXAMPLE. Goppa [4, pp. 139–145] considers the plane quintic $y^4z+yz^4+x^5-x^2z^3$ over the field $\mathbf{F}_4$. This curve has 17 points over $\mathbf{F}_4$. Put $P=(0,1,0)$. Then the gap sequence at P is shown to be 1,2,3,6,7,11. Goppa shows (p. 144) that the code $C^*(D,16P)$, where D is the sum of the remaining 16 $\mathbf{F}_4$-rational points, has minimum distance at least 8. This follows from Theorem (2.3). Our proofs of Theorems (2.1) and (2.3) are essentially generalizations of Goppa's argument.

(3.2) EXAMPLE. Let X denote the curve $y^q+y=x^{q+1}$ defined over $\mathbf{F}_{q^2}$, where $q\geq 5$. This is an example of a Hermitian curve. There are q^3+1 $\mathbf{F}_{q^2}$-rational points on X, the maximum possible (by the Hasse-Weil bound) on a smooth curve of genus $g=(q-1)q/2$ over $\mathbf{F}_{q^2}$. Let P be an $\mathbf{F}_{q^2}$-rational point on X and let D denote the sum of the remaining q^3 $\mathbf{F}_{q^2}$-rational points. The codes $C(D,sP)$ have been studied extensively by Tiersma [11] and Stichtenoth [9], and recently Yang and Kumar [12] have determined the exact minimum distances of these codes. In this example, we show how to use Theorem (2.7) to give an alternate derivation of some of the results of Yang and Kumar.

Let $P=(a,b)$ be an $\mathbf{F}_{q^2}$-rational point on X. The canonical morphism $\varphi:X\to\mathbf{P}^{g-1}$ given by a Hermite basis at P of the regular differentials (cf. [2,8]) is $\varphi((x,y))=$

$$(1:(x-a):\cdots:\tilde{P}^{q-4}:(x-a)\tilde{P}^{q-4}:(x-a)^2\tilde{P}^{q-4}:\tilde{P}^{q-3}:(x-a)\tilde{P}^{q-3}:\tilde{P}^{q-2}),$$

where $\tilde{P}=(y-b)-a^q(x-a)$ is the tangent line at P. We note that $\tilde{P}$ meets X only at P (with multiplicity $q+1$). Denoting by P_∞ the point on X at infinity, we have that

$\varphi(P_\infty) = (0 : 0 : \cdots : 0 : 1)$. By [8] (or see [2]), the Weierstrass nongaps at P smaller than $2g$ are given by:

$$\begin{array}{l} 0 \\ q, q+1 \\ 2q, 2q+1, 2(q+1) \\ \vdots \\ (q-4)q, (q-4)q+1, \ldots, (q-4)(q+1) \\ (q-3)q, (q-3)q+1, \ldots, (q-3)(q+1) \\ (q-2)q, (q-2)q+1, \ldots, (q-2)(q+1) \end{array}$$

Take $j = g-5$, so that $\gamma_j = (q-4)(q+1)+1$. Note that γ_j, γ_j+1, and γ_j+2 are three consecutive gaps.

Let G_k denote the divisor $(\gamma_j + \gamma_k - 1)P = (\gamma_k + (q-4)(q+1))P$ and let D denote the sum of the other q^3 $\mathbf{F}_{q^2}$-rational points on X. The usual lower bound on Goppa codes would show that the minimum distance of $C^*(D, G_k)$ is at least $\deg G_k - (2g-2) = \gamma_k - 2(q+1)$. We claim that the conditions of Theorem (2.7) are satisfied here, so that the minimum distance of $C^*(D, G_k)$ is at least $\gamma_k - 2(q+1)+3$.

To establish this claim, first note that $L_P^{(g-2)}X \cap X = \{P\}$, since $\gamma_g = 2g-1$. Next, we will see that $L_P^{(j-1)}X = L_P^{(g-6)}X$ (i.e., the osculating space of codimension 5 at P) does not meet any line determined by two (distinct) $\overline{\mathbf{F}}_q$-rational points (different from P). Since φ is given by the Hermite basis associated to the point P, the osculating space $L_P^{(g-6)}X$ in $\mathbf{P}^{g-1}$ is given by $X_{g-1} = X_{g-2} = \cdots = X_{g-5} = 0$ (where the X_i are homogeneous coordinates). Suppose $\varphi(x_1, y_1)$ and $\varphi(x_2, y_2)$ are distinct points on $\varphi(X)$, both different from $\varphi(P)$. The line determined by these points is

$$\{\alpha\varphi(x_1, y_1) + \beta\varphi(x_2, y_2) : \alpha, \beta \in \overline{\mathbf{F}_q}\}.$$

Put

$$\tilde{P}_i = \tilde{P}(x_i, y_i) = (y_i - b) - a^q(x_i - a) \text{ for } i = 1, 2.$$

If there is a point on the above line that also lies in $L_P^{(g-6)}X$, then we have

$$(*) \qquad \begin{cases} \alpha\tilde{P}_1^{q-3} + \beta\tilde{P}_2^{q-3} = 0 \\ \alpha(x_1 - a)\tilde{P}_1^{q-3} + \beta(x_2 - a)\tilde{P}_2^{q-3} = 0 \\ \alpha\tilde{P}_1^{q-2} + \beta\tilde{P}_2^{q-2} = 0 \end{cases}$$

Note that $\tilde{P}_1 \neq 0$ and $\tilde{P}_2 \neq 0$; otherwise, there would be another point on the curve (the point (x_i, y_i)) lying on the tangent line to the plane curve X at the point (a, b).

Multiplying the first equation in $(*)$ by $\tilde{P}_1$ and comparing the resulting equation with the third equation in $(*)$, we have $\tilde{P}_1 = \tilde{P}_2$. Multiplying the first equation in $(*)$ by $(x_1 - a)$ and comparing the resulting equation with the second equation in $(*)$, we see that $x_1 = x_2$. From $\tilde{P}_1 = \tilde{P}_2$ and $x_1 = x_2$, we may conclude that $y_1 = y_2$.

This contradicts the assumption that $\varphi(x_1, y_1)$ and $\varphi(x_2, y_2)$ were distinct points. One argues similarly if one of the two points (x_i, y_i) is P_∞.

Finally, suppose (x_1, y_1) is an $\mathbf{F}_{q^2}$-rational point distinct from the point (a, b). In order to determine the tangent line to $\varphi(X)$ at $\varphi(x_1, y_1)$, we take derivatives with respect to x:

$$\frac{dy}{dx}(x_1, y_1) = x_1^q \text{ and } \frac{d\tilde{P}}{dx}(x_1, y_1) = (x_1 - a)^q.$$

Put $\tilde{P}_1 = \tilde{P}(x_1, y_1)$. From the definition of φ, the tangent line to $\varphi(X)$ at $\varphi(x_1, y_1)$ is $\{\alpha\,\varphi(x_1, y_1) + \beta V : \alpha, \beta \in \overline{\mathbf{F}_q}\}$, where $V =$

$$(0 : 1 : \cdots : (q-3)(x_1 - a)^q \tilde{P}_1^{q-4} : \tilde{P}_1^{q-3} + (q-3)(x_1 - a)^{q+1}\tilde{P}_1^{q-4} : (q-2)(x_1 - a)^q \tilde{P}_1^{q-3}).$$

Suppose there is a point Q on this tangent line that also lies in $L_P^{(g-6)}X$. Since no rational point on X except (a, b) lies in this osculating space, we see that $\beta \neq 0$ at Q, so we may assume $\beta = 1$ at Q. Then at Q we must have

$$\begin{cases} \alpha \tilde{P}_1^{q-3} + (q-3)(x_1 - a)^q \tilde{P}_1^{q-4} = 0 \\ \alpha(x_1 - a)\tilde{P}_1^{q-3} + \tilde{P}_1^{q-3} + (q-3)(x_1 - a)^{q+1}\tilde{P}_1^{q-4} = 0 \end{cases}$$

Multiplying the first equation by $(x_1 - a)$ and comparing the resulting equation with the second equation, we see that $\tilde{P}_1 = 0$. But this says that (x_1, y_1) belongs to the tangent line to X at (a, b), a contradiction. Again, one can argue similarly if (x_1, y_1) is P_∞.

Since the projective change of coordinates $(x : y : z) \mapsto (x : z : y)$ induces an automorphism of X that takes $(0 : 1 : 0)$ to $(0 : 0 : 1)$, we see that the conditions of Theorem (2.7) are also satisfied at P_∞. Put d_k equal to the minimum distance of $C^*(D, G_k)$. If we take $\gamma_k = (2 + m)q - 1$, where $1 \le m \le q - 3$, then by Theorem (2.7), we have $d_k \ge mq$. Put $z = (x - a_1)(x - a_2) \cdots (x - a_m)$, where $a_1, a_2, \ldots, a_m$ are distinct elements in $\mathbf{F}_{q^2}$. Then the differential dx/z has a zero of order $(q^2 - q - 2) + mq > \deg G_k$ at P_∞ and has exactly mq simple poles, so it is a codeword of weight mq. Thus, the minimum distance of $C^*(D, G_k)$ will be exactly mq when $\gamma_k = (2 + m)q - 1$. Using Riemann-Roch to compute the dimension of this code, we see that $C^*(D, G_k)$ is a

$$[q^3, q^3 + 4 - q(q + 2m - 1)/2, mq]\text{-code}$$

when $\gamma_k = (2 + m)q - 1$.

4. Although our theorems show that the minimum distance of the code $C^*(D, sP)$ is related to the Weierstrass gap sequence at P, we close with an example to demonstrate that this minimum distance does not depend solely on the gaps. Specifically, we give a curve with two points with the same Weierstrass gaps such that corresponding Goppa codes have different minimum distances.

(4.1) EXAMPLE. Let X denote the nonsingular plane quartic

$$x^4 + y^3 z + x^2 z^2 + x z^3 + 2 x y^2 z = 0$$

over the field $\mathbf{F}_5$. Then X has the following eight $\mathbf{F}_5$-rational points:

$$P_1 = (0,1,0),\ P_2 = (3,1,1),\ P_3 = (4,1,1)$$
$$P_4 = (3,2,1),\ P_5 = (4,3,1),\ P_6 = (2,4,1)$$
$$P = (0,0,1),\ \text{and } P' = (2,3,1).$$

The points P and P' are (ordinary) flexes of X, so the Weierstrass gap sequence at each of these two points is 1,2,4. Put $D = P_1+P_2+\cdots+P_6+P'$ and $D' = P_1+P_2+\cdots+P_6+P$.

The functions 1 and z/x are a basis for $L(3P)$ and a generator matrix for the code $C(D,3P)$ is

$$\begin{pmatrix} 1 & 1 & 1 & 1 & 1 & 1 & 1 \\ 0 & 2 & 4 & 2 & 4 & 3 & 3 \end{pmatrix}.$$

It can then be seen that the weight enumerator (cf. [5, p. 38]) of $C(D,3P)$ is

$$1 + 12z^5 + 4z^6 + 8z^7.$$

By the MacWilliams identity (cf. [5, p. 39]), the weight enumerator of the dual code $C^*(D,3P)$ is

$$1 + 12z^2 + 80z^3 + 400z^4 + 804z^5 + 1180z^6 + 648z^7.$$

The functions 1 and $(x+z)/(y+2z)$ form a basis of $L(3P')$. A generator matrix for $C(D',3P')$ is then

$$\begin{pmatrix} 1 & 1 & 1 & 1 & 1 & 1 & 1 \\ 0 & 3 & 0 & 1 & 0 & 3 & 3 \end{pmatrix}.$$

The weight enumerator of $C(D',3P')$ is then

$$1 + 8z^4 + 4z^6 + 12z^7.$$

Thus $C(D,3P)$ and $C(D',3P')$ have different minimum distances. However, the weight enumerator of $C^*(D',3P')$ is

$$1 + 24z^2 + 60z^3 + 360z^4 + 924z^5 + 1080z^6 + 676z^7,$$

and so $C^*(D',3P')$ has the same minimum distance as $C^*(D,3P)$.

But now consider the divisors $5P$ and $5P'$. The functions $1, z/x$, and $(x+y)z/x^2$ form a basis for $L(5P)$. A generator matrix for $C(D,5P)$ is

$$\begin{pmatrix} 1 & 1 & 1 & 1 & 1 & 1 & 1 \\ 0 & 2 & 4 & 2 & 4 & 3 & 3 \\ 0 & 1 & 0 & 0 & 2 & 4 & 0 \end{pmatrix}.$$

The weight enumerator of $C(D,5P)$ is then

$$1 + 8z^3 + 36z^5 + 64z^6 + 16z^7$$

and the weight enumerator of the dual code $C^*(D,5P)$ is

$$1 + 32z^3 + 52z^4 + 156z^5 + 272z^6 + 112z^7.$$

The functions $1, (x+z)/(y+2z)$, and $(x+z)(x+3z)/(y+2z)^2$ form a basis for $L(5P')$. A generator matrix for $C(D', 5P')$ is

$$\begin{pmatrix} 1 & 1 & 1 & 1 & 1 & 1 & 1 \\ 0 & 3 & 0 & 1 & 0 & 3 & 3 \\ 0 & 1 & 0 & 4 & 2 & 0 & 2 \end{pmatrix}.$$

The weight enumerator of $C(D', 5P')$ is then

$$1 + 4z^3 + 16z^4 + 32z^5 + 40z^6 + 32z^7$$

and the weight enumerator of $C^*(D', 5P')$ is

$$1 + 4z^2 + 12z^3 + 72z^4 + 176z^5 + 232z^6 + 128z^7.$$

In particular, the minimum distance of $C^*(D, 5P)$ is 3, while the minimum distance of $C^*(D', 5P')$ is only 2.

REFERENCES

[1] E. Arbarello, M. Cornalba, P.A. Griffiths, J. Harris, Geometry of algebraic curves, Volume I, Springer-Verlag, New York, 1985.

[2] A. Garcia and P. Viana, Weierstrass points on certain non-classical curves, Arch. Math. 46 (1986), 315–322.

[3] V. D. Goppa, Algebraico-geometric codes, Math. USSR-Izv. 21, no. 1, (1983), 75–91.

[4] V. D. Goppa, Geometry and codes, Kluwer, Dordrecht,1988.

[5] J. H. van Lint, Introduction to coding theory, Springer-Verlag, New York, 1982.

[6] J. H. van Lint and G. van der Geer, Introduction to coding theory and algebraic geometry, Birkhäuser, Basel,1988.

[7] G. Oliveira, Weierstrass semigroups and the canonical ideal of non-trigonal curves, Manus. Math. 71 (1991), 431–450.

[8] F.K. Schmidt, Zur arithmetischen Theorie der algebraischen Funktionen II. Allgemeine Theorie der Weierstrasspunkte, Math. Z. 45 (1939), 75–96.

[9] H. Stichtenoth, A note on Hermitian codes over $GF(q^2)$, IEEE Trans. Inform. Theory 34, no. 5 (1988), 1345–1348.

[10] K.-O. Stöhr and J.F. Voloch, Weicrstrass points and curves over finite fields, Proc. London Math. Soc. (3), 52 (1986), 1–19.

[11] H.J. Tiersma, Remarks on codes from Hermitian curves, IEEE Trans. Inform. Theory, IT–33 (1987), 605–609.

[12] K. Yang and P.V. Kumar, On the true minimum distance of Hermitian codes, these proceedings.

On a Characterization of Some Minihypers in PG(t,q) (q=3 or 4) and its Applications to Error-Correcting Codes

Noboru Hamada[1]
Department of Applied Mathematics
Osaka Women's University
Sakai, Osaka, Japan 590

Tor Helleseth[2]
Department of Informatics
University of Bergen
N-5020 Bergen, Norway

Abstract

A set F of f points in a finite projective geometry $PG(t,q)$ is an $\{f,m;t,q\}$–minihyper if m (≥ 0) is the largest integer such that all hyperplanes in $PG(t,q)$ contain at least m points in F where $t \geq 2$, $f \geq 1$ and q is a prime power. Hamada and Deza [9], [11] characterized all $\{2v_{\alpha+1}+2v_{\beta+1}, 2v_\alpha+2v_\beta; t,q\}$–minihypers for any integers t, q, α and β such that $q \geq 5$ and $0 \leq \alpha < \beta < t$ where $v_l = (q^l-1)/(q-1)$ for any integer $l \geq 0$. Recently, Hamada [5], [6] and Hamada, Helleseth and Ytrehus [18] characterized all $\{2v_1+2v_2, 2v_0+2v_1; t,q\}$–minihypers for the case $t \geq 2$ and $q \in \{3,4\}$. The purpose of this paper is to characterize all $\{2v_{\alpha+1}+2v_{\beta+1}, 2v_\alpha+2v_\beta; t,q\}$–minihypers for any integers t, q, α and β such that $q \in \{3,4\}$, $0 \leq \alpha < \beta < t$ and $\beta \neq \alpha+1$ using several results in Hamada and Helleseth [12], [13], [14], [16], [17].

1. Introduction

Let F be a set of f points in a finite projective geometry $PG(t,q)$ of t dimensions where $t \geq 2$ and q is a prime power. If $|F \cap H| \geq m$ for any hyperplane H in $PG(t,q)$ and $|F \cap H| = m$ for some hyperplane H in $PG(t,q)$, then F is called an $\{f,m;t,q\}$–minihyper where $m \geq 0$ and $|A|$ denotes the number of elements in the set A. In the special case $t=2$ and $m \geq 2$, an $\{f,m;2,q\}$–minihyper F is also called an m–blocking set if F contains no 1–flat in $PG(2,q)$. The concept of a minihyper (or a min•hyper) has been introduced by Hamada and Tamari [19].

[1] Partially supported by Grant-in-aid for Scientific Research of the Ministry of Education, Science and Culture under Contract Numbers 403-4005-02640182.

[2] Partially supported by the Scandinavia Japan Sasakawa Foundation. Partially supported by the Norwegian Research Council for Science and the Humanities

In the case $k \geq 3$ and $1 \leq d \leq q^{k-1} - q$, d can be expressed uniquely as follows:

$$d = q^{k-1} - \left(\varepsilon + \sum_{i=1}^{h} q^{\mu_i}\right)$$

using some ordered set $(\varepsilon, \mu_1, \mu_2, \cdots, \mu_h)$ in $U(k-1, q)$ where $U(t, q)$ denotes the set of all ordered sets $(\varepsilon, \mu_1, \mu_2, \cdots, \mu_h)$ of integers ε, h and μ_i such that (a) $0 \leq \varepsilon \leq q-1$, $1 \leq h \leq (t-1)(q-1)$, $1 \leq \mu_1 \leq \mu_2 \leq \cdots \leq \mu_h < t$ and (b) at most $q-1$ of the μ_i's take the same value.

Hamada [4] showed that in order to characterize all $[n, k, d; q]$–codes meeting the Grismer bound (cf. [2], [23]) for the case $k \geq 3$ and $d = q^{k-1} - \left(\varepsilon + \sum_{i=1}^{h} q^{\mu_i}\right)$ it is sufficient to solve Problem 1.1 below. The connection between codes meeting the Griesmer bound and minihypers is explained in detail in Appendix IV.

Problem 1.1. (1) Find a necessary and sufficient condition on an ordered set $(\varepsilon, \mu_1, \mu_2, \cdots, \mu_h)$ in $U(t, q)$ such that there exists a $\left\{\varepsilon v_1 + \sum_{i=1}^{h} v_{\mu_i+1}, \varepsilon v_0 + \sum_{i=1}^{h} v_{\mu_i}; t, q\right\}$–minihyper where $v_l = (q^l - 1)/(q-1)$ for any integer $l \geq 0$.

(2) Characterize all $\left\{\varepsilon v_1 + \sum_{i=1}^{h} v_{\mu_i+1}, \varepsilon v_0 + \sum_{i=1}^{h} v_{\mu_i}; t, q\right\}$– minihypers in the case where there exist such minihypers.

Let $\Lambda(t, q)$ be the set of all ordered sets $(\lambda_1, \lambda_2, \cdots, \lambda_\eta)$ of integers η and λ_i $(i = 1, 2, \cdots, \eta)$ such that $1 \leq \eta \leq t(q-1)$, $0 \leq \lambda_1 \leq \lambda_2 \leq \cdots \leq \lambda_\eta \leq t-1$, $\lambda_\eta \neq 0$ and $0 \leq n_l(\underline{\lambda}) \leq q-1$ for $l = 0, 1, \cdots, t-1$ where $n_l(\underline{\lambda})$ denotes the number of integers λ_i in $\underline{\lambda} = (\lambda_1, \lambda_2, \cdots, \lambda_\eta)$ such that $\lambda_i = l$ for the given integer l. Note that there is a one-to-one correspondence between the set $U(t, q)$ and the set $\Lambda(t, q)$ as follows : (i) in the case $\varepsilon = 0$, $\eta = h$ and $\lambda_i = \mu_i$ for $i = 1, 2, \cdots, h$ and (ii) in the case $\varepsilon \neq 0$, $\eta = \varepsilon + h$, $\lambda_1 = \lambda_2 = \cdots = \lambda_\varepsilon = 0$ and $\lambda_{\varepsilon+i} = \mu_i$ for $i = 1, 2, \cdots, h$. As occasion demands, we shall consider the following problem instead of Problem 1.1. Of course, Problem 1.2 is equivalent to Problem 1.1.

Problem 1.2. (1) Find a necessary and sufficient condition on an ordered set $(\lambda_1, \lambda_2, \cdots, \lambda_\eta)$ in $\Lambda(t, q)$ such that there exists a $\left\{\sum_{i=1}^{\eta} v_{\lambda_i+1}, \sum_{i=1}^{\eta} v_{\lambda_i}; t, q\right\}$– minihyper.

(2) Characterize all $\left\{\sum_{i=1}^{\eta} v_{\lambda_i+1}, \sum_{i=1}^{\eta} v_{\lambda_i}; t, q\right\}$–minihypers in the case where there exist such minihypers.

Problem 1.2 was solved completely by Helleseth [21] in the case $q = 2$ and by Hamada [3] in the case $q \geq 3$, $1 \leq \eta \leq t$ and $0 < \lambda_1 < \lambda_2 < \cdots < \lambda_\eta \leq t-1$. Hence it is sufficient to solve Problem 1.2 for the case $q \geq 3$, $\eta \geq 2$ and $\lambda_i = \lambda_j$ for some distinct integers i and j.

In the case $\eta = 2$ and $\lambda_1 = \lambda_2$, Problem 1.2 was solved by Hamada [7]. In the case $\eta = 3$, Problem 1.2 was solved completely by Hamada [6], [7], Hamada and Deza [8] and

Hamada and Helleseth [12]-[17]. In the case $\eta = 4$, $q \geq 5$ and $\lambda_1 = \lambda_2 < \lambda_3 = \lambda_4$, Problem 1.2 was solved by Hamada and Deza [9], [11].

Recently, Hamada, Helleseth and Ytrehus [18] solved Problem 1.2 for the case $\eta = 4$, $q \in \{3,4\}$ and $(\lambda_1, \lambda_2, \lambda_3, \lambda_4) = (0,0,1,1)$ (cf. Propositions $II.3$ and $III.4$ in Appendix). The purpose of this paper is to solve Problem 1.2 for the case $\eta = 4$, $q \in \{3,4\}$, $\lambda_1 = \lambda_2 < \lambda_3 = \lambda_4$ and $\lambda_3 \neq \lambda_1 + 1$ using the results in Appendices I, II and III. The main result is as follows.

Theorem 1.1. Let t, q, α and β be any integers such that $q = 3$ or 4, $0 \leq \alpha < \beta < t$ and $\beta \neq \alpha + 1$.

(1) In the case $t \leq 2\beta$, there is no $\{2v_{\alpha+1} + 2v_{\beta+1}, 2v_\alpha + 2v_\beta; t, q\}$– minihyper.

(2) In the case $t \geq 2\beta + 1$, F is a $\{2v_{\alpha+1} + 2v_{\beta+1}, 2v_\alpha + 2v_\beta; t, q\}$– minihyper if and only if F is a union of two α–flats and two β–flats in $PG(t, q)$ which are mutually disjoint.

Remark 1.1. Hamada and Deza [9], [11] showed that in the case $q \geq 5$ Theorem 1.1 holds for any integers t, α and β such that $0 \leq \alpha < \beta < t$. But Hamada, Helleseth and Ytrehus [18] showed that Theorem 1.1 does not hold in the case $q \in \{3,4\}$, $\alpha = 0$ and $\beta = 1$. From Remark $I.3$ in Appendix I and Theorem 1.2 in Hamada and Helleseth [14], it follows that Theorem 1.1 does not hold for any integers t, q, α and β such that $q = 4$, $\alpha \geq 0$, $\beta = \alpha + 1$ and $t \geq 2\alpha + 2$.

Corollary 1.1. Let $n = v_k - 2v_{\alpha+1} - 2v_{\beta+1}$ and $d = q^{k-1} - 2q^\alpha - 2q^\beta$ where $q = 3$ or 4, $0 \leq \alpha < \beta < t$ and $\beta \neq \alpha + 1$.

(1) In the case $k \leq 2\beta + 1$, there is no $[n, k, d; q]$–code meeting the Griesmer bound.

(2) In the case $k \geq 2\beta + 2$, C is a $[n, k, d; q]$–code meeting the Griesmer bound if and only if C is congruent to some $[n, k, d; q]$–code constructed by using a union of two α–flats and two β–flats in $PG(t, q)$ which are mutually disjoint.

Remark 1.2. It is unknown whether or not there exists a $[93, 5, 61; 3]$–code meeting the Griesmer bound. Since $n = 93, d = 61, v_1 = 1, v_3 = 13, v_5 = 121$ in the case $q = 3, k = 5, \alpha = 0$ and $\beta = 2$, Corollary 1.1 shows that there is no $[93, 5, 61; 3]$–code meeting the Griesmer bound.

2. The proof of Theorem 1.1

Let $\mathcal{F}(\alpha, \beta, \gamma, \delta; t, q)$ denote the family of all unions $\bigcup_{i=1}^{4} V_i$ of an α–flat V_1, a β–flat V_2, a γ–flat V_3, a δ–flat V_4 in $PG(t, q)$ which are mutually disjoint where $0 \leq \alpha \leq \beta \leq \gamma \leq \delta < t$. In order to prove Theorem 1.1, we prepare the following two theorems whose proofs will be given in Sections 3 and 4 respectively.

Theorem 2.1. Let $v_0 = 0$, $v_1 = 1$, $v_2 = q + 1$ and $v_3 = q^2 + q + 1$ where $q = 3$ or 4.

(1) In the case $t = 3$ or 4, there is no $\{2v_1 + 2v_3, 2v_0 + 2v_2; t, q\}$– minihyper.

(2) In the case $t \geq 5$, $F \in \mathcal{F}(0,0,2,2;t,q)$ for any $\{2v_1+2v_3, 2v_0+ 2v_2;t,q\}-$ minihyper F.

Theorem 2.2. Let t, β and q be integers such that $t \geq 2\beta \geq 6$ and $q = 3$ or 4. If F is a $\{2v_2+2v_{\beta+1}, 2v_1+2v_\beta;t,q\}-$ minihyper such that (a) $|F \cap G| = 2v_{\beta-1}$ for some $(t-2)-$flat G in $PG(t,q)$ and (b) $F \cap H_j \in \mathcal{F}(0,0,\beta-1,\beta-1;t,q)$ for any hyperplane H_j $(1 \leq j \leq q+1)$ in $PG(t,q)$ which contain G, then $F \in \mathcal{F}(1,1,\beta,\beta;t,q)$.

Remark 2.1. Since $\mathcal{F}(1,1,\beta,\beta;t,q) = \emptyset$ in the case $t = 2\beta$ (cf. Remark I.1), Theorem 2.2 shows that in the case $t = 2\beta$, there is no $\{2v_2+2v_{\beta+1}, 2v_1+2v_\beta;t,q\}-$ minihyper F which satisfies the conditions (a) and (b) in Theorem 2.2.

(Proof of Theorem 1.1) It follows from Proposition I.1 and Remark I.1 that if $F \in \mathcal{F}(\alpha,\alpha,\beta,\beta;t,q)$ in the case $t \geq 2\beta+1$, then F is a $\{2v_{\alpha+1}+2v_{\beta+1}, 2v_\alpha+2v_\beta;t,q\}-$ minihyper.

Conversely, suppose there exists a $\{2v_{\alpha+1}+2v_{\beta+1}, 2v_\alpha+2v_\beta;t,q\}-$ minihyper F for some integers t, q, α and β such that $q = 3$ or 4, $0 \leq \alpha < \beta < t$ and $\beta \neq \alpha+1$. Then it follows from Proposition I.2 ((i) in the case $\alpha = 0$, $\varepsilon = 2$, $h = 2$ and $\mu_1 = \mu_2 = \beta$ and (ii) in the case $\alpha \neq 0$, $\varepsilon = 0$, $h = 4$, $\mu_1 = \mu_2 = \alpha$ and $\mu_3 = \mu_4 = \beta$) that there exists a $(t-2)-$flat G in $PG(t,q)$ such that $|F \cap G| = 2v_{\alpha-1}+2v_{\beta-1}$ where $v_{-1} = v_0 = 0$. Let H_j $(j = 1,2,\cdots,q+1)$ be $q+1$ hyperplanes in $PG(t,q)$ which contain G.

Case I : (α=0 and β=2) It follows from Theorem 2.1 that Theorem 1.1 holds.

Case II : (α=0 and β≥3) It follows from Proposition I.2 that $F \cap H_j$ is a $\{\delta_j+2v_\beta, 2v_{\beta-1};t,q\}-$ minihyper in H_j for $j = 1,2,\cdots,q+1$ where the δ_j are non-negative integers such that $\sum_{j=1}^{q+1} \delta_j = 2$. Without loss of generality, we can assume that either (a) $\delta_1 = \delta_2 = \cdots = \delta_{q-1} = 0$ and $\delta_q = \delta_{q+1} = 1$ or (b) $\delta_1 = \delta_2 = \cdots = \delta_q = 0$ and $\delta_{q+1} = 2$.

(A) In the case $\delta_1 = \delta_2 = \cdots = \delta_{q-1} = 0$ and $\delta_q = \delta_{q+1} = 1$, it follows that $F \cap H_i$ is a $\{2v_\beta, 2v_{\beta-1};t,q\}-$ minihyper in the $(t-1)-$flat H_i for $i = 1,2,\cdots,q-1$ and $F \cap H_j$ is a $\{v_1+2v_\beta, v_0+2v_{\beta-1};t,q\}-$ minihyper in H_j for $j = q, q+1$. From Remark I.2, Proposition I.5 , II.1 and III.1 in Appendices, it follows that (i) in the case $t-1 \leq 2(\beta-1)$, there is no $\{2v_\beta, 2v_{\beta-1};t,q\}-$ minihyper in any $(t-1)-$flat H, a contradiction, and (ii) in the case $t-1 \geq 2(\beta-1)+1$, $F \cap H_i \in \mathcal{F}(\beta-1,\beta-1;t,q)$ for $i = 1,2,\cdots,q-1$ and $F \cap H_j \in \mathcal{F}(0,\beta-1,\beta-1;t,q)$ for $j = q, q+1$. Hence it follows from (i) and Proposition I.3 ($\varepsilon = 2$, $h = 2$, $\mu_1 = \mu_2 = \beta$) that (1) in the case $t \leq 2\beta-1$, there is no $\{2v_1+2v_{\beta+1}, 2v_0+2v_\beta;t,q\}-$ minihyper F which satisfies the condition (a) and (2) in the case $t \geq 2\beta$, $F \in \mathcal{F}(0,0,\beta,\beta;t,q)$. Since $\mathcal{F}(0,0,\beta,\beta;t,q) \neq \emptyset$ if and only if $t \geq 2\beta+1$ (cf. Remark I.1), there is no $\{2v_1+2v_{\beta+1}, 2v_0+2v_{\beta-1};t,q\}-$ minihyper in the case $t = 2\beta$.

(B) In the case $\delta_1 = \delta_2 = \cdots = \delta_q = 0$ and $\delta_{q+1} = 2$, it follows from Remark I.2, Proposition I.5, Case I and induction on β that (i) in the case $t-1 \leq 2(\beta-1)$, there is no $\{2v_\beta, 2v_{\beta-1};t,q\}-$ minihyper in H_1, a contradiction, and (ii) in the case $t-1 \geq 2(\beta-1)+1$, $F \cap H_i \in \mathcal{F}(\beta-1,\beta-1;t,q)$ for $i = 1,2,\cdots,q$ and $F \cap H_{q+1} \in \mathcal{F}(0,0,\beta-1,\beta-1;t,q)$. Hence it follows from (i), Proposition I.3 and Remark I.1 that (1) in the case $t \leq 2\beta$, there is no $\{2v_1 + 2v_{\beta+1}, 2v_0 + 2v_\beta; t, q\}-$ minihyper F which satisfies the condition (b) and (2) in the case $t \geq 2\beta + 1$, $F \in \mathcal{F}(0,0,\beta,\beta;t,q)$.

From (A) and (B), it follows that Theorem 1.1 holds in Case II.

Case III : (α=1 and β≥3) It follows from Proposition $I.2$ ($\varepsilon = 0$, $h = 4$, $\mu_1 = \mu_2 = 1$, $\mu_3 = \mu_4 = \beta$) that $F \cap H_j$ is a $\{2v_1 + 2v_\beta, 2v_0 + 2v_{\beta-1}; t, q\}-$ minihyper in H_j for $j = 1, 2, \cdots, q+1$. Hence it follows from Remark $I.2$, Cases I and II that (i) in the case $t - 1 \leq 2(\beta - 1)$, there is no $\{2v_1 + 2v_\beta, 2v_0 + 2v_{\beta-1}; t, q\}-$ minihyper in H_1, a contradiction, and (ii) in the case $t - 1 \geq 2(\beta - 1) + 1$, $F \cap H_j \in \mathcal{F}(0, 0, \beta - 1, \beta - 1; t, q)$ for $j = 1, 2, \cdots, q+1$. Using Theorem 2.2 and Remark $I.1$, it can be shown that Theorem 1.1 holds in Case III.

Case IV : (α≥2 and β≥α+2) It follows from Proposition $I.2$ that $F \cap H_j$ is a $\{2v_\alpha + 2v_\beta, 2v_{\alpha-1} + 2v_{\beta-1}; t, q\}-$ minihyper in H_j for $j = 1, 2, \cdots, q+1$. Hence it follows from Remark $I.2$, Cases I, II, III, induction on α and β, Proposition $I.3$ and Remark $I.1$ that Theorem 1.1 holds in Case IV. This completes the proof of Theorem 1.1.

3. The proof of Theorem 2.1

Let $\mathcal{F}(\lambda_1, \lambda_2, \cdots, \lambda_\eta; t, q)$, $\overline{\mathcal{F}}(0,1,1;t,3)$, $\overline{\mathcal{F}_i}(0,0,1,1;t,3)$ and $\overline{\mathcal{F}_i}(0,0,1,1;t,4)$ ($i = 1,2,3,4$) denote the families given in Appendix I, Definition $II.1$, $II.2$, $III.2$ respectively. In order to prove Theorem 2.1, we prepare the following four lemmas whose proofs will be given in Sections 5, 6, 7 and 8 respectively.

Lemma 3.1. In the case $t \geq 4$, there is no $\{2v_1 + 2v_3, 2v_0 + 2v_2; t, 3\}-$ minihyper F such that (a) $|F \cap G| = 2$ for some $(t-2)-$flat G in $PG(t,3)$ and (b) $F \cap H_1 \in \mathcal{F}(1,1;t,3)$, $F \cap H_2 \in \mathcal{F}(1,1;t,3)$, $F \cap H_3 \in \mathcal{F}(0,1,1;t,3)$ and $F \cap H_4 \in \overline{\mathcal{F}}(0,1,1;t,3)$ where $v_0 = 0$, $v_1 = 1$, $v_2 = 4$, $v_3 = 13$ and the H_i denote four hyperplanes in $PG(t,3)$ which contain G.

Lemma 3.2. In the case $t \geq 4$, there is no $\{2v_1 + 2v_3, 2v_0 + 2v_2; t, 3\}-$ minihyper F such that (a) $|F \cap G| = 2$ for some $(t-2)-$flat G in $PG(t,3)$ and (b) $F \cap H_1 \in \mathcal{F}(1,1;t,3)$, $F \cap H_2 \in \mathcal{F}(1,1;t,3)$, $F \cap H_3 \in \overline{\mathcal{F}}(0,1,1;t,3)$ and $F \cap H_4 \in \overline{\mathcal{F}}(0,1,1;t,3)$ where the H_i denote four hyperplanes in $PG(t,3)$ which contain G.

Lemma 3.3. In the case $t \geq 4$, there is no $\{2v_1 + 2v_3, 2v_0 + 2v_2; t, 3\}-$ minihyper F such that (a) $|F \cap G| = 2$ for some $(t-2)-$flat G in $PG(t,3)$ and (b) $F \cap H_1 \in \mathcal{F}(1,1;t,3)$, $F \cap H_2 \in \mathcal{F}(1,1;t,3)$, $F \cap H_3 \in \mathcal{F}(1,1;t,3)$ and $F \cap H_4 \in \overline{\mathcal{F}_\theta}(0,0,1,1;t,3)$ for some θ in $\{1,2,3,4\}$ where the H_i denote four hyperplanes in $PG(t,3)$ which contain G.

Lemma 3.4. In the case $t \geq 4$, there is no $\{2v_1 + 2v_3, 2v_0 + 2v_2; t, 4\}-$ minihyper F such that (a) $|F \cap G| = 2$ for some $(t-2)-$flat G in $PG(t,4)$ and (b) $F \cap H_i \in \mathcal{F}(1,1;t,4)$ for $i = 1,2,3,4$ and $F \cap H_5 \in \overline{\mathcal{F}_\theta}(0,0,1,1;t,4)$ for some θ in $\{1,2,3,4\}$ where $v_0 = 0$, $v_1 = 1$, $v_2 = 5$, $v_3 = 21$ and the H_i denote five hyperplanes in $PG(t,4)$ which contain G.

(Proof of Theorem 2.1) Suppose there exists a $\{2v_1 + 2v_3, 2v_0 + 2v_2; t, q\}-$ minihyper F for some integer $t \geq 3$. Then it follows from Proposition $I.2$ ($\varepsilon = 2$, $h = 2$, $\mu_1 = \mu_2 = 2$)

that (a) $|F \cap G| = 2$ for some $(t-2)$–flat G in $PG(t,q)$ and (b) $F \cap H_j$ is a $\{\delta_j + 2v_2, 2v_1; t, q\}$–minihyper in H_j for any hyperplane H_j $(1 \le j \le q+1)$ in $PG(t,q)$ which contain G where the δ_j are nonnegative integers such that $\sum_{j=1}^{q+1} \delta_j = 2$. Without loss of generality, we can assume that either (α) $\delta_1 = \delta_2 = \cdots = \delta_{q-1} = 0$ and $\delta_q = \delta_{q+1} = 1$ or (β) $\delta_1 = \delta_2 = \cdots = \delta_q = 0$ and $\delta_{q+1} = 2$.

Case I : (q=3, $\delta_1 = \delta_2 = 0$ and $\delta_3 = \delta_4 = 1$). It follows from Remark $I.2$, Propositions $I.5$ and $II.2$ that (i) in the case $t-1=2$, there is no $\{2v_2, 2v_1; t, 3\}$–minihyper in H_1, a contradiction, and (ii) in the case $t-1 \ge 3$, $F \cap H_1 \in \mathcal{F}(1,1;t,3)$, $F \cap H_2 \in \mathcal{F}(1,1;t,3)$ and either $F \cap H_i \in \mathcal{F}(0,1,1;t,3)$ or $F \cap H_i \in \overline{\mathcal{F}}(0,1,1;t,3)$ for $i = 3,4$. It follows from Lemmas 3.1 and 3.2 that $F \cap H_i \in \mathcal{F}(0,1,1;t,3)$ for $i = 3,4$. Hence it follows from (i), Proposition $I.3$ ($\varepsilon = 2$, $h = 2$, $\mu_1 = \mu_2 = 2$) and Remark $I.1$ that Theorem 2.1 holds in Case I.

Case II : (q=4, $\delta_1 = \delta_2 = \delta_3 = 0$ and $\delta_4 = \delta_5 = 1$). It follows from Remark $I.2$, Propositions $I.5$ and $III.1$ ($\alpha = 0, \beta = \gamma = 1$) that (i) in the case $t-1=2$, there is no $\{2v_2, 2v_1; t, 4\}$–minihyper in H_1, a contradiction, and (ii) in the case $t-1 \ge 3$, $F \cap H_i \in \mathcal{F}(1,1;t,4)$, for $i = 1,2,3$ and $F \cap H_j \in \mathcal{F}(0,1,1;t,4)$ for $j = 4,5$. Hence it follows from (i), Proposition $I.3$ and Remark $I.1$ that Theorem 2.1 holds in Case II.

Case III : (q=3, $\delta_1 = \delta_2 = \delta_3 = 0$ and $\delta_4 = 2$). It follows from Remark $I.2$, Propositions $I.5$ and $II.3$ that (i) in the case $t-1=2$, there is no $\{2v_2, 2v_1; t, 3\}$–minihyper in H_1, a contradiction, and (ii) in the case $t-1 \ge 3$, $F \cap H_i \in \mathcal{F}(1,1;t,3)$, for $i = 1,2,3$ and either $F \cap H_4 \in \mathcal{F}(0,0,1,1;t,3)$ or $F \cap H_4 \in \overline{\mathcal{F}_\theta}(0,0,1,1;t,3)$ for some θ in $\{1,2,3,4\}$. Hence it follows from (i), Lemma 3.3, Proposition $I.3$ and Remark $I.1$ that Theorem 2.1 holds in Case III.

Case IV : (q=4, $\delta_1 = \delta_2 = \delta_3 = \delta_4 = 0$ and $\delta_5 = 2$). It follows from Remark $I.2$, Propositions $I.5$ and $III.4$ that (i) in the case $t-1=2$, there is no $\{2v_2, 2v_1; t, 4\}$–minihyper in H_1, a contradiction, and (ii) in the case $t-1 \ge 3$, $F \cap H_i \in \mathcal{F}(1,1;t,4)$ for $i = 1,2,3,4$ and either $F \cap H_5 \in \mathcal{F}(0,0,1,1;t,4)$ or $F \cap H_5 \in \overline{\mathcal{F}_\theta}(0,0,1,1;t,4)$ for some θ in $\{1,2,3,4\}$. Hence it follows from (i), Lemma 3.4, Proposition $I.3$ and Remark $I.1$ that Theorem 2.1 holds in Case IV. This completes the proof of Theorem 2.1.

4. The proof of Theorem 2.2

Lemma 4.1. Let t, β and q be integers such that $t \ge 2\beta + 1 \ge 7$ and $q \ge 3$. If F is a $\{2v_2 + 2v_{\beta+1}, 2v_1 + 2v_\beta; t, q\}$–minihyper such that $F = X \cup Y_1 \cup Y_2$ for some set X (of $2v_2$ points in $PG(t,q)$) and some β–flats Y_1, Y_2 in $PG(t,q)$ which are mutually disjoint, then $X \in \mathcal{F}(1,1;t,q)$ and $F \in \mathcal{F}(1,1,\beta,\beta;t,q)$.

Proof. If $|X \cap H| \ge 2$ for any hyperplane H in $PG(t,q)$ and $|X \cap H| = 2$ for some hyperplane hyperplane H in $PG(t,q)$, it follows from $|X| = 2v_2$ and Proposition $I.5$ that $X \in \mathcal{F}(1,1;t,q)$ and $F \in \mathcal{F}(1,1,\beta,\beta;t,q)$. Since $|F \cap H| = |X \cap H| + \sum_{i=1}^{2} |Y_i \cap H|$ and $|Y_i \cap H| = v_\beta$ or $v_{\beta+1}$ for any hyperplane H in $PG(t,q)$, it is sufficient to show that there is no hyperplane H in $PG(t,q)$ such that $|F \cap H| = v_\beta + v_{\beta+1}$, $1 + v_\beta + v_{\beta+1}$, $2v_{\beta+1}$ or $1 + 2v_{\beta+1}$.

Suppose there exists a hyperplane H in $PG(t,q)$ such that $|F \cap H| = 3 + 2v_\beta$, $4 + 2v_\beta$, $v_\beta + v_{\beta+1}$, $1 + v_\beta + v_{\beta+1}$, $2v_{\beta+1}$ or $1 + 2v_{\beta+1}$.

Case I : ($|F \cap H| = 3v_1 + 2v_\beta$). Suppose there exists a $(t-2)$–flat G in H such that $|F \cap G| \le 2v_{\beta-1}$. Let H_i $(i = 1, 2, \cdots, q)$ be q hyperplanes in $PG(t,q)$, except for H, which contain G. Since $|F| = 2v_2 + 2v_{\beta+1}$ and $|F \cap H_i| \ge 2v_1 + 2v_\beta$ for $i = 1, 2, \cdots, q$, it follows from $qv_{\beta-1} = v_\beta - 1$ and $qv_\beta = v_{\beta+1} - 1$ that $|F| = |F \cap H| + \sum_{i=1}^{q} \{|F \cap H_i| - |F \cap G|\} \ge 2v_{\beta+1} + 2q + 3 > |F|$, a contradiction. Hence $|F \cap G| \ge v_1 + 2v_{\beta-1}$ for any $(t-2)$–flat G in H. Using Proposition *I*.6 ($\theta = t-1, \varepsilon_1 = 1, \varepsilon_{\beta-1} = 2$), we have $|F \cap H| \ge v_2 + 2v_\beta > 3 + 2v_\beta = |F \cap H|$, a contradiction. Therefore, there is no hyperplane H in $PG(t,q)$ such that $|F \cap H| = 3 + 2v_\beta$.

Case II : ($q \ge 4$ *and* $|F \cap H| = 4v_1 + 2v_\beta$). Using a method similar to Case I, it can be shown that there is no hyperplane H in $PG(t,q)$ such that $|F \cap H| = 4 + 2v_\beta$ in the case $q \ge 4$.

Case III : ($q=3$ *and* $|F \cap H| = v_2 + 2v_\beta$). Suppose there exists a $(t-2)$–flat G in H such that $|F \cap G| \le -1 + v_1 + 2v_{\beta-1}$. Let H_i $(i = 1, 2, 3)$ be three hyperplanes in $PG(t,3)$, except for H, which contain G. Then $|F| = |F \cap H| + \sum_{i=1}^{3} \{|F \cap H_i| - |F \cap G|\} \ge 2v_{\beta+1} + v_2 + 6 > |F|$, a contradiction. Hence $|F \cap G| \ge v_1 + 2v_{\beta-1}$ for any $(t-2)$–flat G in H. Since $|F \cap H| = v_2 + 2v_\beta$, it follows from Proposition *I*.6 that there exists a $(t-2)$–flat G in H such that $|F \cap G| = v_1 + 2v_{\beta-1}$. Let Π_i $(i = 1, 2, 3)$ be three hyperplanes in $PG(t,3)$, except for H, which contain G.

Since $\sum_{i=1}^{3} |F \cap (\Pi_i \setminus G)| = |F| - |F \cap H| = 3(2 \cdot 3^{\beta-1} + 1) + 1$ and $|F \cap (\Pi_i \setminus G)| = |F \cap \Pi_i| - |F \cap G| \ge 2 \cdot 3^{\beta-1} + 1$ for $i = 1, 2, 3$, there exists a hyperplane Π in $\{\Pi_1, \Pi_2, \Pi_3\}$ such that $|F \cap \Pi| = (2v_1 + 2v_\beta) + 1 = 3 + 2v_\beta$, a contradiction. Hence there is no hyperplane H in $PG(t,3)$ such that $|F \cap H| = 4 + 2v_\beta$ in the case $q = 3$.

Case IV : ($|F \cap H| = v_\beta + v_{\beta+1}$). Using a method similar to Case III, it can be shown that there exists a $(t-2)$–flat G in H such that $|F \cap G| = v_{\beta-1} + v_\beta$. Let Π_i $(i = 1, 2, \cdots, q)$ be q hyperplanes in $PG(t,q)$, except for H, which contain G.

Since $\sum_{i=1}^{q} |F \cap (\Pi_i \setminus G)| = |F| - |F \cap H| = q(q^{\beta-1} + 2) + 2$ and $|F \cap (\Pi_i \setminus G)| = |F \cap \Pi_i| - |F \cap G| \ge q^{\beta-1} + 2$ for $i = 1, 2, \cdots, q$, there exists a hyperplane Π in $\{\Pi_1, \Pi_2, \cdots, \Pi_q\}$ such that $|F \cap \Pi| = 3 + 2v_\beta$ or $4 + 2v_\beta$, a contradiction. Hence there is no hyperplane H in $PG(t,q)$ such that $|F \cap H| = v_\beta + v_{\beta+1}$.

Similarly, it can be shown that there is no hyperplane H in $PG(t,q)$ such that $|F \cap H| = 2v_{\beta+1}$.

Case V : ($|F \cap H| = 1 + v_\beta + v_{\beta+1}$ *or* $1 + 2v_{\beta+1}$). Using a method similar to Case III, it can be shown that there exists a hyperplane Π in $PG(t,q)$ such that $|F \cap H| = 3 + 2v_\beta$, a contradiction. This completes the proof.

(Proof of Theorem 2.2) Let F be any $\{2v_2+2v_{\beta+1}, 2v_1+2v_\beta; t, q\}$– minihyper which satisfies the conditions (a) and (b) in Theorem 2.2 where $t \geq 2\beta \geq 6$. Then $F \cap H_i = \{P_{i1}, P_{i2}\} \cup W_{i1} \cup W_{i2}$ for some points P_{i1}, P_{i2} and some $(\beta-1)$–flats W_{i1}, W_{i2} in H_i which are mutually disjoint for $i = 1, 2, \cdots, q+1$. Since $|G \cap W| = v_{\beta-1}$ or v_β for any $(\beta-1)$–flat W and any $(t-2)$–flat G in H_i, it follows from $|F \cap G| = 2v_{\beta-1}$ that $P_{ij} \notin G$ and $G \cap W_{ij} = V_j$ $(i = 1, 2, \cdots, q+1,\ j = 1, 2)$ for some $(\beta-2)$–flat V_j in G. Let $X = \bigcup_{i=1}^{q+1} \{P_{i1}, P_{i2}\}$, $Y_1 = \bigcup_{i=1}^{q+1} W_{i1}$ and $Y_2 = \bigcup_{i=1}^{q+1} W_{i2}$.

Let $E_i = H_i \cap (W_{11} \oplus W_{21})$ for $i = 3, 4, \cdots, q+1$. Then E_i is a $(\beta-1)$–flat in H_i such that $G \cap E_i = V_1$ for $i = 3, 4, \cdots, q+1$. Note that (i) $W_{11} \cup W_{21} \cup E_3 \cup \cdots \cup E_{q+1}$ is a β–flat in $PG(t,q)$ and (ii) either $W_{i1} = E_i$ or $W_{i1} \cap E_i = V_1$ for each i. Hence if $W_{i1} = E_i$ for $i = 3, 4, \cdots, q+1$, then Y_1 is a β–flat in $PG(t,q)$.

Suppose $W_{i1} \cap E_i = V_1$ for some i in $\{3, 4, \cdots, q+1\}$. Without loss of generality, we can assume that $W_{21} = E_2$, $W_{31} = E_3$, $\cdots$, $W_{\theta-1,1} = E_{\theta-1}$ and $W_{\theta,1} \cap E_\theta = \cdots = W_{q+1,1} \cap E_{q+1} = V_1$ for some θ in $\{3, 4, \cdots, q+1\}$ where $E_2 = W_{21}$.

Let Ω be a $(t-2)$–flat in H_{q+1} such that $E_{q+1} \subset \Omega$, $\Omega \cap W_{q+1,1} = V_1$ and $\Omega \cap V_2$ is a $(\beta-3)$–flat in G. Let Π_l $(l = 1, 2, \cdots, q)$ be q hyperplanes in $PG(t,q)$, except for H_{q+1}, which contain Ω. Then $\Pi_l \cap W_{q+1,1} = \Omega \cap W_{q+1,1} = V_1$ and $\Pi_l \cap W_{i2}$ is a $(\beta-2)$–flat in H_i (i.e., $|\Pi_l \cap W_{i2}| = v_{\beta-1}$) for $l = 1, 2, \cdots, q$ and $i = 1, 2, \cdots, q+1$.

Since $W_{11} \subset \Pi_\alpha$, $W_{21} \subset \Pi_\alpha$ and $W_{i1} \subset \Pi_{\theta_i}$ for some integers $\alpha, \theta_3, \theta_4, \cdots, \theta_q$ in $\{1, 2, \cdots, q\}$, there exists a hyperplane Π in $\{\Pi_1, \Pi_2, \cdots, \Pi_q\}$ such that $\Pi \cap W_{i1} = V_1$ for $i = 1, 2, \cdots, q$. Hence it follows from $F = X \cup Y_1 \cup Y_2$ that $|F \cap \Pi| = |X \cap \Pi| + |Y_1 \cap \Pi| + |Y_2 \cap \Pi| = |X \cap \Pi| + |V_1| + (q+1)v_{\beta-1} = |X \cap \Pi| + (q+2)v_{\beta-1} < 2v_1 + 2v_\beta$ unless $q = 3, \beta = 3$ and $|X \cap \Pi| = 8$.

Case I : ($q \geq 4$ or $q=3$ and $\beta \geq 4$). It follows that $|F \cap \Pi| < 2v_1 + 2v_\beta$, a contradiction. Hence $W_{i1} = E_i$ for $i = 3, 4, \cdots, q+1$. This implies that Y_1 is a β–flat in $PG(t,q)$. Similarly, it can be shown that Y_2 is a β–flat in $PG(t,q)$. Hence it follows from Lemma 4.1 that $F \in \mathcal{F}(1, 1, \beta, \beta; t, q)$ in the case $t \geq 2\beta + 1$.

In the case $t = 2\beta$, there do not exist two β–flats Y_1 and Y_2 in $PG(t,q)$ such that $Y_1 \cap Y_2 = \emptyset$. Hence in the case $t = 2\beta$, there is no $\{2v_2+2v_{\beta+1}, 2v_1+2v_\beta; t, q\}$–minihyper F which satisfies the conditions (a) and (b) in Theorem 2.2. This implies that Theorem 2.2 holds in Case I.

Case II : ($q=3$ and $\beta=3$). It follows that $\theta = 3$ or 4, i.e., either (α) $W_{31} \cap E_3 = W_{41} \cap E_4 = V_1$ or (β) $W_{31} = E_3$ and $W_{41} \cap E_4 = V_1$.

(A) In the case $W_{31} = E_3$, we can assume without loss of generality that $W_{11} \subset \Pi_1$, $W_{21} \subset \Pi_1$ and $W_{31} \subset \Pi_1$. Hence there exists a hyperplane Π in $\{\Pi_2, \Pi_3\}$ such that $P_{11} \notin \Pi$ and $\Pi \cap W_{i1} = V_1$ for $i = 1, 2, 3$. This implies that $|X \cap \Pi| \leq 7$, i.e., $|F \cap \Pi| < 2v_1 + 2v_\beta$, a contradiction. Using a method similar to Case I, it can be shown that Theorem 2.2 holds in the case (A).

(B) In the case $W_{31} \cap E_3 = V_1$ and $P_{31} \in E_3$ there exists a hyperplane Π in $\{\Pi_1, \Pi_2, \Pi_3\}$ such that $P_{31} \notin \Pi$ and $\Pi \cap W_{i1} = V_1$ for $i = 1, 2, 3$. This implies that $|X \cap \Pi| \leq 7$, i.e., $|F \cap \Pi| < 2v_1 + 2v_\beta$, a contradiction. Hence Theorem 2.2 holds in the case (B).

(C) In the case $W_{31} \cap E_3 = V_1$ and $P_{31} \notin E_3$, let Ω be a $(t-2)$–flat in H_3 such that $P_{31} \notin \Omega$, $E_3 \subset \Omega$, $\Omega \cap W_{31} = V_1$ and $\Omega \cap V_2$ is a $(\beta - 3)$–flat in G. Let Π_l $(l = 1,2,3)$ be three hyperplanes in $PG(t,3)$, except for H_3, which contain Ω. Since $P_{31} \notin \Pi_l$ for $l = 1,2,3$, there exists a hyperplane Π in $\{\Pi_1, \Pi_2, \Pi_3\}$ such that $|F \cap \Pi| < 2v_1 + 2v_\beta$, a contradiction. Hence Theorem 2.2 holds in the case (C). This completes the proof of Theorem 2.2.

5. The proof of Lemma 3.1

Suppose there exists a $\{2v_1 + 2v_3, 2v_0 + 2v_2; t, 3\}$–minihyper F which satisfies the conditions (a) and (b) in Lemma 3.1. Then $F \cap G = \{P_1, P_2\}$, $F \cap H_1 = L_{11} \cup L_{12}$, $F \cap H_2 = L_{21} \cup L_{22}$, $F \cap H_3 = L_{31} \cup L_{32} \cup \{P_3\}$ and $F \cap H_4 = V \backslash \{Q_1, Q_2, Q_3, Q_4\}$ for some 2–flat V in H_4 and some 4–arc $\{Q_1, Q_2, Q_3, Q_4\}$ in V where L_{i1} and L_{i2} are 1–flats in H_i such that $L_{i1} \cap L_{i2} = \emptyset$, $G \cap L_{i1} = \{P_1\}$ and $G \cap L_{i2} = \{P_2\}$ for $i = 1,2,3$ and P_3 is a point in $H_3 \backslash G$ and $G \cap V$ is a 1–flat in G which contains two points P_1 and P_2. Without loss of generality, we can assume that $G \cap V = \{P_1, P_2, Q_1, Q_2\}$.

In order to prove Lemma 3.1, it is sufficient to show that there exists a hyperplane Π in $PG(t,3)$ such that $|F \cap \Pi| < 2v_2 = 8$. Let $E_i = H_i \cap (L_{21} \oplus L_{31})$ for $i = 1,4$. Then E_i is a 1–flat in H_i such that $G \cap E_i = \{P_1\}$ for $i = 1,4$. Note that either $L_{11} = E_1$ or $L_{11} \cap E_1 = \{P_1\}$.

Case I : ($L_{11} \cap E_1 = \{P_1\}$). Let $M = G \cap (L_{11} \oplus E_1)$. Then M is a 1–flat in G passing through P_1. Let Σ be a $(t-3)$–flat in G such that $\Sigma \cap M = \{P_1\}$ and $P_2 \notin \Sigma$. Let $\Pi_j (j = 1,2,3)$ be three hyperplanes in $PG(t,3)$, except for H_1, which contain the $(t-2)$–flat $\Sigma \oplus E_1$ in H_1. Since $P_2 \in V$, $P_2 \in L_{i2}$, $P_2 \notin \Pi_j$ and $M \not\subset \Pi_j$ for $i = 1,2,3$ and $j = 1,2,3$, we have $|V \cap \Pi_j| = 4$, $|L_{i2} \cap \Pi_j| = 1$ and $L_{11} \cap \Pi_j = \{P_1\}$ for $i = 1,2,3$ and $j = 1,2,3$.

Since $L_{21} \subset \Pi_\alpha$, $L_{31} \subset \Pi_\alpha$ and $P_3 \in \Pi_\beta$ for some integers α and β in $\{1,2,3\}$, there exists a hyperplane Π in $\{\Pi_1, \Pi_2, \Pi_3\}$ such that $P_3 \notin \Pi$ and $\Pi \cap L_{i1} = \{P_1\}$ for $i = 2,3$. This implies that there exists a hyperplane Π in $PG(t,3)$ such that $|F \cap \Pi| = \sum_{i=1}^{3} |L_{i2} \cap \Pi| + |(F \cap H_4) \cap \Pi| \le 3 + |V \cap \Pi| = 7 < 2v_2$, a contradiction. Hence $L_{11} = E_1$.

Case II : ($L_{11} = E_1$). Let Σ be a $(t-3)$–flat in G such that $P_1 \in \Sigma$ and $P_2 \notin \Sigma$. Let Π_j $(j = 1,2,3)$ be three hyperplanes in $PG(t,3)$, except for H_4, which contain the $(t-2)$–flat $\Sigma \oplus E_4$ in H_4. Since $L_{11} \subset \Pi_\alpha$, $L_{21} \subset \Pi_\alpha$, $L_{31} \subset \Pi_\alpha$ and $P_3 \in \Pi_\beta$ for some integers α and β in $\{1,2,3\}$, there exists a hyperplane Π in $\{\Pi_1, \Pi_2, \Pi_3\}$ such that $P_3 \notin \Pi$ and $\Pi \cap L_{i1} = \{P_1\}$ for $i = 1,2,3$, i.e., $|F \cap \Pi| = \sum_{i=1}^{3} |L_{i2} \cap \Pi| + |(F \cap H_4) \cap \Pi| < 2v_2$, a contradiction. Hence there is no $\{2v_1 + 2v_3, 2v_0 + 2v_2; t, 3\}$–minihyper F which satisfies the conditions (a) and (b) in Lemma 3.1.

6. The proof of Lemma 3.2

Suppose there exists a $\{2v_1 + 2v_3, 2v_0 + 2v_2; t, 3\}$–minihyper F which satisfies the conditions (a) and (b) in Lemma 3.2. Then $F \cap G = \{P_1, P_2\}$, $F \cap H_1 = L_{11} \cup L_{12}$, $F \cap H_2 = L_{21} \cup L_{22}$ and $F \cap H_j = V_j \backslash \{Q_{j1}, Q_{j2}, Q_{j3}, Q_{j4}\}$ for some 2–flat V_j in H_j and some 4–arc $\{Q_{j1}, Q_{j2}, Q_{j3}, Q_{j4}\}$ in V_j ($j = 3, 4$) where L_{i1} and L_{i2} are 1–flats in H_i such that $L_{i1} \cap L_{i2} = \emptyset$, $G \cap L_{i1} = \{P_1\}$ and $G \cap L_{i2} = \{P_2\}$ for $i = 1, 2$.

Since $G \cap V_j$ is a 1–flat in G which contains two points P_1 and P_2, we can assume without loss of generality that $G \cap V_3 = G \cap V_4 = \{P_1, P_2, Q_1, Q_2\}$ where $Q_{j1} = Q_1$ and $Q_{j2} = Q_2$ for $j = 3, 4$. Since $\{Q_1, Q_2, Q_{43}, Q_{44}\}$ and $P_1 \oplus P_2$ are a 4–arc and a 1–flat, respectively, in the 2–flat V_4, it follows that $(P_1 \oplus P_2) \cap (Q_{43} \oplus Q_{44}) = \{P_1\}$ or $\{P_2\}$. Without loss of generality, we can assume that $(P_1 \oplus P_2) \cap (Q_{43} \oplus Q_{44}) = \{P_1\}$.

Let Σ be a $(t-3)$–flat in G such that $P_1 \in \Sigma$ and $P_2 \notin \Sigma$. Let Π_l ($l = 1, 2, 3$) be three hyperplanes in $PG(t, 3)$, except for H_4, which contain the $(t-2)$–flat $\Sigma \oplus (Q_{43} \oplus Q_{44})$ in H_4. Then $|L_{12} \cap \Pi_l| = 1$, $|L_{22} \cap \Pi_l| = 1$, $|V_3 \cap \Pi_l| = 4$ and $V_4 \cap \Pi_l = \{Q_{43}, Q_{44}, P_1, R\}$ (i.e., $|(F \cap V_4) \cap \Pi_l| = 2$) for $l = 1, 2, 3$ where R denotes the point in $Q_{43} \oplus Q_{44}$ except for three points P_1, Q_{43} and Q_{44}.

Since $L_{11} \subset \Pi_\alpha$ and $L_{21} \subset \Pi_\beta$ for some integers α and β in $\{1, 2, 3\}$, there exists a hyperplane Π in $\{\Pi_1, \Pi_2, \Pi_3\}$ such that $\Pi \cap L_{11} = \{P_1\}$ and $\Pi \cap L_{21} = \{P_1\}$ i.e., $|F \cap \Pi| = |L_{12} \cap \Pi| + |L_{22} \cap \Pi| + |(F \cap V_3) \cap \Pi| + |(F \cap V_4) \cap \Pi| - 1 \leq 7 < 2v_2$, a contradiction. Hence there is no $\{2v_1 + 2v_3, 2v_0 + 2v_2; t, 3\}$–minihyper F which satisfies the conditions (a) and (b) in Lemma 3.2.

7. The proof of Lemma 3.3

Suppose there exists a $\{2v_1 + 2v_3, 2v_0 + 2v_2; t, 3\}$–minihyper F which satisfies the conditions (a) and (b) in Lemma 3.3. Then $F \cap G = \{P_1, P_2\}$, $F \cap H_1 = L_{11} \cup L_{12}$, $F \cap H_2 = L_{21} \cup L_{22}$, $F \cap H_3 = L_{31} \cup L_{32}$ where L_{i1} and L_{i2} are 1–flats in H_i such that $L_{i1} \cap L_{i2} = \emptyset$, $G \cap L_{i1} = \{P_1\}$ and $G \cap L_{i2} = \{P_2\}$ for $i = 1, 2, 3$.

Case I : (θ=1). It follows from Definition *II*.2 that $F \cap H_4 = V \backslash \{Q_1, Q_2, Q_3\}$ for some 2–flat V in H_4 and some 3–arc $\{Q_1, Q_2, Q_3\}$ in V. Let $E_4 = H_4 \cap (L_{11} \oplus L_{21})$ and let Σ be a $(t-3)$–flat in G such that $P_1 \in \Sigma$ and $P_2 \notin \Sigma$. Let Π_j ($j = 1, 2, 3$) be three hyperplanes in $PG(t, 3)$, except for H_4, which contain the $(t-2)$–flat $\Sigma \oplus E_4$ in H_4. Then there exists a hyperplane Π in $\{\Pi_1, \Pi_2, \Pi_3\}$ such that $|F \cap \Pi| < 2v_2$, a contradiction.

Case II : (θ=2). It follows that $F \cap H_4 = (V \backslash S) \cup \{P\}$ for some 2–flat V in H_4, some 4–arc $S = \{Q_1, Q_2, Q_3, Q_4\}$ in V and some point P in $H_4 \backslash V$. Let $E_4 = H_4 \cap (L_{11} \oplus L_{21})$ and $G \cap V = \{P_1, P_2, Q_1, Q_2\}$.

(A) In the case $P \notin E_4$ let $M = G \cap (E_4 \oplus P)$ and let Σ be a $(t-3)$–flat in G such that $\Sigma \cap M = \{P_1\}$ and $P_2 \notin \Sigma$. Let Π_j ($j = 1, 2, 3$) be three hyperplanes in $PG(t, 3)$, except for H_4, which contain $\Sigma \oplus E_4$. Since $P \notin \Pi_j$ and $|V \cap \Pi_j| = 4$ for $j = 1, 2, 3$, there exists a hyperplane Π in $\{\Pi_1, \Pi_2, \Pi_3\}$ such that $|F \cap \Pi| < 2v_2$, a contradiction.

(B) In the case $P \in E_4$, let $N = G \cap (E_4 \oplus Q_3)$ and let Σ be a $(t-3)$–flat in G such that $N \subset \Sigma$ and $P_2 \notin \Sigma$. Let Π_j ($j = 1, 2, 3$) be three hyperplanes in $PG(t, 3)$, except for H_4, which contain $\Sigma \oplus E_4$. Since $V \cap \Pi_j = P_1 \oplus Q_3$ (i.e., $|(V \backslash S) \cap \Pi_j| \leq 3$) for $j = 1, 2, 3$, there exists a hyperplane Π in $\{\Pi_1, \Pi_2, \Pi_3\}$ such that $|F \cap \Pi| < 2v_2$, a contradiction.

Case III : (θ=3). It follows that $F \cap H_4 = L \cup K^*$ for some 1–flat L in H_4 and some minihyper K^* in $\overline{\mathcal{F}}(0, 0, 1; t, 3)$ such that $L \cap K^* = \emptyset$, $G \cap L = \{P_1\}$ and $G \cap K^* = \{P_2\}$. Let

V be the 2–flat in H_4 which contains K^* and let $M = G \cap V$. Let $E_4 = H_4 \cap (L_{12} \oplus L_{22})$ and let Σ be a $(t-3)$–flat in G such that $P_1 \notin \Sigma$ and $\Sigma \cap M = \{P_2\}$. Let Π_j $(j = 1,2,3)$ be three hyperplanes in $PG(t,3)$, except for H_4, which contain $\Sigma \oplus E_4$.

Since $M \not\subset \Sigma$ and $|K^* \cap L^*| \leq 3$ for any 1–flat L^* in V, $V \cap \Pi_j$ is a 1–flat in V and $|K^* \cap (V \cap \Pi_j)| \leq 3$ for $j = 1,2,3$. Hence it follows from $|L_{i1} \cap \Pi_j| = 1$ and $|L \cap \Pi_j| = 1$ $(i = 1,2,3,\ j = 1,2,3)$ that there exists a hyperplane Π in $\{\Pi_1, \Pi_2, \Pi_3\}$ such that $|F \cap \Pi| < 2v_2$, a contradiction.

Case IV : (θ=4). It follows from Definition $II.2$ that $F \cap H_4 = K$ for some minihyper K in $\overline{\mathcal{F}_4}(0,0,1,1;t,3)$ where $K = \{(\xi_0),(\xi_1),(\xi_2),(\xi_3),(2\xi_0+\xi_1),(2\xi_0+\xi_2),(2\xi_0+\xi_3),(2\xi_1+\xi_2),(2\xi_1+\xi_3),(2\xi_2+\xi_3)\}$ for some four linearly independent points $(\xi_0),(\xi_1),(\xi_2)$ and (ξ_3) in $PG(t,3)$.

Let W be the 3–flat in H_4 generated by four points $(\xi_0),(\xi_1),(\xi_2)$ and (ξ_3) in $PG(t,3)$. Since G is a $(t-2)$–flat in the $(t-1)$–flat H_4 such that $K \cap G = (F \cap H_4) \cap G = F \cap G = \{P_1, P_2\}$, $W \cap G$ must be a 2–flat (denoted by Δ) in G such that $K \cap \Delta = \{P_1, P_2\}$. Without loss of generality, we can assume that $P_1 = (2\xi_0 + \xi_1)$ and $P_1 = (2\xi_2 + \xi_3)$.

Let $J = P_1 \oplus P_2$, $\Delta_1 = J \oplus (2\xi_1 + \xi_2)$, $\Delta_2 = J \oplus (\xi_0)$, $\Delta_3 = J \oplus (\xi_2)$ and $\Delta_4 = J \oplus (\xi_0 + \xi_1 + \xi_2 + \xi_3)$. Then Δ_i $(i = 1,2,3,4)$ are four 2–flats in W which contain J and $K \cap \Delta_1 = \{(2\xi_0+\xi_1),(2\xi_0+\xi_2),(2\xi_0+\xi_3),(2\xi_1+\xi_2),(2\xi_1+\xi_3),(2\xi_2+\xi_3)\}$, $K \cap \Delta_2 = \{(\xi_0),(\xi_1),(2\xi_0+\xi_1),(2\xi_2+\xi_3)\}$, $K \cap \Delta_3 = \{(\xi_2),(\xi_3),(2\xi_0+\xi_1),(2\xi_2+\xi_3)\}$ and $K \cap \Delta_4 = \{(2\xi_0+\xi_1),(2\xi_2+\xi_3)\}$ where $\Delta = \Delta_4$. Let $E_4 = H_4 \cap (L_{11} \oplus L_{21})$. Then E_4 is a 1–flat in H_4 such that $G \cap E_4 = \{P_1\}$. Note that either $E_4 \subset W$ or $E_4 \cap W = \{P_1\}$.

(A) In the case $E_4 \subset W$ (i.e., $E_4 \subset \Delta_i$) for some i in $\{1,2,3\}$), there exists a 2–flat V in W such that $E_4 \subset W$, $P_2 \notin V$ and $|K \cap V| \leq 4$. Let $N = G \cap V$. Then N is a 1–flat in G such that $P_1 \in N$ and $P_2 \notin N$.

Let Σ be a $(t-3)$–flat in G such that $N \subset \Sigma$ and $P_2 \notin \Sigma$. Let Π_j $(j = 1,2,3)$ be three hyperplanes in $PG(t,3)$, except for H_4, which contain $\Sigma \oplus E_4$. Since $|L_{i2} \cap \Pi_j| = 1$ and $W \cap \Pi_j = V$ for $i = 1,2,3$ and $j = 1,2,3$, there exists a hyperplane Π in $\{\Pi_1, \Pi_2, \Pi_3\}$ such that $|F \cap \Pi| = 3 + |K \cap V| \leq 7 < 2v_2$, a contradiction.

(B) In the case $E_4 \cap W = \{P_1\}$, let V be a 2–flat in W such that $P_1 \in V$, $P_2 \notin V$ and $|K \cap V| \leq 4$. Let Ω be a $(t-2)$–flat in H_4 such that $E_4 \subset \Omega$ and $\Omega \cap W = V$. Let Π_j $(j = 1,2,3)$ be three hyperplanes in $PG(t,3)$, except for H_4, which contain Ω. Since $W \cap \Pi_j = W \cap \Omega = V$ for $j = 1,2,3$, there exists a hyperplane Π in $\{\Pi_1, \Pi_2, \Pi_3\}$ such that $|F \cap \Pi| < 2v_2$, a contradiction. This completes the proof.

8. The proof of Lemma 3.4

Suppose there exists a $\{2v_1 + 2v_3, 2v_0 + 2v_2; t, 4\}$–minihyper F which satisfies the conditions (a) and (b) in Lemma 3.4 where $v_0 = 0$, $v_1 = 1$, $v_2 = 5$ and $v_3 = 21$. Then $F \cap G = \{P_1, P_2\}$ and $F \cap H_i = M_{i1} \cup M_{i2}$ for some 1–flats M_{i1} and M_{i2} in H_i such that $M_{i1} \cap M_{i2} = \emptyset$, $G \cap M_{i1} = \{P_1\}$ and $G \cap M_{i2} = \{P_2\}$ for $i = 1,2,3,4$.

Case I : (θ=1). It follows from Definition $III.2$ that $F \cap H_5 = K$ for some minihyper K in $\overline{\mathcal{F}_1}(0,0,1,1;t,4)$. Let V be the 2–flat which contains K. Let $E = H_5 \cap (M_{11} \oplus M_{21})$.

Then E is a 1–flat in H_5 such that $G \cap E = \{P_1\}$. Note that (i) $K \cap G = (F \cap H_5) \cap G = F \cap G = \{P_1, P_2\}$ and (ii) either $E \subset V$ or $E \cap V = \{P_1\}$.

(A) In the case $E \subset V$, let Ω be a $(t-2)$–flat in H_5 such that $\Omega \cap V = E$. Let Π_l $(l = 1,2,3,4)$ be four hyperplanes in $PG(t,4)$, except for H_5, which contain Ω. Since $P_2 \in M_{i2}$, $P_2 \in V$, $P_2 \notin \Omega$ and $P_2 \notin \Pi_l$ for $i = 1,2,3,4$ and $l = 1,2,3,4$, it follows that that $|M_{i2} \cap \Pi_l| = 1$ and $V \cap \Pi_l = V \cap \Omega = E$. Since $M_{11} \subset \Pi_a$, $M_{21} \subset \Pi_a$, $M_{31} \subset \Pi_b$ and $M_{41} \subset \Pi_c$ for some integers a, b and c in $\{1,2,3,4\}$, there exists at least one hyperplane Π in $\{\Pi_1, \Pi_2, \Pi_3, \Pi_4\}$ such that $\Pi \cap M_{i1} = \{P_1\}$ for $i = 1,2,3,4$. This implies that $|F \cap \Pi| = \sum_{i=1}^{4} |M_{i2} \cap \Pi| + |(F \cap H_5) \cap \Pi| \le 4 + |V \cap \Omega| = 9 < 2v_2$, a contradiction. Hence $V \cap E = \{P_1\}$. Note that $2v_2 = 10$ in Lemma 3.4.

(B) In the case $V \cap E = \{P_1\}$, let Ω be a $(t-2)$–flat in H_5 such that $E \subset \Omega$ and $\Omega \cap V = N$ where N is a 1–flat in V such that $P_1 \in N$ and $P_2 \notin N$. Using a method similar to (A), we have a contradiction. Hence $\theta \ne 1$.

Case II : (θ=2 or 3). Using a method similar to Case I, it can be shown that there exists a hyperplane Π in $PG(t,4)$ such that $|F \cap \Pi| < 2v_2$, a contradiction. Hence $\theta \ne 2, 3$.

Case III : (θ=4). It follows from Definition $III.2$ that $F \cap H_5 = L \cup K^*$ for some 1–flat L in H_5 and some minihyper K^* in $\overline{\mathcal{F}}(0,0,1;t,4)$ such that $L \cap K^* = \emptyset$ where $K^* = \{(\omega_0), (\omega_1), (\omega_2), (\omega_0 + \omega_1), (\omega_0 + \omega_2), (\omega_1 + \omega_2)(\omega_0 + \omega_1 + \omega_2)\}$ for some noncollinear points $(\omega_0), (\omega_1)$ and (ω_2) in H_5. Note that $|K^* \cap E| \le 3$ for any 1–flat E in V where V denotes the 2–flat generated by the three points $(\omega_0), (\omega_1)$ and (ω_2).

Since $|L \cap G| = 1$ or 5 and $(L \cup K^*) \cap G = (F \cap H_5) \cap G = F \cap G = \{P_1, P_2\}$, we can assume without loss of generality that $K^* \cap G = \{P_1\}$ and $L \cap G = \{P_2\}$. Let $E = H_5 \cap (M_{11} \oplus M_{21})$. Then either $E \subset V$ or $E \cap V = \{P_1\}$

(A) In the case $E \subset V$, let Ω be a $(t-2)$–flat in H_5 such that $P_2 \notin \Omega$ and $\Omega \cap V = E$. Let Π_l $(l = 1,2,3,4)$ be four hyperplanes in $PG(t,4)$, except for H_5, which contain Ω. Then $|M_{i2} \cap \Pi_l| = 1$, $|L \cap \Pi_l| = 1$ and $V \cap \Pi_l = E$ for $i = 1,2,3,4$ and $l = 1,2,3,4$. Hence there exists a hyperplane Π in $\{\Pi_1, \Pi_2, \Pi_3, \Pi_4\}$ such that $|F \cap \Pi| = \sum_{i=1}^{4} |M_{i2} \cap \Pi| + |L \cap \Pi| + |K^* \cap E| \le 8 < 2v_2$, a contradiction. Hence $V \cap E = \{P_1\}$.

(B) In the case $V \cap E = \{P_1\}$, let Ω be a $(t-2)$–flat in H_5 such that $E \subset \Omega$ and $\Omega \cap V = N$ where N is a 1–flat in V such that $P_1 \in N$ and $P_2 \notin N$. Using a method similar to (A), we have a contradiction. Hence $\theta \ne 4$. This completes the proof of Lemma 3.4.

Appendix I. Preliminary results in the general case

Let $U(t,q)$ denote the set of all ordered sets $(\varepsilon, \mu_1, \mu_2, \cdots, \mu_h)$ of integers ε, h and μ_i $(i = 1,2,\cdots,h)$ such that $0 \le \varepsilon \le q-1$, $1 \le h \le (t-1)(q-1)$, $1 \le \mu_1 \le \mu_2 \le \cdots \le \mu_h < t$ and $0 \le n_l(\underline{\mu}) \le q-1$ for $l = 1,2,\cdots,t-1$ where $n_l(\underline{\mu})$ denotes the number of integers μ_i in $\underline{\mu} = (\mu_1, \mu_2, \cdots, \mu_h)$ such that $\mu_i = l$ for the given integer l. In the case $k \ge 3$ and

$1 \le d < q^{k-1} - q$, d and the Griesmer bound can be expressed as follows.

$$d = q^{k-1} - \left(\varepsilon + \sum_{i=1}^{h} q^{\mu_i}\right) \quad and \quad n \ge v_k - \left(\varepsilon + \sum_{i=1}^{h} v_{\mu_i+1}\right) \tag{I.1}$$

using some ordered set $(\varepsilon, \mu_1, \mu_2, \cdots, \mu_h)$ in $U(k-1, q)$ where $v_l = (q^l - 1)/(q-1)$ for any integer $l \ge 0$.

Let $\mathcal{F}_U(\varepsilon, \mu_1, \mu_2, \cdots, \mu_h; t, q)$ denote the family of all unions of ε points, a μ_1–flat, a μ_2–flat, $\cdots$, a μ_h–flat in $PG(t,q)$ which are mutually disjoint where $(\varepsilon, \mu_1, \mu_2, \cdots, \mu_h) \in U(t,q)$. As occasion demands, we shall denote $\mathcal{F}_U(\varepsilon, \mu_1, \mu_2, \cdots, \mu_h; t, q)$ by $\mathcal{F}(\lambda_1, \lambda_2, \cdots, \lambda_\eta; t, q)$ where $\eta = h + \varepsilon$, $\lambda_i = 0$ $(i = 1, 2, \cdots, \varepsilon)$ and $\lambda_{\varepsilon+j} = \mu_j$ $(j = 1, 2, \cdots, h)$. For example, $\mathcal{F}(\alpha, \beta, \gamma, \delta; t, q)$ denotes the family of all unions $\bigcup_{i=1}^{4} V_i$ of an α–flat V_1, a β–flat V_2, a γ–flat V_3 and a δ–flat V_4 in $PG(t,q)$ which are mutually disjoint where $0 \le \alpha \le \beta \le \gamma \le \delta < t$.

In order to prove Theorem 1.1, we prepare the following propositions which play an important role in solving Problem 1.1 or 1.2.

Proposition I.1. (Hamada [4]). (1) If $F \in \mathcal{F}_U(\varepsilon, \mu; t, q)$ in the case $t \ge \mu + 1$, then F is a $\{\varepsilon v_1 + v_{\mu+1}, \varepsilon v_0 + v_\mu; t, q\}$–minihyper.

(2) If $F \in \mathcal{F}_U(\varepsilon, \mu_1, \mu_2, \cdots, \mu_h; t, q)$ in the case $h \ge 2$ and $t \ge \mu_{h-1} + \mu_h + 1$, then F is a $\left\{\varepsilon v_1 + \sum_{i=1}^{h} v_{\mu_i+1}, \varepsilon v_0 + \sum_{i=1}^{h} v_{\mu_i}; t, q\right\}$–minihyper.

Remark I.1. It is known (cf. Theorems 2.2 and 2.3 in Hamada and Tamari [20]) that $\mathcal{F}_U(\varepsilon, \mu_1, \mu_2, \cdots, \mu_h; t, q) \ne \emptyset$ if and only if either (a) $h = 1$ or (b) $h \ge 2$ and $t \ge \mu_{h-1} + \mu_h + 1$ where $(\varepsilon, \mu_1, \mu_2, \cdots, \mu_h) \in U(t,q)$.

Proposition I.2. (Hamada [3]). If there exists a $\{\varepsilon v_1 + \sum_{i=1}^{h} v_{\mu_i+1}, \varepsilon v_0 + \sum_{i=1}^{h} v_{\mu_i}; t, q\}$–minihyper F for some ordered set $(\varepsilon, \mu_1, \mu_2, \cdots, \mu_h)$ in $U(t,q)$, then $| F \cap \Delta | \ge \sum_{i=1}^{h} v_{\mu_i - 1}$ for any $(t-2)$–flat Δ in $PG(t,q)$ and $| F \cap G | = \sum_{i=1}^{h} v_{\mu_i - 1}$ for some $(t-2)$–flat G in $PG(t,q)$. Let H_j $(j = 1, 2, \cdots, q+1)$ be $q+1$ hyperplanes in $PG(t,q)$ which contain G. Then $F \cap H_j$ is a $\left\{\delta_j v_1 + \sum_{i=1}^{h} v_{\mu_i}, \delta_j v_0 + \sum_{i=1}^{h} v_{\mu_i - 1}; t, q\right\}$–minihyper in H_j for $j = 1, 2, \cdots, q+1$ where the δ_j are nonnegative integers such that $\sum_{j=1}^{q+1} \delta_j = \varepsilon$.

Remark I.2. (1) For any $(t-2)$–flat G in $PG(t,q)$, there are $q+1$ hyperplanes in $PG(t,q)$ which contain G.

(2) There exists an $\{f,m;t,q\}$−minihyper F such that $F \subset \Omega$ for some θ−flat Ω in $PG(t,q)$ if and only if there exists an $\{f,m;\theta,q\}$−minihyper where $2 \le \theta < t$ and $0 \le m < f < v_{\theta+1}$.

Proposition I.3. (Hamada [4]). Let $(\varepsilon,\mu_1,\mu_2,\cdots,\mu_h)$ be an ordered set in $U(t,q)$ such that either (α) $h=1$ and $\mu_1 \ge 2$, (β) $h \ge 2$, $\varepsilon = 0$, $\mu_1 = 1$, $\mu_2 \ge 2$ and $t \ge \mu_{h-1}+\mu_h$ or (γ) $h \ge 2$, $\mu_1 \ge 2$ and $t \ge \mu_{h-1}+\mu_h$. Let δ_j $(j=1,2,\cdots,q+1)$ be nonnegative integers such that $\sum_{j=1}^{q+1}\delta_j = \varepsilon$. If there exists a $\{\varepsilon v_1 + \sum_{i=1}^{h} v_{\mu_i+1}, \varepsilon v_0 + \sum_{i=1}^{h} v_{\mu_i}; t,q\}$−minihyper F such that (a) $|F \cap G| = \sum_{i=1}^{h} v_{\mu_i-1}$ for some $(t-2)$−flat G in $PG(t,q)$ and (b) $F \cap H_j \in \mathcal{F}_U(\delta_j,\mu_1-1,\mu_2-1,\cdots,\mu_h-1;t,q)$ for any hyperplane H_j $(1 \le j \le q+1)$ which contains G, then $F \in \mathcal{F}_U(\varepsilon,\mu_1,\mu_2,\cdots,\mu_h;t,q)$.

Proposition I.4. (Hamada [3]). Let t, q, ε, h and μ_i $(i=1,2,\cdots,h)$ be any integers such that $t \ge 2$, $q \ge 3$, $\varepsilon = 0$ or 1, $1 \le h < t$ and $1 \le \mu_1 < \mu_2 < \cdots < \mu_h < t$.

(1) In the case $h=1$, F is a $\{\varepsilon v_1 + v_{\mu+1}, \varepsilon v_0 + v_\mu; t,q\}$−minihyper if and only if $F \in \mathcal{F}_U(\varepsilon,\mu;t,q)$.

(2) In the case $h \ge 2$ and $t \ge \mu_{h-1}+\mu_h+1$, F is a $\left\{\varepsilon v_1 + \sum_{i=1}^{h} v_{\mu_i+1}, \varepsilon v_0 + \sum_{i=1}^{h} v_{\mu_i}; t,q\right\}$−minihyper if and only if $F \in \mathcal{F}_U(\varepsilon,\mu_1,\mu_2,\cdots,\mu_h;t,q)$.

(3) In the case $h \ge 2$ and $t \le \mu_{h-1}+\mu_h$, there is no $\left\{\varepsilon v_1 + \sum_{i=1}^{h} v_{\mu_i+1}, \varepsilon v_0 + \sum_{i=1}^{h} v_{\mu_i}; t,q\right\}$−minihyper.

Proposition I.5. (Hamada [7]). Let $t \ge 2$, $q \ge 3$ and $1 \le \mu < t$.

(1) In the case $t \le 2\mu$, there is no $\{2v_{\mu+1}, 2v_\mu; t,q\}$−minihyper.

(2) In the case $t \ge 2\mu+1$, F is a $\{2v_{\mu+1}, 2v_\mu; t,q\}$−minihyper if and only if $F \in \mathcal{F}(\mu,\mu;t,q)$.

Definition I.1. Let V be a θ−flat in $PG(t,q)$ where $2 \le \theta \le t$. A set S of m points in V is called an m−arc in V if no θ+1 points in S are linearly dependent where $m \ge \theta+1$. For convenience sake, a set S of θ points in V is also called an θ−arc in V if θ points in S are linearly independent.

Remark I.3. Let ε_{i1} and ε_{i2} $(i = 0,1,\cdots,t-1)$ be nonnegative integers such that $0 \le \varepsilon_{i1}+\varepsilon_{i2} \le q-1$. If F_j is a $\left\{\sum_{i=0}^{t-1}\varepsilon_{ij}v_{i+1}, \sum_{i=1}^{t-1}\varepsilon_{ij}v_i; t,q\right\}$−minihyper for $j=1,2$ and $F_1 \cap F_2 = \emptyset$, then $F_1 \cup F_2$ is a $\left\{\sum_{i=0}^{t-1}(\varepsilon_{i1}+\varepsilon_{i2})v_{i+1}, \sum_{i=1}^{t-1}(\varepsilon_{i1}+\varepsilon_{i2})v_i; t,q\right\}$−minihyper (cf. (1), (2) and (3) in Definition *II*.2 and (4) in Definition *III*.2).

Definition I.2. Let V and W be a μ–flat and a ν–flat in $PG(t,q)$, respectively, such that $V \cap W$ is an m–flat in $PG(t,q)$ where $0 \leq m \leq \mu \leq \nu < t$. Let $V \oplus W$ denote the $(\mu+\nu-m)$–flat in $PG(t,q)$ which contain two flats V and W. In the special case $V \cap W = \emptyset$ (i.e., $m = -1$), $V \oplus W$ denotes the $(\mu+\nu+1)$–flat in $PG(t,q)$ which contains V and W.

Proposition I.6. (Hamada [7]). Let ε_i $(i = 1, 2, \cdots, t-1)$ be integers such that either (a) $0 \leq \varepsilon_i \leq q-1$ for $i = 1, 2, \cdots, t-1$ or (b) $\varepsilon_1 = \varepsilon_2 = \cdots = \varepsilon_{\lambda-1} = 0$, $\varepsilon_\lambda = q$, $0 \leq \varepsilon_{\lambda+1} \leq q-1$, $0 \leq \varepsilon_{t-1} \leq q-1$ for some integer λ in $\{1, 2, \cdots, t-1\}$. Let H be a θ–flat in $PG(t,q)$ where $2 \leq \theta \leq t$. If F is a set of points in H such that $|F \cap G| \geq \sum_{i=1}^{t-1} \varepsilon_i v_i$ for any $(\theta-1)$–flat G in H, then $|F \cap H| \geq \sum_{i=1}^{t-1} \varepsilon_i v_{i+1}$.

Appendix II. Preliminary results in the case q=3

In this appendix, let $q = 3$ and $v_l = (3^l - 1)/(3-1)$ for any integer $l \geq 0$.

Proposition II.1. (Hamada and Helleseth [16], [17]). Let t, α, β and γ be integers such that either (a) $0 \leq \alpha = \beta < \gamma < t$ and $\gamma \neq \alpha + 1$ or (b) $0 \leq \alpha < \beta = \gamma < t$ and $\gamma \neq \alpha + 1$.

(1) In the case $t \leq \beta + \gamma$, there is no $\{v_{\alpha+1} + v_{\beta+1} + v_{\gamma+1}, v_\alpha + v_\beta + v_\gamma; t, 3\}$–minihyper.

(2) In the case $t \geq \beta+\gamma+1$, F is a $\{v_{\alpha+1} + v_{\beta+1} + v_{\gamma+1}, v_\alpha + v_\beta + v_\gamma; t, 3\}$– minihyper if and only if $F \in \mathcal{F}(\alpha, \beta, \gamma; t, 3)$.

Definition II.1. (1) Let $\overline{\mathcal{F}}(0,1,1;t,3)$ denote the family of all sets K in $PG(t,3)$ such that $K = V \setminus S$ for some 2–flat V in $PG(t,3)$ and some 4–arc S in V where $t \geq 2$ (cf. Definition $I.1$ in Appendix I).

(2) Let $\overline{\mathcal{F}}(0,0,1;t,3)$ denote the family of all sets K in $PG(t,3)$ such that $K = \{(\nu_1), (\nu_0+\nu_1), (2\nu_0+\nu_1), (\nu_2), (\nu_1+\nu_2), (\nu_0+2\nu_1+\nu_2)\}$ for some noncollinear points (ν_0), (ν_1) and (ν_2) in $PG(t,3)$.

Remark II.1. In this appendix, (ν) and (2ν) represent the same point in $PG(t,3)$ for any nonzero element ν in the Galois field $GF(3^{t+1})$.

Remark II.2. In (2) of Definition $II.1$, let $\xi_0 = \nu_1$, $\xi_1 = 2\nu_0 + 2\nu_1$ and $\xi_2 = \nu_1 + \nu_2$. Then $K = \{(\xi_0), (\xi_1), (\xi_2), (2\xi_0+\xi_1), (2\xi_0+\xi_2), (2\xi_1+\xi_2)\}$ (cf. (4) in Definition $II.2$).

Proposition II.2. (Hamada [6], [7]). (1) In the case $t \geq 2$, F is a $\{2v_1 + v_2, 2v_0 + v_1; t, 3\}$– minihyper if and only if either $F \in \mathcal{F}(0,0,1;t,3)$ or $F \in \overline{\mathcal{F}}(0,0,1;t,3)$.

(2) In the case $t = 2$, F is a $\{v_1 + 2v_2, v_0 + 2v_1; 2, 3\}$–minihyper if and only if $F \in \overline{\mathcal{F}}(0,1,1;2,3)$.

(3) In the case $t \geq 3$, F is a $\{v_1 + 2v_2, v_0 + 2v_1; t, 3\}$–minihyper if and only if either $F \in \mathcal{F}(0,1,1;t,3)$ or $F \in \overline{\mathcal{F}}(0,1,1;t,3)$.

Definition II.2. (1) Let $\overline{\mathcal{F}_1}(0,0,1,1;t,3)$ denote the family of all sets K in $PG(t,3)$ such that $K = V \setminus S$ for some 2−flat V in $PG(t,3)$ and some 3−arc S in V where $t \geq 2$.

(2) Let $\overline{\mathcal{F}_2}(0,0,1,1;t,3)$ denote the family of all sets K in $PG(t,3)$ such that $K = (V \setminus S) \cup \{P\}$ for some 2−flat V in $PG(t,3)$, some 4−arc S in V and some point $P \notin V$ where $t \geq 3$.

(3) Let $\overline{\mathcal{F}_3}(0,0,1,1;t,3)$ denote the family of all sets K in $PG(t,3)$ such that $K = L \cup K^*$ for some 1−flat L in $PG(t,3)$ and some minihyper K^* in $\overline{\mathcal{F}}(0,0,1;t,3)$ such that $L \cap K^* = \emptyset$ where $t \geq 2$.

(4) Let $\overline{\mathcal{F}_4}(0,0,1,1;t,3)$ denote the family of all sets K in $PG(t,3)$ such that $K = \{(\xi_0),(\xi_1),(\xi_2),(\xi_3),(2\xi_0+\xi_1),(2\xi_0+\xi_2),(2\xi_0+\xi_3),(2\xi_1+\xi_2), (2\xi_1+\xi_3),(2\xi_2+\xi_3)\}$ for some four linearly independent points (ξ_0), (ξ_1), (ξ_2) and (ξ_3) in $PG(t,3)$ where $t \geq 3$.

Remark II.3. In Theorem 2.2 of Hamada, Helleseth and Ytrehus [18], let $\xi_0 = 2\nu_0$, $\xi_1 = c_1\nu_1$, $\xi_2 = c_2\nu_2$ and $\xi_3 = c_3\nu_3$. Then the set of 10 points in Theorem 2.2 can be expressed as K in (4) of Definition $II.2$.

Proposition II.3. (Hamada [6] and Hamada, Helleseth and Ytrehus [18]). (1) In the case $t = 2$, F is a $\{2v_1+2v_2, 2v_0+2v_1; 2,3\}$− minihyper if and only if $F \in \overline{\mathcal{F}_1}(0,0,1,1;2,3)$.

(2) In the case $t \geq 3$, F is a $\{2v_1+2v_2, 2v_0+2v_1; t,3\}$−minihyper if and only if either $F \in \mathcal{F}(0,0,1,1;t,3)$ or $F \in \overline{\mathcal{F}_i}(0,0,1,1;t,3)$ for some i in $\{1,2,3,4\}$.

Appendix III. Preliminary results in the case q=4

In this appendix, let $q = 4$ and $v_l = (4^l - 1)/(4-1)$ for any integer $l \geq 0$.

Proposition III.1. (Hamada and Helleseth [13]). Let t, α, β and γ be integers such that either (a) $0 \leq \alpha < \beta \leq \gamma < t$ or (b) $0 \leq \alpha = \beta < \gamma < t$ and $\gamma \neq \alpha + 1$.

(1) In the case $t \leq \beta + \gamma$, there is no $\{v_{\alpha+1}+v_{\beta+1}+v_{\gamma+1}, v_\alpha+v_\beta+v_\gamma; t,4\}$−minihyper.

(2) In the case $t \geq \beta+\gamma+1$, F is a $\{v_{\alpha+1}+v_{\beta+1}+v_{\gamma+1}, v_\alpha+v_\beta+v_\gamma; t,4\}$− minihyper if and only if $F \in \mathcal{F}(\alpha,\beta,\gamma;t,4)$.

Definition III.1. Let $\overline{\mathcal{F}}(0,0,1;t,4)$ denote the family of all sets K in $PG(t,4)$ such that $K = \{(w_0),(w_1),(w_2),(w_0+w_1),(w_0+w_2),(w_1+w_2),(w_0+w_1+w_2)\}$ for some noncollinear points (w_0), (w_1) and (w_2) in $PG(t,4)$ where $t \geq 2$.

Proposition III.2. (Hamada [6]). In the case $t \geq 2$, F is a $\{2v_1+v_2, 2v_0+v_1; t,4\}$−minihyper if and only if either $F \in \mathcal{F}(0,0,1;t,4)$ or $F \in \overline{\mathcal{F}}(0,0,1;t,4)$.

Definition III.2. (1) Let $\overline{\mathcal{F}_1}(0,0,1,1;t,4)$ denote the family of all sets K in $PG(t,4)$ such that $K = L_0 \cup L_1 \cup \{(c_0w_0+w_1+w_2),(c_1w_0+\alpha w_1+w_2),(c_2w_0+\alpha^2w_1+w_2)\}$ for some noncollinear points (w_0), (w_1) and (w_2) in $PG(t,4)$ and some elements c_0, c_1 and c_2

in $\{0,1,\alpha,\alpha^2\}$ where $t \geq 2$, $L_0 = (w_0) \oplus (w_1)$, $L_1 = (w_0) \oplus (w_2)$ and α is a primitive element in $GF(2^2)$ such that $\alpha^2 = \alpha + 1$.

(2) Let $\overline{\mathcal{F}_2}(0,0,1,1;t,4)$ denote the family of all sets K in $PG(t,4)$ such that $K = L_0 \cup \{(w_2),(w_1+w_2),(w_0+w_1+w_2),(w_0+\alpha w_1+w_2),(\alpha^2 w_0+\alpha w_1+w_2),(w_0+\alpha^2 w_1+w_2),(\alpha w_0+\alpha^2 w_1+w_2)\}$ for some noncollinear points (w_0), (w_1) and (w_2) in $PG(t,4)$ where $t \geq 2$ and $L_0 = (w_0) \oplus (w_1)$.

(3) Let $\overline{\mathcal{F}_3}(0,0,1,1;t,4)$ denote the family of all sets K in $PG(t,4)$ such that $K = (L_0\backslash\{(w_1)\}) \cup (L_1\backslash\{(w_2)\}) \cup (L_2\backslash\{(w_1+w_2)\}) \cup \{(\alpha w_1+w_2),(\alpha^2 w_1+w_2)\}$ for some noncollinear points (w_0), (w_1) and (w_2) in $PG(t,4)$ where $t \geq 2$, $L_0 = (w_0) \oplus (w_1)$, $L_1 = (w_0) \oplus (w_2)$ and $L_2 = (w_0) \oplus (w_1+w_2)$.

(4) Let $\overline{\mathcal{F}_4}(0,0,1,1;t,4)$ denote the family of all sets K in $PG(t,4)$ such that $K = L \cup K^*$ for some $1-$flat L in $PG(t,4)$ and some minihyper K^* in $\overline{\mathcal{F}}(0,0,1;t,4)$ such that $L \cap K^* = \emptyset$ where $t \geq 3$.

Remark III.1. If $c_2 = c_0\alpha + c_1\alpha^2$ in (1) of Definition $III.2$, then K in (1) contains three $1-$flats L_0, L_1 and L^* where $L^* = (c_0 w_0 + w_1 + w_2) \oplus (c_1 w_0 + \alpha w_1 + w_2)$ (cf. Definition $I.2$ with respect to the notation $\oplus$).

Proposition III.3. (Hamada [5], [6]). In the case $t = 2$, F is a $\{2v_1 + 2v_2, 2v_0 + 2v_1; 2, 4\}-$minihyper if and only if $F \in \overline{\mathcal{F}_i}(0,0,1,1;2,4)$ for some i in $\{1,2,3\}$.

Proposition III.4. (Hamada, Helleseth and Ytrehus [18]). In the case $t \geq 3$, F is a $\{2v_1 + 2v_2, 2v_0 + 2v_1; t, 4\}-$minihyper if and only if either $F \in \mathcal{F}(0,0,1,1;t,4)$ or $F \in \overline{\mathcal{F}_i}(0,0,1,1;t,4)$ for some i in $\{1,2,3,4\}$.

Appendix IV. The correspondence between minihypers and codes meeting the Griesmer bound

Let $S(k,q)$ be the set of all column vectors $\mathbf{c}$, $\mathbf{c}' = (c_0, c_1, \cdots, c_{k-1})$, in $W(k,q)$ such that either $c_{k-1} = 1$ or $c_i = 1$, $c_{i+1} = c_{i+2} = \cdots = c_{k-1} = 0$, for some integer i in $\{0,1,\cdots,k-2\}$ where $k \geq 3$ and $W(k,q)$ denotes a $k-$dimensional vector space consisting of column vectors over $GF(q)$. Then $S(k,q)$ consists of $(q^k-1)/(q-1)$ nonzero vectors in $W(k,q)$ and there is no element σ in $GF(q)$ such that $\mathbf{x}_2 = \sigma \mathbf{x}_1$ for any two vectors $\mathbf{x}_1$ and $\mathbf{x}_2$ in $S(k,q)$. Hence the $(q^k-1)/(q-1)$ nonzero vectors in $S(k,q)$ may be regarded as $(q^k-1)/(q-1)$ points in $PG(k-1,q)$.

Proposition IV.1. (Hamada[4]). Let F be a set of f vectors in $S(k,q)$ and let C be the subspace of $V(n,q)$ generated by a $k \times n$ matrix (denoted by G) whose column vectors are all the vectors in $S(k,q) \setminus F$ where $n = v_k - f$, $1 \leq f < v_k - 1$ and $v_i = (q^i-1)/(q-1)$ for any integer $i \geq 0$.

(1) Let $H_{\mathbf{z}} = \{\mathbf{y} \epsilon S(k,q) | \ \mathbf{z}'\mathbf{y} = 0 \text{ over } GF(q)\}$ for a nonzero vector $\mathbf{z}$ in $W(k,q)$. Then $H_{\mathbf{z}}$ is a hyperplane in $PG(k-1,q)$ and the weight of the code vector $\mathbf{z}'G$ in C is equal

to $|F \cap H_z| + q^{k-1} - f$, i.e.,

$$w(z'G) = |F \cap H_z| + q^{k-1} - f, \tag{IV.1}$$

where $w(\mathbf{x})$ and $\mathbf{z}'$ denote the number of nonzero elements in the vector $\mathbf{x}$ and the transpose of the vector $\mathbf{z}$, respectively.

(2) In the case $k \geq 3$ and $1 \leq d < q^{k-1}$, C is an $[n, k, d; q]$−code meeting the Griesmer bound if and only if F is a $\{v_k - n, v_{k-1} - n + d; k-1, q\}$−minihyper.

Remark IV.1. For any two $k \times n$ matrices G_1 and G_2 whose column vectors are all the vectors in $S(k,q) \setminus F$, there exists an $n \times n$ permutation matrix P such that $G_2 = G_1 P$.

Let C be an $[n, k, d; q]$−code meeting the Griesmer bound for some integers k, d and q such that $k \geq 3$ and $1 \leq d < q^{k-1}$. Let $A = [\mathbf{a}_1, \mathbf{a}_2, \cdots, \mathbf{a}_n]$ be a $k \times n$ generator matrix of C. Then there exists a unique vector $\mathbf{b}_i$ in $S(k,q)$ for each vector $\mathbf{a}_i$ in A such that $\mathbf{b}_i = \sigma_i \mathbf{a}_i$ for some nonzero element σ_i in $GF(q)$. This implies that $B = AD$ for a nonsingular diagonal matrix D=diag$(\sigma_1, \sigma_2, \cdots, \sigma_n)$ where $B = [\mathbf{b}_1, \mathbf{b}_2, \cdots, \mathbf{b}_n]$. Hence we introduce an equivalence relation among $[n, k, d; q]$−codes as follows:

Definition IV.1. Two $[n, k, d; q]$−codes C_1 and C_2 are said to be equivalent if there exists a $k \times n$ matrix G_2 of the code C_2 such that $G_2 = G_1 PD$ (or $G_2 = G_1 DP$) for some permutation matrix P and some nonsingular diagonal matrix D with entries from $GF(q)$, where G_1 is a $k \times n$ generator matrix of C_1.

From Proposition IV.1 and Definition IV.1, we have

Proposition IV.2. In the case $k \geq 3$ and $1 \leq d < q^{k-1}$, there is a one-to-one correspondence between the set of all nonequivalent $[n, k, d; q]$−codes meeting the Griesmer bound and the set of all $\{v_k - n, v_{k-1} - n + d; k-1, q\}$−minihypers.

Corollary IV.3. In the case $k \geq 3$ and $d = q^{k-1} - \left(\varepsilon + \sum_{i=1}^{h} q^{\mu_i}\right)$, there is a one-to-one correspondence between the set of all nonequivalent $[n, k, d; q]$−codes meeting the Griesmer bound and the set of all $\left\{\varepsilon v_1 + \sum_{i=1}^{h} v_{\mu_i+1}, \varepsilon v_0 + \sum_{i=1}^{h} v_{\mu_i}; k-1, q\right\}$−minihypers where $n = v_k - \left(\varepsilon v_1 + \sum_{i=1}^{h} v_{\mu_i+1}\right)$.

Corollary IV.4. In the case $k \geq 3$ and $d = q^{k-1} - \sum_{i=1}^{\eta} q^{\lambda_i}$, there is a one-to-one correspondence between the set of all nonequivalent $[n, k, d; q]$−codes meeting the Griesmer bound and the set of all $\left\{\sum_{i=1}^{\eta} v_{\lambda_i+1}, \sum_{i=1}^{\eta} v_{\lambda_i}; k-1, q\right\}$−minihypers where $n = v_k - \sum_{i=1}^{\eta} v_{\lambda_i+1}$.

References

1. A.A. Bruen and R. Silverman, Arcs and blocking sets II, Europ. J. Combin. 8 (1987), 351–356.
2. J.H. Griesmer, A bound for error-correcting codes, IBM J. Res. Develop. 4 (1960), 532–542.
3. N. Hamada, Characterization resp. nonexistence of certain q−ary linear codes attaining the Griesmer bound, Bull. Osaka Women's Univ. 22 (1985), 1–47.
4. N. Hamada, Characterization of min·hypers in a finite projective geometry and its applications to error-correcting codes, Bull. Osaka Women's Univ. 24 (1987), 1–24.
5. N. Hamada, Characterization of $\{12,2;2,4\}$−min·hypers in a finite projective geometry $PG(2,4)$, Bull. Osaka Women's Univ. 24 (1987), 25–31.
6. N. Hamada, Characterization of $\{(q+1)+2,1;t,q\}$−min·hypers and $\{2(q+1)+2,2;2,q\}$−min·hypers in a finite projective geometry, Graphs and Combin. 5 (1989), 63–81.
7. N. Hamada, A characterization of some $(n,k,d;q)$−codes meeting the Griesmer bound using a minihyper in a finite projective geometry, to appear in Discrete Math., In Chapter 4 in "Combinatorial Aspect of Design Experiments".
8. N. Hamada and M. Deza, A characterization of some $(n,k,d;q)$−codes meeting the Griesmer bound for given integers $k \geq 3$, $q \geq 5$ and $d = q^{k-1} - q^{\alpha} - q^{\beta} - q^{\gamma}$ $(0 \leq \alpha \leq \beta < \gamma < k-1$ or $0 \leq \alpha < \beta \leq \gamma < k-1)$, In: First Sino-Franco Conference on Combinatorics, Algorithms, and Coding Theory, Bull. Inst. Math. Academia Sinica 16 (1988), 321–338.
9. N. Hamada and M. Deza, Characterization of $\{2(q+1)+2,2;t,q\}$−min·hyper in $PG(t,q)$ $(t \geq 3,\ q \geq 5)$ and its applications to error-correcting codes, Discrete Math. 71 (1988), 219–231.
10. N. Hamada and M. Deza, A characterization of $\{v_{\mu+1}+\varepsilon, v_{\mu}; t, q\}$−min·hyper and its applications to error-correcting codes and factorial designs, J. Statist. Plann. Inference 22 (1989), 323–336.
11. N. Hamada and M. Deza, A characterization of $\{2v_{\alpha+1}+2v_{\beta+1}, 2v_{\alpha}+2v_{\beta}; t, q\}$−minhypers in $PG(t,q)$ $(t \geq 2,\ q \geq 5$ and $0 \leq \alpha < \beta < t)$ and its applications to error-correcting codes, Discrete Math. 91 (1991), xxx-xxx.
12. N. Hamada and T. Helleseth, A characterization of some $\{3v_2, 3v_1; t, q\}$− minihypers and some $\{2v_2+v_{\gamma+1}, 2v_1+v_{\gamma}; t, q\}$−minihypers $(q = 3$ or $4,\ 2 \leq \gamma < t)$ and its applications to error-correcting codes, Bull. Osaka Women's Univ. 27 (1990), 49–107.
13. N. Hamada and T. Helleseth, A characterization of some minihypers in a finite projective geometry $PG(t,4)$, Europ. J. Combin. 11 (1990) 541–548.
14. N. Hamada and T. Helleseth, A characterization of some linear codes over $GF(4)$ meeting the Griesmer bound, to appear in Mathematica Japonica 37 (1992).
15. N. Hamada and T. Helleseth, A characterization of some $\{3v_{\mu+1}, 3v_{\mu}; k-1, q\}$− minihypers and some $(n, k, q^{k-1} - 3q^{\mu}; q)$−codes $(k \geq 3,\ q \geq 5,\ 1 \leq \mu < k-1)$ meeting the Griesmer bound, submitted for publication.
16. N. Hamada and T. Helleseth, A characterization of some $\{2v_{\alpha+1}+v_{\gamma+1}, 2v_{\alpha}+v_{\gamma}; k-1, 3\}$−minihypers and some $(n, k, 3^{k-1} - 2\cdot 3^{\alpha} - 3^{\gamma}; 3)$−codes $(k \geq 3,\ 0 \leq \alpha < \gamma < k-1)$ meeting the Griesmer bound, to appear in Discrete Math.

17. N. Hamada and T. Helleseth, A characterization of some minihypers in $PG(t,3)$ and some ternary codes meeting the Griesmer bound, submitted for publication.
18. N. Hamada, T. Helleseth and Ø. Ytrehus, Characterization of $\{2(q+1)+2,2;t,q\}$-minihypers in $PG(t,q)$ ($t \geq 3$, $q \in \{3,4\}$), to appear in Discrete Math.
19. N. Hamada and F. Tamari, On a geometrical method of construction of maximal t-linearly independent sets, J. Combin. Theory 25 (A) (1978), 14–28.
20. N. Hamada and F. Tamari, Construction of optimal linear codes using flats and spreads in a finite projective geometry, Europ. J. Combin. 3 (1982), 129–141.
21. T. Helleseth, A characterization of codes meeting the Griesmer bound, Inform. and Control 50 (1981), 128–159.
22. R. Hill, Caps and codes, Discrete Math. 22 (1978), 111–137.
23. G. Solomon and J.J. Stiffler, Algebraically punctured cyclic codes, Inform. and Control 8 (1965), 170–179.
24. F. Tamari, A note on the construction of optimal linear codes, J. Statist. Plann. Inference 5 (1981), 405–411.
25. F. Tamari, On linear codes which attain the Solomon-Stiffler bound, Discrete Math. 49 (1984), 179–191.

Deligne-Lusztig Varieties and Group Codes

by

Johan P. Hansen[1]

Matematisk Institut, Aarhus Universitet, Ny Munkegade, 8000 Aarhus C, Denmark

e-mail: matjph@mi.aau.dk

Abstract

We construct algebraic geometric codes using the Deligne-Lusztig varieties [De-Lu] associated to a connected reductive algebraic group G defined over a finite field $\mathbb{F}_q$, with Frobenius map F. The codes are obtained as geometric Goppa codes, that is linear error-correcting codes constructed from algebraic varieties [Go1] and [Go2]. The finite group G^F of Lie type acts as $\mathbb{F}_q$-rational automorphisms on the codes and they become modules over the group algebra $\mathbb{F}_q[G^F]$. Algebraic geometric codes with a group algebra structure induced from automorphisms of the underlying variety have been constructed and studied in [Ha1], [Ha2], [Ha-St] and [V].

The Deligne-Lusztig varieties used in the construction of the codes have in some cases many $\mathbb{F}_q$-rational points, which ensures that the codes have a large word length. In case G is of type ${}^2A_2$ the Deligne-Lusztig curve considered have $1+q^3$ points over $\mathbb{F}_{q^2}$. In case G is a Suzuki group ${}^2B_2$, respectively a Ree group ${}^2G_2$, the Deligne-Lusztig curves considered have $1+q^2$, respectively $1+q^3$, points over $\mathbb{F}_q$. In relation to their genera these numbers are maximal as determined by the "explicit formulas" of Weil.

Tabel of Contents

[1]Supported by the Danish National Science Foundation.

Introduction

Many classical linear error-correcting codes can be realized as ideals in group-algebras or as modules over group-algebras. S. D. Berman [Be] etablished that the Reed-Muller codes over $\mathbb{F}_2$ are ideals in a group-algebra over an elementary abelian 2-group. This was generalized by P. Charpin [Ch1], [Ch2], [Ch3] and P. Landrock and O. Manz [La-Ma], who showed that any Generalized Reed-Muller code is an ideal in a group-algebra over an elementary abelian p-group.

In [Ha1], [Ha2] and [Ha-St] H. Stichtenoth and the author construct algebraic geometric error-correcting codes with a group algebra structure. In [V] S. G. Vladut shows that the asymptotically good codes on classical and Drinfeld modular curves [T-V-Z], [M-V] constructed by Yu. I. Manin, M. A. Tsfasman, Th. Zink and himself can be realized as group codes.

The general setup, which is treated in section 1, is to consider an algebraic variety X, defined over a finite field $\mathbb{F}_q$, with a group G of $\mathbb{F}_q$-rational automorphisms. On the variety X we consider an $\mathbb{F}_q$-rational and G-invariant divisor D together with a G-stable set $P_1, P_2, \ldots, P_n$ of $\mathbb{F}_q$-rational points on X, none in the support of D. The associated Goppa code [Go1] and [Go2] is the image $C = \phi(\mathrm{L}(D)) \subseteq \mathbb{F}_q^n$ of the $\mathbb{F}_q$-linear map:

$$\phi: \mathrm{L}(D) \rightarrow \mathbb{F}_q^n$$
$$f \mapsto (f(P_1), \ldots, f(P_n)).$$

The group G acts on the G-stable set $P_1, P_2, \ldots, P_n$ of $\mathbb{F}_q$-rational points giving $\mathbb{F}_q^n$ a $\mathbb{F}_q[G]$-module structure. As the divisor D is G-stable, G acts on $\mathrm{L}(D)$ and it becomes a $\mathbb{F}_q[G]$-module. The $\mathbb{F}_q$-linear map:

$$\phi: \mathrm{L}(D) \rightarrow \mathbb{F}_q^n$$

becomes a $\mathbb{F}_q[G]$-morphism, and the geometric Goppa code

$$C = \phi(\mathrm{L}(D)) \subseteq \mathbb{F}_q$$

a $\mathbb{F}_q[G]$-module.

In [Ha1] a series of geometric Goppa codes is obtained from the Klein quartic, codes which are ideals in the group-algebra $\mathbb{F}_8[G]$, where G is a Frobenius group of order 21. In [Ha2], [St2] and [Ti] series of geometric Goppa codes are constructed from Hermitian curves, among these there are codes that are ideals in the group-algebra $\mathbb{F}_{q^2}[G]$, G being a non-abelian p-group of order q^3, $q = p^n$, p prime. In [Ha-St] another series of group codes are presented. The codes are ideals in $\mathbb{F}_q[S]$, where S is a Sylow-2-subgroup of order q^2 of the Suzuki-group of order $q^2(q-1)(q^2+1)$ and $q = 2^{2m+1}$. The codes are geometric Goppa codes over $\mathbb{F}_q$ with good parameters.

Section 2 introduces another series of varieties with large groups of automorphisms. The varieties are Deligne-Lusztig varieties associated to a connected reductive algebraic groups G defined over a finite field $\mathbb{F}_q$, with Frobenius map $F: G \rightarrow G$ [De-Lu]. Specifically, let G be a connected, reductive algebraic group G and let X_G be the $\mathbb{F}_q$-scheme of all Borel subgroups of G with Frobenius morphism $F: X_G \rightarrow X_G$. For $w \in W$ in the Weyl group

$$X(w) \subseteq X_G$$

is the subshceme of X_G of all Borel subgroups B of G such that B and $F(B)$ are in relative position w. Let $w = s_1 \cdot \ldots \cdot s_n$ be a minimal expression for w. Then

$$\overline{X}(s_1, \ldots, s_n)$$

is the space of sequences $(B_0, \ldots, B_n)$ of Borel subgroups of G such that $B_n = FB_0$ and B_{i-1} and B_i are in relative position e or s_i. The scheme

$\overline{X}(s_1,\ldots,s_n)$ is of dimension n and it is a compactification of $X(w)$. The $\mathbb{F}_q$-rational points of $\overline{X}(s_1,\ldots,s_n)$ is $X(e)$ and the finite group G^F of Lie type acts as $\mathbb{F}_q$-rational automorphisms on $\overline{X}(s_1,\ldots,s_n)$, $X(w)$ and the $\mathbb{F}_q$-rational points $X(e)$.

Section 3 treats the case where G is of type ${}^2A_2$. Then the finite group G^F has order $q^3(q^2-1)(q^3+1)$. For a simple reflection $s \in W$ in the Weyl group, the curve $\overline{X}(s)$ is irreducible with genus

$$g = \frac{(q^2-q)}{2}$$

and it has $1+q^2+2gq$ $\mathbb{F}_{q^2}$-rational points. This is the maximal number of $\mathbb{F}_{q^2}$-rational points a curve of that genus can have according to the Weil bound.

Section 4 treats the case where G is a Suzuki-group. Then the finite group $G^F = {}^2B_2(q)$, $q = 2^{2m+1}$, has order $q^2(q-1)(q^2+1)$. For a simple reflection $s \in W$ in the Weyl group, the curve $\overline{X}(s)$ is irreducible with genus

$$g = \frac{\sqrt{q}^3 - \sqrt{q}}{\sqrt{2}}$$

and it has $1+q^2$ $\mathbb{F}_q$-rational points. This is the maximal number of $\mathbb{F}_q$-rational points a curve of that genus can have according to Weil's explicit bound discussed in the appendix. In [Ha-St] plane models have been given and resulting codes have been studied.

Section 5 treats the case where G is a Ree group. Then the finite group $G^F = {}^2G_2(q)$, $q = 3^{2m+1}$, has order $q^3(q-1)(q^3+1)$. For a simple reflection $s \in W$ in the Weyl group, the curve $\overline{X}(s)$ is irreducible with genus

$$g = \frac{\sqrt{3}(\sqrt{q}^5 - \sqrt{q})}{2} + \frac{(q^2-q)}{2}$$

and it has $1+q^3$ $\mathbb{F}_q$-rational points, which is the maximal number a curve of that genus can have according to Weil's explicit bound discussed in the appendix.

The appendix discusses Weil's explicit formulas bounding the number of $\mathbb{F}_q$-rational points on the curves and thereby the length of the resulting geometric Goppa codes.

I am grateful to J.P. Serre who suggested me to study Deligne-Lusztig varieties in search of curves with large groups of automorphisms in relation to their genera.

1. Geometric Goppa codes as group codes

1.1 Let X be a projective curve of genus g defined over a finite field $\mathbb{F}_q$. Let D be a $\mathbb{F}_q$-rational and positive divisor and let $L(D)$ denote the $\mathbb{F}_q$-vectorspace of rational functions defined over $\mathbb{F}_q$ such that $f=0$ or $\mathrm{div}(f) \geq -D$. Finally let $P_1, P_2, \ldots, P_n$ be a set of $\mathbb{F}_q$-rational points on X, none in the support of D. The associated Goppa code [Go1] and [Go2] is the image $C = \phi(L(D)) \subseteq \mathbb{F}_q^n$ of the $\mathbb{F}_q$-linear map:

(1.1.1)
$$\phi: L(D) \rightarrow \mathbb{F}_q^n$$
$$f \mapsto (f(P_1), \ldots, f(P_n)).$$

Theorem 1.2 (V. D. Goppa, cf. [Go1],[Go2]). *Assume* $0 \leq \deg D \leq n$. *The length* n *and the minimal distance* d *of the code* $C = \phi(L(D)) \subseteq \mathbb{F}_q^n$ *satisfies:*

(1.2.1) $$d \geq n - \deg D .$$

If X *is smooth, the dimension* k *of the code* C *satisfies:*

(1.2.2) $$k = \deg D + 1 - g \qquad \text{for } \deg D > 2g-2$$

(1.2.3) $$k \geq \deg D + 1 - g \qquad \text{for } \deg D \leq 2g-2$$

In particular

(1.2.4) $$\frac{k}{n} + \frac{d}{n} \geq 1 + \frac{1}{n} - \frac{g}{n} .$$

1.3 From (1.2.4) it's clear that geometric Goppa codes with good parameters are to be found on curves where $\frac{g}{n}$ is small, that is on curves with a large number n of $\mathbb{F}_q$-rational points compared to the genus g. As for the number N of all $\mathbb{F}_q$-rational points on a curve, Weil's bound asserts, that

(1.3.1) $$|N - (1+q)| \leq 2 \cdot g\sqrt{q}$$

With the "explicit formula" of Weil this general bound can in concrete cases be improved. This technique is presented in the appendix and applied in the last 3 sections of this paper.

1.4 Let G be a group of $\mathbb{F}_q$-rational automorphisms on the curve X. The action of G on X induces an action on the divisors on X. Assume that the divisor D is G-invariant and assume that the set $P_1, P_2, \ldots, P_n$ of $\mathbb{F}_q$-rational points on X is G-stable. The group G acts on the G-stable set $P_1, P_2, \ldots, P_n$ of $\mathbb{F}_q$-rational points giving $\mathbb{F}_q^n$ a $\mathbb{F}_q[G]$-module structure. As the divisor D is G-invariant, G acts on $L(D)$ by $f^g = f \circ g^{-1}$ for $f \in L(D)$ and $g \in G$, and it becomes a left $\mathbb{F}_q[G]$-module. The $\mathbb{F}_q$-linear map:

$$\phi : L(D) \quad \rightarrow \quad \mathbb{F}_q^n$$

of (1.1.1) becomes a $\mathbb{F}_q[G]$-morphism, and the geometric Goppa code

$$C = \phi(L(D)) \subseteq \mathbb{F}_q^n$$

a left $\mathbb{F}_q[G]$-module. In case G acts freely and transitively on $P_1, P_2, \ldots, P_n$ then the $\mathbb{F}_q$-vectorspace on the points can be identified with the group algebra $\mathbb{F}_q[G]$ and the geometric Goppa code $C = \phi(L(D)) \subseteq \mathbb{F}_q^n$ becomes a left ideal in $\mathbb{F}_q[G]$.

1.5 Let X be a projective curve of genus g defined over a finite field $\mathbb{F}_q$ of characteristic p. Let G denote the automorphism group of X. In case $p=0$ Hurwitz showed that G is finite and that

(1.5.1) $$|G| \leq 84(g-1), \qquad p=0$$

In case $p>0$ H. L. Schmidt showed that G is finite, but $|G|$ is not bounded as above. Stichtenoth [St1] obtains

(1.5.2) $$|G| \leq 16g^4, \qquad p \geq 0$$

except in the case where X is defined by the affine equation

(1.5.2.1) $$y^{p^n} + y = x^{p^n+1}, \qquad p^n > 3$$

H.-W. Henn [He] obtains

(1.5.3) $$|G| < 8g^3, \qquad p \geq 0$$

excluding 4 cases. In a footnote he asserts that this bound can be strengthen to

(1.5.3.1) $$|G| < 3(2g)^2\sqrt{2g}$$

excluding 2 more cases. However in section 5 we will construct curves not on his list of excluded curves with more automorphisms than allowed by the proclaimed bound (1.5.3.1).

2. Deligne-Lusztig varieties

Let k be an algebraically closed field of characteristic p.

2.1 The basic properties of affine algebraic groups can be found in [Ca1] and [Ca2]. Here we recollect what is needed for our purpose. An *affine algebraic group* G over k is an affine variety defined over k which is also a group such that the multiplication map $G \times G \to G$ and the inversion map $G \to G$ are morphisms of varieties. Every affine algebraic group is isomorphic to a closed subgroup of the general linear group $GL_n(k)$ for some n. An affine algebraic group is called *simple* if it has no non-trivial closed connected normal subgroup.

2.1.1 The multiplicative group $k^* \simeq GL_1(k)$ is an algebraic group. An algebraic group isomorphic to $k^* \times \ldots \times k^*$ is called a *torus*. A *Borel subgroup* B of a connected affine algebraic group G is a maximal connected solvable subgroup of G. Any two Borel subgroups of G are conjugate in G. A maximal torus lies in some Borel subgroup of G and two maximal tori in G are conjugate. The group G has a maximal closed connected normal subgroup all of whose elements are unipotent. This is the *unipotent radical*. The group G is called *reductive* if the unipotent radical is trivial. Let G be a connected, reductive algebraic groups G defined over the field k. Let B be a Borel subgroup of G, let T be a maximal torus of G in B and let U be the unipotent radical of B. The *Weyl group* of G is the finite group $W = N(T)/T$ where $N(T)$ is the normalizer T in G.

2.1.2 Let $X = \mathrm{Hom}(T, k^*)$ be the character group of T, that is the group of algebraic group homomorphisms from T to the multiplicative group k^* and let likewise $Y = \mathrm{Hom}(k^*, T)$ be the group of cocharacters of T. X and Y are free abelian groups of the same finite rank. Let $\chi \in X$ and $\gamma \in Y$. By composition we obtain a morphism

$$k^* \xrightarrow{\gamma} T \xrightarrow{\chi} k^*$$

so that $\chi \circ \gamma \in \mathrm{Hom}(k^*, k^*)$ and therefore of the form

$$(\chi \circ \gamma)(\lambda) = \lambda^m \qquad \lambda \in k^*$$

for some integer m.

This gives a nondegenerate pairing

$$X \times Y \to \mathbb{Z}$$
$$(\chi, \gamma) \mapsto <\chi, \gamma>$$

where

$$(\chi \circ \gamma)(\lambda) = \lambda^{<\chi,\gamma>} \qquad \lambda \in k^*$$

The groups X and Y are in duality and there is a bijection between them

$$X \to Y$$
$$\alpha \mapsto \alpha^\vee$$

such that $<\alpha, \alpha^\vee> = 2$.

2.1.3 Consider the finitely many minimal closed subgroups of U normalised by T. Each of these are isomorphic to the additive group k. An element $t \in T$ acts on any of these subgroups by conjugation, the corresponding automorhism $k \to k$ is multiplication by an element in k^*. Hence the action of T by conjugation gives an element in $X = \mathrm{Hom}(T, k^*)$ for any minimal closed subgroups of U normalised by T. These elements are called *positive roots*. There is a unique Borel subgroup B^- of G containing T such that $B \cap B^- = T$. Let U^- be the unipotent radical of B^-. As before we consider the minimal closed subgroups of U^- normalised by T. The action of T by conjugation gives an element in $X =$

Hom(T, k^*) for any of these groups. These elements are called *negative roots*. The action of the Weyl group W on T gives rise to actions of W on X and Y defined by

(2.1.3.1)
$$\begin{aligned} (w\chi)\,t &= \chi(w^{-1}(t)), & w\in W, \chi\in X, t\in T \\ (w\gamma)\,\lambda &= w(\gamma(\lambda)), & w\in W, \gamma\in Y, \lambda\in k^* \end{aligned}$$

For each root α there is an element $w_\alpha \in W$ such that
$$w_\alpha = w_{-\alpha} \text{ and } w_\alpha^2 = 1$$
Let $\theta \subseteq X$ be the set of roots. Then $\{ w_\alpha \mid \alpha\in\theta \}$ generates W and such that the action of the Weyl group on X and Y is determined by

(2.1.3.2)
$$\begin{aligned} w_\alpha(\chi) &= \chi - <\chi, \alpha^\vee> \alpha & \chi \in X \\ w_\alpha(\gamma) &= \gamma - <\alpha,\gamma> \alpha^\vee & \gamma \in Y \end{aligned}$$

2.1.4 We have seen (2.1.3) that every connected reductive group G has a root datum
$$(X, \theta, Y, \theta^\vee)$$
associated to it, where X, Y are the character and cocharacter groups of a maximal torus of G and θ is the set of roots and $\theta^\vee$ the set of coroots. The root datum $(X, \theta, Y, \theta^\vee)$ uniguely determines the connected reductive group G.

2.1.5 Let G be a simple algebraic group. Let θ^+ denote the positive roots (2.1.3). A positive root is called *simple* if it can not be expressed as the sum of two positive roots. Let $\{\alpha_1,\ldots,\alpha_l\}$ be the simple roots for G. *The Cartan integers*

(2.1.5.1)
$$A_{ij} = <\alpha_j, \alpha_i{}^\vee>$$
takes on the value 2 if $i=j$ and the values 0, -1, -2, -3 if $i\neq j$ in such a way that the integers

(2.1.5.2)
$$n_{ij} = A_{ij}\,A_{ji}$$
takes on one of the values 0, 1, 2, 3. Let

(2.1.5.3)
$$s_i = w_{\alpha_i} \in W \qquad i=1,\ldots,l$$
and let the order of $s_i s_j$ be m_{ij}. Then
$$W = <s_1,\ldots,s_l \mid (s_i)^2 = 1\ ,\ (s_i s_j)^{m_{ij}} = 1 \text{ for } i\neq j>$$
and

(2.1.5.4)
$$\begin{aligned} n_{ij} &= 0 & \Leftrightarrow \quad m_{ij} &= 2 \\ n_{ij} &= 1 & \Leftrightarrow \quad m_{ij} &= 3 \\ n_{ij} &= 2 & \Leftrightarrow \quad m_{ij} &= 4 \\ n_{ij} &= 3 & \Leftrightarrow \quad m_{ij} &= 6 \end{aligned}$$

The *Dynkin diagram* of G is a graph with l nodes corresponding to the simple roots α_i. The nodes corresponding to different simple roots α_i and α_j are connected by $n_{ij} = A_{ij}\,A_{ji}$ bounds. An arrow is pointing from the node corresponding to α_i to the node corresponding to α_j if $A_{ji} \neq -1$.

The Dynkin diagrams interesting for our purpose are the following:

A_2 

B_2 

G_2

2.2 A connected reductive group G over k is isomorphic to a closed subgroup af $GL_n(k)$ for some n (2.1). A map
$$Fr_p{}^e : G \to G$$

is called a *standard Frobenius map* if Fr_{p^e} is the restriction of the morphism

$$Fr_{p^e}[a_{ij}] = [a_{ij}^{p^e}]$$

on $\mathrm{GL}_n(k)$ for some embedding of G into some $\mathrm{GL}_n(k)$. A *Frobenius map* is a morphism

$$F : G \to G$$

such that some power F^s is a standard Frobenius map. The finite groups G^F where G is a connected reductive group and F is a Frobenius map are called *finite groups of Lie type.* The real number Q defined by

(2.2.0.1) $$Q^s = p^e \qquad \text{where } F^s = Fr_{p^e}$$

will be of later importance.

The fundamental theorem of Lang-Steinberg [L] and [Ste] asserts that the map

(2.2.0.2) $$L: G \to G \ , \quad L(g) = g^{-1}F(g)$$

is surjective. This result has important consequences; in the following we recollect what is needed for our purpose.

2.2.1 Let G be a connected reductive group, then G *has a F-stable Borel subgroup.* Namely let B be a fixed Borel subgroup, any other Borel subgroup is of the form gBg^{-1} for some $g \in G$ (2.1.1). The group $F(B)$ is also a Borel subgroup, so there is an element $g_0 \in G$ such that $g_0 F(B)(g_0)^{-1} = B$. From (2.2.0.2) we find a $g \in G$ such that $L(g) = g^{-1}F(g) = g_0$. Now gBg^{-1} is a F-stable Borel subgroup as $F(gBg^{-1}) = F(g)F(B)F(g^{-1}) = (gg_0)F(B)(gg_0)^{-1} = gBg^{-1}$.

2.2.2 Let G be a connected reductive group, then *any two F-stable Borel subgroups are conjugate by an element in* G^F. For let B_0, B_1 be two F-stable Borel subgroups, then $B_1 = gB_0g^{-1}$ for some $g \in G$ (2.2.1). As both groups are F-stable we have $g^{-1}F(g) \in N_G(B_0) = B_0$. By the theorem of Lang-Steinberg applied to B_0 there is a $b \in B_0$ such that $L(b) = b^{-1}F(b) = g^{-1}F(g)$, that is $gb^{-1} \in G^F$ and $B_1 = (gb^{-1})B_0(gb^{-1})$.

2.2.3 A Frobenius morphism induces a graph automorphism ρ of the Dynkin diagram (2.1.5) when the arrows are disregarded. Let T be a maximally split torus. The Frobenius morphism induces an action on the character and cocharacter groups of T:

$$F : X \to X$$
$$F(\chi)t = \chi(F(t)) \qquad \chi \in X$$

and

$$F : Y \to Y$$
$$F(\gamma)\lambda = F(\gamma(\lambda)) \qquad \gamma \in Y$$

The action of F on the roots is related to the graph automorphism ρ of the Dynkin diagram. Specifically $F(\rho(\alpha))$ is a positive multiple of α each positive root α.

The Dynkin diagrams with F-action interesting for our purpose are the following twisted groups where the F-action permutes the two roots:

$^2A_2$

$^2B_2$

$^2G_2$

2.2.4 If G has type ${}^2A_2$, the real number Q (2.2.0.1) can take any value p^e, $e \in \mathbb{N}$, $e \neq 0$. These groups are defined with respect to a non-degenerate Hermitian form in 3 variables on $\mathbb{F}_{Q^2}$ corresponding to the involution $\lambda \mapsto \lambda^Q$.

2.2.5 If G has type ${}^2B_2$, then then characteristic p has to be 2 and the real number Q (2.2.0.1) must satiesfy $Q^2 = 2^{2n+1}$, $n \in \mathbb{N}$. The finite groups $G^F = {}^2B_2(Q^2)$ are the Suzuki groups.

2.2.6 If G has type ${}^2G_2$, then then characteristic p has to be 3 and the real number Q (2.2.0.1) must satiesfy $Q^2 = 3^{2n+1}$, $n \in \mathbb{N}$. The finite groups $G^F = {}^2G_2(Q^2)$ are the Ree groups.

2.3 We now introduce the Deligne-Lusztig varieties associated to a connected reductive algebraic groups G defined over a finite field $\mathbb{F}_q$, with Frobenius map $F: G \to G$ [De-Lu]. Specifically, let G be a connected, reductive algebraic group G and let X_G be the $\mathbb{F}_q$-scheme of all Borel subgroups of G with Frobenius morphism $F: X_G \to X_G$. The group G acts on X_G by conjugation and for each Borel subgroup B of G the stabiliser of the corresponding point in X_G is B , and their is a natural isomorphism $G/B \to X_G$, $g \mapsto gBg^{-1}$.

2.3.1 The set of orbits of G in $X_G \times X_G$ can be identified with the Weyl group (2.2.1). For T a maximal torus and B a Borel subgroup containing it we have isomorphisms:

$$W \xrightarrow{\sim} N(T)/T \xrightarrow{\sim} B\backslash G/B \xrightarrow{\sim} G\backslash(G/B \times G/B) \xrightarrow{\sim} G\backslash(X_G \times X_G)$$

For $w \in W$ in the Weyl group the orbit of G in $X_G \times X_G$ is denoted by $O(w)$ and

(2.3.1.1) $$X(w) \subseteq X_G$$

is the subshceme of X_G of all Borel subgroups B of G such that $(B, F(B)) \in O(w)$ are in relative position w.

2.3.2 The subscheme $X(w) \subseteq X_G$ (2.3.1.1) is smooth of pure dimension

(2.3.2.1) $$\dim X(w) = l(w) = n$$

where $w = s_1 \cdot \ldots \cdot s_n$ is a minimal expression for w as simple reflections [De-Lu, 1.3].

2.3.3 The subscheme $X(w) \subseteq X_G$ (2.3.1.1) is G^F-stable.

2.3.4 Let $w = s_1 \cdot \ldots \cdot s_n$ be a minimal expression for w. Then

(2.3.4.1) $$\overline{X}(s_1, \ldots, s_n)$$

is the space of sequences $(B_0, \ldots, B_n)$ of Borel subgroups of G such that $B_n = FB_0$ and B_{i-1} and B_i are in relative position e or s_i. The scheme $\overline{X}(s_1, \ldots, s_n)$ is of dimension n and it is a compactification of $X(w)$ [De-Lu, 9.1].

2.3.5 The $\mathbb{F}_q$-rational points of $\overline{X}(s_1, \ldots, s_n)$ is $X(e)$ and the finite group G^F of Lie type acts as $\mathbb{F}_q$-rational automorphisms on $\overline{X}(s_1, \ldots, s_n)$, $X(w)$ and the $\mathbb{F}_q$-rational points $X(e)$.

2.3.6 Let $w = s_1 \cdot \ldots \cdot s_n$ be a minimal expression for w as simple reflections. Then $\overline{X}(s_1, \ldots, s_n)$ is irreducible if and only if any simple reflection $s \in W$ is in the F-orbit of some s_i where $i = 1, \ldots, n$. [Lu1, 3.10 d]

2.3.7 The Euler characteristic of $X(w)$ is according to [De-Lu, Theorem 7.1] determined by

(2.3.6.1) $$\chi(X(w)) = (-1)^{\sigma(G)-\sigma(T)} \frac{|G^F|}{\mathrm{St}_G(e)\ |T^F|}$$

where T is a F-stable maximal torus contained in $B \in X(w)$, St_G is the Steinberg representation of G^F and $\sigma(G)$ (resp. $\sigma(T)$) is the $\mathbb{F}_q$-rank of G (resp. T). The order $|T^F|$ is calculated by the formula

(2.3.6.2) $$|T^F| = |\det_{Y_0 \otimes \mathbb{R}}(w^{-1} \circ F - 1)|,$$

cf. [Ca2], where $Y_0 = \mathrm{Hom}(k^*, T_0)$ be the group of cocharacters of T_0 (2.2.1), where T_0 is a F-stable maximal torus contained in a F-stable Borel subgroup and the action of F and w on Y is described in (2.2.3) and (2.1.3).

2.4 In case $s \in W$ is a simple reflection we obtain from (2.3) that the Deligne-Lusztig variety

$$\overline{X}(s) = X(s) \cup X(e)$$

is a curve with the group G^F of Lie type acting as $\mathbb{F}_q$-rational automorphisms. The $\mathbb{F}_q$-rational points on $\overline{X}(s)$ is $X(e)$ and the curve $\overline{X}(s)$ is irreducible if and only if any simple reflection $s \in W$ is in the F-orbit of s.

2.4.1 The genus and Euler characteristic of $\overline{X}(s) = X(s) \cup X(e)$ is determined by

(2.4.1.1) $$2 - 2g = \chi(X(w)) + \chi(X(e))$$

which is calculated using (2.3.6.1) and the number of $\mathbb{F}_q$-rational points on $\overline{X}(s)$ is $\chi(X(e))$, which is also calculated using (2.3.6.1).

3. Groups of type $^2A_2$ — Hermite curves. There are two simple roots α_1, α_2. The corresponding Dynkin diagram has two nodes with 1 bound between them. The Cartan matrix (2.1.5.1) is

(3.0.1) $$\begin{bmatrix} +2 & -1 \\ -1 & +2 \end{bmatrix}$$

The Frobenius morphism interchanges the two nodes. If G has type $^2A_2$, the real number Q (2.2.0.1) can take any value p^e, $e \in \mathbb{N}$, $e \neq 0$. Let $q = Q$. These groups are defined with respect to a non-degenerate Hermitian form $x^q + y^q + z^q = 0$ on $\mathbb{F}_{q^2}$ corresponding to the involution $\lambda \mapsto \lambda^q$. The finite groups $G^F = {}^2A_2(q^2)$ have order $q^3(q^2-1)(q^3+1)$.
The element $w_{\alpha_1} \in W$ acts on the simple coroots according to (2.1.3.2) and the Cartan matrix (3.0.1)

$$w_{\alpha_1}(\alpha_1^\vee) = \alpha_1^\vee - \langle \alpha_1, \alpha_1^\vee \rangle \alpha_1^\vee = -\alpha_1^\vee$$
$$w_{\alpha_1}(\alpha_2^\vee) = \alpha_2^\vee - \langle \alpha_1, \alpha_2^\vee \rangle \alpha_1^\vee = \alpha_2^\vee + \alpha_1^\vee$$

The endomorphism of $Y_0 \otimes \mathbb{R}$ induced by the element $w_{\alpha_1} \in W$ therefore has matrix

(3.0.2) $$\begin{bmatrix} -1 & 1 \\ 0 & 1 \end{bmatrix}$$

The endomorphism of $Y_0 \otimes \mathbb{R}$ induced by the Frobenius has matrix

(3.0.3)
$$q\begin{bmatrix} 0 & 1 \\ 1 & 0 \end{bmatrix}$$

St_G is the Steinberg representation of ${}^2A_2(q^2)$ and assumes the value $\mathrm{St}_G(e) = q^3$.

3.1 Now let T be a F-stable maximal torus contained in $B \in X(w_{\alpha_1})$ and let T_o be a F-stable maximal torus contained in $B_o \in X(e)$. Then according to (2.3.6.2) we obtain

(3.1.1)
$$|T^F| = \det{}_{Y_0 \otimes \mathbf{R}}(w_{\alpha_1}^{-1} \circ F - 1)| =$$

$$\left| \begin{bmatrix} -1 & 1 \\ 0 & 1 \end{bmatrix} \circ q \begin{bmatrix} 0 & 1 \\ 1 & 0 \end{bmatrix} - \begin{bmatrix} 1 & 0 \\ 0 & 1 \end{bmatrix} \right| = q^2 - q + 1$$

(3.1.2)
$$|T_0^F| = |\det{}_{Y_0 \otimes \mathbf{R}}(e \circ F - 1)| =$$

$$\left| q \begin{bmatrix} 0 & 1 \\ 1 & 0 \end{bmatrix} - \begin{bmatrix} 1 & 0 \\ 0 & 1 \end{bmatrix} \right| = q^2 - 1$$

PROPOSITION 3.2 *Let G be a connected reductive group of type ${}^2A_2$ over $k = \mathbb{F}_{q^2}$. Let $w_{\alpha_1} \in W$ be a simple reflection and let $\overline{X}(w_{\alpha_1})$ be the corresponding Deligne-Lusztig variety. Then $\overline{X}(w_{\alpha_1})$ is an irreducible curve of genus*

$$g = \frac{q^2 - q}{2}$$

with $1 + q^3$ points over $\mathbb{F}_{q^2}$. The finite group $G^F = {}^2A_2(q^2)$ has order $q^3(q^2-1)(q^3+1)$ and it acts as a group of $\mathbb{F}_{q^2}$-rational automorphisms on $\overline{X}(w_{\alpha_1})$.

The variety is a curve according to (2.3.2) and irreducible according to (2.3.6) as the Frobenius interchanges the two simple roots. From (2.3.7) we determine the Euler characteristica using (3.1.1) and (3.1.2):

(3.2.1)
$$\chi(X(w_{\alpha_1})) = (-1)^{\sigma(G) - \sigma(T)} \frac{|G^F|}{\mathrm{St}_G(e)\ |T^F|} =$$

$$- \frac{q^3(q^2-1)(q^3+1)}{q^3(q^2 - q + 1)} = -(q^2-1)(q+1)$$

(3.2.2)
$$\chi(X(e)) = (-1)^{\sigma(G) - \sigma(T_0)} \frac{|G^F|}{\mathrm{St}_G(e)\ |T_0^F|} =$$

$$\frac{q^3(q^2-1)(q^3+1)}{q^3(q^2-1)} = 1+q^3$$

and the genus using 2.4.1.1

(3.2.3)
$$2-2g = \chi(X(e)) + \chi(X(w_{\alpha_1})) =$$
$$1+q^3-(q^2-1)(q+1) \quad \Leftrightarrow$$
$$g = \frac{q^2-q}{2}$$

The finite group $G^F = {}^2A_2(q)$ order $q^3(q^2-1)(q^3+1)$ and it acts as a group of $\mathbb{F}_{q^2}$-rational automorphisms on $\overline{X}(w_{\alpha_1})$ by (2.3.3).

PROPOSITION 3.3 *The irreducible curve* $\overline{X}(w_{\alpha_1})$ *of genus*

$$g = \frac{q^2-q}{2}$$

with $1+q^3$ *points over* $\mathbb{F}_{q^2}$ *has the maximal number of rational points allowed by the Weil-bound. The Zeta-function of the curve is*

$$Z(X,\mathbb{F}_q)(t) = \frac{(1+qt^2)^g}{(1-t)(1-qt)}$$

and the number N_m *of* $\mathbb{F}_{q^m}$*-rational points is determined by the formula*

$$N_m = 1+q^m-(i^m+(-i)^m)\,g\,q^{\frac{m}{2}}, \qquad m\in\mathbb{N}$$

The claim follows from the general theory of Zeta-functions (see the appendix), as the formula for N_2 is seen to be true by inspection.

REMARK 3.4 It is possible to construct geometric Goppa codes over $\mathbb{F}_{q^2}$ such that

$$\textit{dimension} + \textit{minimal distance} \geq 1+q^3 - \frac{q^2-q}{2}$$

The codes are modules over the group-ring $\mathbb{F}_{q^2}[{}^2A_2(q^2)]$. These codes have been studied in detail in [Ha2].

4. **Groups of type ${}^2B_2$ — Suzuki groups.** There are two simple roots α_1, α_2. The corresponding Dynkin diagram has two nodes with 2 bounds between them and an arrow from the node corresponding to α_1 to that of α_2 (2.2.4). The Cartan matrix (2.1.5.1) is

(4.0.1)
$$\begin{bmatrix} +2 & -1 \\ -2 & +2 \end{bmatrix}$$

The Frobenius morphism interchanges the two nodes. If G has type ${}^2B_2$, then then characteristic p has to be 2 and the real number Q (2.2.0.1) must satiesfy $Q^2 = 2^{2n+1}$, $n\in\mathbb{N}$. Let $q = Q^2$. The finite groups $G^F = {}^2B_2(q)$ are the

Suzuki groups and they have order $q^2(q-1)(q^2+1)$.
The element $w_{\alpha_1} \in W$ acts on the simple coroots according to (2.1.3.2) and the Cartan matrix (4.0.1)

$$\begin{aligned} w_{\alpha_1}(\alpha_1^\vee) &= \alpha_1^\vee - \langle \alpha_1, \alpha_1^\vee \rangle \alpha_1^\vee = -\alpha_1^\vee \\ w_{\alpha_1}(\alpha_2^\vee) &= \alpha_2^\vee - \langle \alpha_1, \alpha_2^\vee \rangle \alpha_1^\vee = \alpha_2^\vee + 2\alpha_1^\vee \end{aligned}$$

The endomorphism of $Y_0 \otimes \mathbb{R}$ induced by the element $w_{\alpha_1} \in W$ therefore has matrix

$$\begin{bmatrix} -1 & 1 \\ 0 & 2 \end{bmatrix} \tag{4.0.2}$$

The endomorphism of $Y_0 \otimes \mathbb{R}$ induced by the Frobenius has matrix

$$\sqrt{q} \begin{bmatrix} 0 & \sqrt{2} \\ \frac{1}{\sqrt{2}} & 0 \end{bmatrix} \tag{4.0.3}$$

St_G is the Steinberg representation of ${}^2B_2(q)$ and assumes the value $\mathrm{St}_G(e) = q^2$.

4.1 Now let T be a F-stable maximal torus contained in $B \in X(w_{\alpha_1})$ and let T_o be a F-stable maximal torus contained in $B_o \in X(e)$. Then according to (2.3.6.2) we obtain

(4.1.1) $$|T^F| = \det_{Y_0 \otimes \mathbb{R}}(w_{\alpha_1}^{-1} \circ F - 1)| =$$

$$\left| \begin{bmatrix} -1 & 1 \\ 0 & 2 \end{bmatrix} \circ \sqrt{q} \begin{bmatrix} 0 & \sqrt{2} \\ \frac{1}{\sqrt{2}} & 0 \end{bmatrix} - \begin{bmatrix} 1 & 0 \\ 0 & 1 \end{bmatrix} \right| = q - \sqrt{2}\sqrt{q} + 1$$

(4.1.2) $$|T_0^F| = |\det_{Y_0 \otimes \mathbb{R}}(e \circ F - 1)| =$$

$$\left| \sqrt{q} \begin{bmatrix} 0 & \sqrt{2} \\ \frac{1}{\sqrt{2}} & 0 \end{bmatrix} - \begin{bmatrix} 1 & 0 \\ 0 & 1 \end{bmatrix} \right| = q - 1$$

PROPOSITION 4.2 *Let G be a connected reductive group of type ${}^2\mathrm{B}_2$ over $k = \mathbb{F}_q$, $q = 2^{2n+1}$, $n \in \mathbb{N}$. Let $w_{\alpha_1} \in W$ be a simple reflection and let $\overline{X}(w_{\alpha_1})$ be the corresponding Deligne-Lusztig variety. Then $\overline{X}(w_{\alpha_1})$ is an irreducible curve of genus*

$$g = \frac{\sqrt{q}^3 - \sqrt{q}}{\sqrt{2}}$$

with $1+q^2$ points over $\mathbb{F}_q$. The finite group $G^F = {}^2\mathrm{B}_2(q)$ is a Suzuki group, it has order $q^2(q-1)(q^2+1)$ and it acts as a group of $\mathbb{F}_q$-rational automorphisms on $\overline{X}(w_{\alpha_1})$

The variety is a curve according to (2.3.2) and irreducible according to (2.3.6) as the Frobenius interchanges the two simple roots. From (2.3.7) we determine the Euler characteristica using (4.1.1) and (4.1.2):

(4.2.1)
$$\chi(X(w_{\alpha_1})) = (-1)^{\sigma(G)-\sigma(T)} \frac{|G^F|}{\mathrm{St}_G(e)\ |T^F|} =$$

$$-\frac{q^2(q-1)(q^2+1)}{q^2(q-\sqrt{2}\sqrt{q}+1)}. = -(q^2+\sqrt{2}(\sqrt{q})^3-\sqrt{2}\sqrt{q}-1)$$

(4.2.2)
$$\chi(X(e)) = (-1)^{\sigma(G)-\sigma(T_0)} \frac{|G^F|}{\mathrm{St}_G(e)\ |T_0^F|} =$$

$$\frac{q^2(q-1)(q^2+1)}{q^2(q-1)} = q^2+1$$

and the genus using 2.4.1.1

(4.2.3)
$$2-2g = \chi(X_{T_0\subseteq B_0}) + \chi(X_{T\subseteq B}) =$$

$$q^2+1 - (q^2+\sqrt{2}(\sqrt{q})^3-\sqrt{2}\sqrt{q}-1) \qquad \Leftrightarrow$$

$$g = \frac{\sqrt{q}^3-\sqrt{q}}{\sqrt{2}}$$

The finite group $G^F = {}^2B_2(q)$ is a Suzuki group, it has order $q^2(q-1)(q^2+1)$ and it acts as a group of $\mathbb{F}_q$-rational automorphisms on $\overline{X}(w_{\alpha_1})$ by (2.3.3). We collect what we know about $\overline{X}(w_{\alpha_1})$ in the following proposition.

PROPOSITION 4.3 *The irreducible curve* $\overline{X}(w_{\alpha_1})$ *of genus*

$$g = \frac{\sqrt{q}^3-\sqrt{q}}{\sqrt{2}}$$

with $1+q^2$ *points over* $\mathbb{F}_q$ *has the maximal number of rational points allowed by the "explicit formulas" of Weil. The Zeta-function of the curve is*

$$Z(X,\mathbb{F}_q)(t) = \frac{(1+\sqrt{2}\sqrt{q}\,t+qt^2)^g}{(1-t)(1-qt)}$$

and the number N_m *of* $\mathbb{F}_{q^m}$*-rational points is determined by the formula*

$$N_m = 1+q^m-g\,q^{\frac{m}{2}}\,\delta$$

where $\delta=-2$ *if* $m=4$ *mod* 8, $\delta=-\sqrt{2}$ *if* $m=1,7$ *mod* 8, $\delta=0$ *if* $m=2$ *mod* 4, $\delta=\sqrt{2}$ *if* $m=3,5$ *mod* 8 *and* $\delta=2$ *if* $m=0$ *mod* 8.

The claims follow from the determination of the eigenvalues of the Frobenius given in [Lu2] and the remarks in the appendix using

$$f(\theta) = 1+2\left(\tfrac{\sqrt{2}}{2}\cos(\theta)+\tfrac{1}{4}\cos(2\theta)\right)$$

in (a.6) having roots $\theta=\pm\frac{3}{4}\pi$, such that the α_i of (a.3) are $\sqrt{q}\left(-\frac{\sqrt{2}}{2}\pm i\frac{\sqrt{2}}{2}\right)$ imidiately giving the Zeta-function and consequentely the formulas for M_m .

REMARK 4.4 It is possible to construct geometric Goppa codes over $\mathbb{F}_q$, $q = 2^{2n+1}$, $n \in \mathbb{N}$, such that

$$\textit{dimension} + \textit{minimal distance} \geq 1 + q^2 - \frac{\sqrt{q}^3 - \sqrt{q}}{\sqrt{2}}$$

The codes are modules over the group-ring $\mathbb{F}_q[{}^2A_2(q)]$. These codes have been studied in detail in [Ha-St].

5. Group codes from Ree groups. There are two simple roots α_1, α_2. The corresponding Dynkin diagram has two nodes with 3 bounds between them and an arrow from the node corresponding to α_1 to that of α_2. The Cartan matrix is

(5.0.1)
$$\begin{bmatrix} +2 & -1 \\ -3 & +2 \end{bmatrix}$$

The Frobenius morphism interchanges the two nodes. If G has type ${}^2G_2$, then then characteristic p has to be 3 and the real number Q (2.2.0.1) must satiesfy $Q^2 = 3^{2n+1}$, $n \in \mathbb{N}$. The finite groups $G^F = {}^2G_2(Q^2)$ are the Ree groups. Let $q = Q^2$, then $G^F = {}^2G_2(q)$, $q = 3^{2m+1}$, has order $q^3(q-1)(q^3+1)$. The element $w_{\alpha_1} \in W$ acts on the simple coroots according to (2.1.3.2) and the Cartan matrix (5.0.1)

$$w_{\alpha_1}(\alpha_1^\vee) = \alpha_1^\vee - \langle \alpha_1, \alpha_1^\vee \rangle \alpha_1^\vee = -\alpha_1^\vee$$
$$w_{\alpha_1}(\alpha_2^\vee) = \alpha_2^\vee - \langle \alpha_1, \alpha_2^\vee \rangle \alpha_1^\vee = \alpha_2^\vee + 3\alpha_1^\vee$$

The endomorphism of $Y_0 \otimes \mathbb{R}$ induced by the element $w_{\alpha_1} \in W$ therefore has matrix

(5.0.2)
$$\begin{bmatrix} -1 & 1 \\ 0 & 3 \end{bmatrix}$$

The endomorphism of $Y_0 \otimes \mathbb{R}$ induced by the Frobenius has matrix

(5.0.3)
$$\sqrt{q} \begin{bmatrix} 0 & \sqrt{3} \\ \frac{1}{\sqrt{3}} & 0 \end{bmatrix}$$

St_G is the Steinberg representation of ${}^2G_2(q)$ and assumes the value $\mathrm{St}_G(e) = q^3$.

5.1 Now let T be a F-stable maximal torus contained in $B \in X(w_{\alpha_1})$ and let $T_\circ$ be a F-stable maximal torus contained in $B_\circ \in X(e)$. Then according to (2.3.6.2) we obtain

(5.1.1)
$$|T^F| = |\det{}_{Y_0 \otimes \mathbb{R}}(w_{\alpha_1}^{-1} \circ F - 1)| =$$

$$\left| \begin{bmatrix} -1 & 1 \\ 0 & 3 \end{bmatrix} \circ \sqrt{q} \begin{bmatrix} 0 & \sqrt{3} \\ \frac{1}{\sqrt{3}} & 0 \end{bmatrix} - \begin{bmatrix} 1 & 0 \\ 0 & 1 \end{bmatrix} \right| = q - \sqrt{3}\sqrt{q} + 1$$

(5.1.2) $$| T_0^F | = | \det_{Y_0 \otimes \mathbb{R}}(e \circ F - 1) | =$$

$$\left| \sqrt{q} \begin{bmatrix} 0 & \sqrt{3} \\ \frac{1}{\sqrt{3}} & 0 \end{bmatrix} - \begin{bmatrix} 1 & 0 \\ 0 & 1 \end{bmatrix} \right| = q - 1$$

PROPOSITION 5.2 *Let G be a connected reductive group of type ${}^2G_2$ over $k = \mathbb{F}_q$, $q = 3^{2n+1}$, $n \in \mathbb{N}$. Let $w_{\alpha_1} \in W$ be a simple reflection and let $\overline{X}(w_{\alpha_1})$ be the corresponding Deligne-Lusztig variety. Then $\overline{X}(w_{\alpha_1})$ is an irreducible curve of genus*

$$g = \frac{\sqrt{3}(\sqrt{q}^5 - \sqrt{q})}{2} + \frac{(q^2 - q)}{2}$$

with $1 + q^3$ points over $\mathbb{F}_q$. The finite group $G^F = {}^2G_2(q)$ is a Ree group, it has order $q^3(q-1)(q^3+1)$ and it acts as a group of $\mathbb{F}_q$-rational automorphisms on $\overline{X}(w_{\alpha_1})$.

The variety is a curve according to (2.3.2) and irreducible according to (2.3.6) as the Frobenius interchanges the two simple roots. From (2.3.7) we determine the Euler characteristica using (5.1.1) and (5.1.2):

(5.2.1) $$\chi(X_{T \subseteq B}) = (-1)^{\sigma(G) - \sigma(T)} \frac{| G^F |}{\mathrm{St}_G(e) \; | T^F |} =$$

$$-\frac{q^3(q-1)(q^3+1)}{q^3(q - \sqrt{3}\sqrt{q} + 1)} = -(q^3 + \sqrt{3}(\sqrt{q})^5 + q^2 - q - \sqrt{3}\sqrt{q} - 1)$$

(5.2.2) $$\chi(X_{T_0 \subseteq B_0}) = (-1)^{\sigma(G) - \sigma(T_0)} \frac{| G^F |}{\mathrm{St}_G(e) \; | T_0^F |} =$$

$$\frac{q^3(q-1)(q^3+1)}{q^3(q - 1)} = q^3 + 1$$

and the genus using 2.4.1.1

$$2 - 2g = \chi(X_{T_0 \subseteq B_0}) + \chi(X_{T \subseteq B}) =$$

$$q^3 + 1 - (q^3 + \sqrt{3}(\sqrt{q})^5 + q^2 - q^1 - \sqrt{3}\sqrt{q} - 1) \Leftrightarrow$$

$$g = \frac{\sqrt{3}(\sqrt{q}^5 - \sqrt{q})}{2} + \frac{(q^2 - q)}{2}$$

The finite group $G^F = {}^2G_2(q)$ is a Ree group, it has order $q^3(q-1)(q^3+1)$ and it acts as a group of $\mathbb{F}_q$-rational automorphisms on $\overline{X}(w_{\alpha_1})$ by (2.3.3). We collect what we know about $\overline{X}(w_{\alpha_1})$ in the following proposition.

PROPOSITION 5.3 *The irreducible curve* $\overline{X}(w_{\alpha_1})$ *of genus*

$$g = \frac{\sqrt{3}(\sqrt{q^5} - \sqrt{q})}{2} + \frac{(q^2 - q)}{2}$$

with $1+q^3$ *points over* $\mathbb{F}_q$ *has the maximal number of rational points allowed by the "explicit formulas" of Weil. Let* $q_0 = \frac{\sqrt{q}}{\sqrt{3}}$ *the Zeta-function of the curve is*

$$Z(X,\mathbb{F}_q)(t) = \frac{(1+3q_0t+qt^2)^{q_0(q^2-1)}(1+qt^2)^{q_0(q-1)(q+3q_0+1)/2}}{(1-t)(1-qt)}$$

and the number N_m *of* $\mathbb{F}_{q^m}$*-rational points is determined by the formula*

$$N_m = 1+q^m - q_0\sqrt{q}(q-1)\left[(q+3q_0+1)\cos m\pi/2 + 2(1+q)\cos 5m\pi/6\right]$$

The claims follow from the determination of the eigenvalues of the Frobenius given in [Lu2] and the remarks in the appendix using

$$f(\theta) = \tfrac{1}{3}\left(1 + \sqrt{3}\cos(\theta) + \cos(2\theta)\right)^2$$

in (a.6) having roots $\theta_i = \pm\frac{1}{2}\pi$ counted with multiplicity a and $\pm\frac{5}{6}\pi$ counted with multiplicity b, such that the α_i of (a.3) are $\pm i\sqrt{q}$, $\left(-\frac{\sqrt{3}}{2} \pm i\frac{1}{2}\right)\sqrt{q}$ imidiately giving the Zeta-function and consequentely the formulas for M_m. (cf. [Pe]).

REMARK 5.4 It is possible to construct geometric Goppa codes over $\mathbb{F}_q$, $q = 3^{2n+1}$, $n \in \mathbb{N}$, such that

$$\textit{dimension} + \textit{minimal distance} \geq 1+q^3 - \left(\frac{\sqrt{3}(\sqrt{q^5} - \sqrt{q})}{2} + \frac{(q^2 - q)}{2}\right)$$

The codes are modules over the group-ring $\mathbb{F}_q[{}^2G_2(q)]$.

REMARK 4.5 In [PE] curves with the same genera, Zeta-function and automorphism groups have been realized as singular curves in $\mathbb{P}^3$.

Appendix. Weil's "explicit formulas".

In [Se1] the "explicit formulas" of Weil for bounding the number N of $\mathbb{F}_q$-rational points on a curve of genus g is treated. Let X be a curve defined over the finite field $\mathbb{F}_q$. Let a_d denote the number of primedivisors on X of degree d and let

(a.1) $$N_m = \sum_{d|m} d\cdot a_d, \qquad m \geq 1$$

be the number of $\mathbb{F}_{q^m}$- rational points on X.
The Zeta-function of X is the formal power series

(a.2) $$Z(X,\mathbb{F}_q)(t) = \exp\Big(\sum_{m\geq 1} N_m \frac{t^m}{m}\Big),$$

which is a rational function of the form

(a.3) $$Z(X,\mathbf{F}_q)(t) = \frac{\prod_{j=1}^{g}(1-\alpha_j t)(1-\bar{\alpha}_j t)}{(1-t)(1-qt)},$$

satiesfying the Riemann hypotheses

(a.4) $$\alpha_j = \sqrt{q}\exp(i\theta_j), \qquad \theta_j \in \mathbf{R}$$

Taking the logarithmic derivative of $Z(X,\mathbf{F}_q)(t)$ we obtain

(a.5) $$N_m = 1 + q^m - (\sqrt{q})^m \sum_{j=1}^{g} 2\cos(m\theta_j)$$

Consider the trigonometric expressions

(a.6) $$f(\theta) = 1 + 2\sum_{n\geq 1} c_n \cos(n\theta) \quad , \quad \Psi_d(t) = \sum_{n\geq 1} c_{nd}\, t^{nd}, \qquad d \geq 1$$

Then it follows from (a.5) that

(a.7) $$\sum_{j=1}^{g} f(\theta_j) + \sum_{d\geq 1} d\cdot a_d \cdot \Psi_d(t) = g + \Psi_1((\sqrt{q})^{-1})\cdot\Psi_1(\sqrt{q}).$$

From this one concludes about the number $N = N_1$ of $\mathbf{F}_q$- rational points on X.

Proposition a.8 Let the notation be as above and assume that $f(\theta) \geq 0$ for all $\theta \in \mathbf{R}$ and that $c_n \geq 0$ for all $n \geq 1$. Then

$$N \leq \frac{1}{\Psi_1((\sqrt{q})^{-1})}\cdot g + \left(\frac{\Psi_1(\sqrt{q})}{\Psi_1((\sqrt{q})^{-1})} + 1\right)$$

and eguality holds if and only if

$$\sum_{j=1}^{g} f(\theta_j) = 0 \quad \text{and} \quad \sum_{d\geq 2} d\cdot a_d \cdot \Psi_d(t) = 0.$$

The Deligne-Lusztig curves associated to the algebraic groups of type ${}^2A_2$, ${}^2B_2$ and ${}^2G_2$ all have the maximal number $\mathbf{F}_q$- rational points on X allowed by the above formulas as we have seen.

References

[Be] S.D. Berman, *On the Theory of Group Codes*, Kibernetika, 3 (1967), 31-39.

[Ca1] R..W. Carter, *Simple Groups of Lie Type*, John Wiley and Sons Ltd, 1972.

[Ca2] R..W. Carter, *Finite Groups of Lie Type*, John Wiley and Sons Ltd, 1985.

[Ch1] P. Charpin, *Codes idéaux de certaines algèbres modulaires*, These de 3ième cycle, Université de Paris VII (1982).

[Ch2] P. Charpin, *The extended Reed-Solomon codes considered as ideals of a modular algebra*, Annals of Discrete Math. 17 (1983), 171-176.

[Ch3] P. Charpin, *A new description of some polynomial Codes: the primitive generalized Reed-Muller codes*, preprint.

[Che] C. Chevalley, *Introduction to the theory of algebraic functions of one variable*, New York 1951.

[Da-La] I. Damgård and P. Landrock, *Ideals and Codes in Group Algebras*, Aarhus Universitet, Preprint Series.

[De-Lu] P. Deligne and G. Lusztig, *Representations of reductive groups over finite fields*, Ann. of Math. 103 (1976), 103-161.

[Go1] V. D. Goppa, *Codes on algebraic Curves*, Dokl. Akad. NAUK, SSSR, 259 (1981), 1289-1290; (Soviet Math. Dokl. 24 (1981), 170-172.)

[Go2] V. D. Goppa, *Algebraico-geometric Codes*, Izv. Akad. NAUK, SSSR, 46 (1982); (Math. U.S.S.R. Izvestiya, 21 (1983), 75-91.)

[Ha1] J. P. Hansen, *Codes on the Klein Quartic, Ideals and decoding*, IEEE Trans. on Inform. Theory 33 (1987).

[Ha2] J. P. Hansen, *Group Codes on Algebraic Curves*, Mathematica Gottingensis, Heft 9, Feb. 1987.

[Ha-St] J. P. Hansen and H. Stichtenoth, *Group Codes on Certain Algebraic Curves with Many Rational Points*, Aarhus Universitet, Preprint Series 1988/89 No. 7. (to appear in Applicable Algebra in Engineering, Communication and Computing).

[Has] H. Hasse, *Theorie der relativ zyklischen algebraischen Funktionenkörper*, J. Reine Angew. Math. 172 (1934), 37-54.

[He] H. W. Henn, *Funktionenkörper mit grosser Automorphismengruppe*, J. Reine u. Angew. Math. 302 (1978), 96-115.

[L] S. Lang, *Algebraic Groups over finite Fields*, Amer. J. Math. 78 (1956), 555-563.

[La-Ma] P. Landrock and O. Manz, *Classical Codes as Ideals in Group Algebras* Aarhus Universitet, Preprint Series, 1986/87 No. 18.

[Lu1] G. Lusztig, *Representations of finite Chevalley groups* Regional conference series in mathematics; no. 39. AMS 1978.

[Lu2] G. Lusztig, *Representations of finite Chevalley groups* Re

[Pe] J. P. Pedersen, *A Function Field related to the Ree group*, to appear in the proceedings.

[Se1] J. P. Serre, *Sur le nombre des points rationnels d'une courbe algébrique sur un corps fini*,

C. R. Acad. Sci. Paris Sér. I Math, 296 (1983), 397-402.

[Se2] J.-P. Serre, *Corps locaux*, Paris 1962.

[Se3] J.-P. Serre, *Annuaire du Collège de France*, 1983-1984.

[St1] H. Stichtenoth, *On Automorphisms of Geometric Goppa Codes*, (to appear in J. of Alg).

[St2] H. Stichtenoth, *A note on Hermitian Codes over* GF(q^2), IEEE Trans. on Inform. Theory 34 (1988), 1345-1348.

[St3] H. Stichtenoth, *Self-dual Goppa Codes*, J. Pure Appl. Algebra 55 (1988), 199-211.

[Ste] R. Steinberg, *Endomorphisms of linear algebraic groups*, Mem. Amer. Math. Soc., 80 (1968).

[Ti] H. J. Tiersma, *Remarks on Codes from Hermitian Curves*, IEEE Trans. on Inform. Theory 33 (1987), 605-609.

[T-V-Z] M. A. Tsfasman, S. G. Vladut and Th. Zink, *Modular curves, Shimura curves and Goppa codes, better than Varshamov-Gilbert bound*, Math. Nachr., 109 (1982), 13-28.

[V] S. G. Vladut, *Modular Codes as Group Codes*, MAT-REPORT NO.1990-25, Matematisk Institut, DanmarksTekniske Højskole.

Spectra of Linear Codes and Error Probability of Decoding

Gregory L. KATSMAN, Michael A. TSFASMAN, Serge G. VLADUŢ

G.K.: Institute of Control Science
65, Profsojuznaja st. Moscow 117342, U.S.S.R.
M.Ts.& S.Vl.: Institute of Information Transmission
19 Ermolovoi st., Moscow GSP-4, U.S.S.R.
e-mail: tsfasman@ippi.msk.su , vladut@ippi.msk.su

Introduction

One of the most important characteristics of an error-correcting code is its spectrum. Therefore for asymptotic families of codes of growing length it is natural to consider the asymptotic behaviour of their spectra. Moreover the knowledge of this behaviour helps one to estimate the probability of error for maximum likelihood decoding for the considered family of codes.

This paper is based on the study of spectra of algebraic-geometric codes - and arbitrary linear codes as well - carried out in [1] (cf. also [2] 1.1.3, 3.1.3). The main idea of that paper was to redevelop the enumerator $W_C(x:y)$ over the basis $\{x^j(x-y)^{n-j}\}$ and to estimate the corresponding coefficients B_j which have a lucid algebraic-geometric interpretation.

Let C be a linear $[n,k,d]_q$-code and A_i be the number of codewords of weight i. The enumerator is given by

$$W_C(x:y) = \sum_{i=0}^{n} A_i x^{n-i} y^i = x^n + \sum_{j=0}^{n-d} B_j (x-y)^j y^{n-j} .$$

For a family F of codes with $n \longrightarrow \infty$ we set

$$\alpha_F(\omega) = \limsup \frac{1}{n} \log_q A_{\lceil n\omega \rceil} ,$$

where $0 \le \omega \le 1$, the limit is taken over all $[n,k,d]_q$-codes $C \in F$, $n \longrightarrow \infty$, and $\lceil \cdot \rceil$ denotes the integer part.

We would like to find out or at least to estimate the function $\alpha_F(\omega)$ in terms of asymptotic parameters of the family F, such as the rate R, the relative distance δ, the relative dual distanse $\delta^{\perp}$, etc.

First (in Section 1) we use the estimate for B_j obtained in [1] and obvious relations between A_i and B_j to estimate A_i and $\alpha_F(\omega)$ for arbitrary linear codes. Then (in Section 2) we give some subtler estimates for algebraic-geometric codes.

Bounds for $\alpha_F(\omega)$ can be used to estimate the error probability of maximum likelihood decoding. Let $x \in C$ be an input of a q-ary symmetric channel with error probability per symbol p, let y be the output of maximum likelihood decoding. Then $P_C(p)$ is defined as the probability for y to be different from x. For a family F we can set

$$E_F(p) = \limsup \frac{1}{n} \log_q P_C(p) .$$

Then for the given rate R let

$$E(p,R) = \liminf E_F(p) ,$$

the limit being taken over all families F whose rate equals R, i.e. $E(p,R)$ corresponds to the error probability of decoding in the best family of codes with rate R.

This is the function to estimate. The best estimate known is due to Gallager [3].

It is but natural that using algebraic-geometric codes and estimates for their spectra we can ameliorate the Gallager bound in many cases (Section 3).

1. Spectra of linear codes

In this section we give some upper bounds for the spectrum of an arbitrary linear q-ary code in terms of its parameters and its dual distance. Here we present some asymptotic results (for sequences of codes of growing length). For the case of a fixed code of a "finite length" see Remark 3 below.

We use the approach introduced in [1] (see also [2] 1.1.3). Let us recall some notation.

Let C be a linear $[n,k,d]_q$-code, $d^{\perp}$ its dual distance. Its spectrum $\{A_j\}$ is defined as

$$A_j = A_j(C) = |\{x \in C \mid \|x\|=j\}| \, .$$

Of course $A_0 = 1$ and $A_j = 0$ for $1 \le j \le d-1$. The enumerator $W_C(x:y)$ is defined as

$$W_C(x:y) = x^n + \sum_{j=d}^{n} A_j x^{n-j} y^j \, .$$

We define B_i from the decomposition

$$W_C(x:y) = x^n + \sum_{i=0}^{n-d} B_i (x-y)^i y^{n-i} .$$

Clearly

$$B_i = \sum_{j=d}^{n-i} \binom{n-j}{i} A_j \ge 0 \, .$$

We get the following obvious

Lemma 1. $A_j \le \min\limits_{0 \le i \le n-j} B_i / \binom{n-j}{i}$. ■

It is easy to check that

$$B_\ell = \sum \left(q^{k(i_1,\ldots,i_\ell)} - 1\right) ,$$

where $k(i_1,\dots,i_\ell) = \dim C(i_1,\dots,i_\ell)$, the subcode $C(i_1,\dots,i_\ell) \subseteq C$ being defined as the space of vectors having zeroes at the positions $i_1,\dots,i_\ell$, and the sum is taken over all subsets $\{i_1,\dots,i_\ell\} \subseteq \{1,\dots,n\}$ of cardinality ℓ, $1 \le i_1 \le \dots \le i_\ell \le n$.

The following statement (see [2] thm. 1.1.26) easily follows from the MacWilliams identity.

Lemma 2. *Let* $i \le d^\perp - 1$. *Then*

$$B_i = \binom{n}{i}(q^{k-i} - 1)\,. \blacksquare$$

Together Lemma 1 and Lemma 2 show that

$$A_j \le \min_{0 \le i \le d^\perp - 1} \binom{n}{i} q^{k-i} \Big/ \binom{n-j}{i}\,.$$

Using $\binom{n}{i}\binom{n-i}{j} = \binom{n}{j}\binom{n-j}{i}$ we get

Lemma 3. $A_j \le \binom{n}{j} q^k \min_{0 \le i \le d^\perp - 1} q^{-i} \Big/ \binom{n-i}{j}\,. \blacksquare$

The next statement is just as simple.

Lemma 4. $A_j \le \binom{n}{j} \min_{0 \le i \le n-d} q^{K(n-i,d)} \Big/ \binom{n-i}{j}\,,$

where $K(n-i, d)$ *is the maximal possible dimension* k *of an* $[n-i, k, d]_q$*-code.*

Proof:

$$B_\ell = \sum (q^{k(i_1,\dots,i_\ell)} - 1) \le \binom{n}{\ell} q^{K(n-\ell,d)}\,.$$

Then use again Lemma 1 and $\binom{n}{\ell}\binom{n-\ell}{j} = \binom{n}{j}\binom{n-j}{\ell}\,. \blacksquare$

Let us establish an asymptotic upper bound for the spectrum. To do this the following notation looks appropriate. Consider an $[n,k,d]_q$-code C with the dual distance $d^\perp$. Let $R = R(C) = k/n$, $\delta = \delta(C) = d/n$, $\delta^\perp = \delta^\perp(C) = d^\perp/n$. For $j = d,\dots,n$ let $\omega = j/n$ and

$$\alpha_q(\omega, C) = \frac{1}{n} \log_q A_j .$$

Consider a family of codes $F = \{C_\ell\}$ of growing length. By an abuse of language we speak then about "a code" of growing length. We shall consider only families F such that the limit $R = R(F) = \lim R(C_\ell)$ does exist. Note that any family contains such a subfamily. Set $\delta = \delta(F) = \liminf \delta(C_\ell)$, and $\delta^\perp = \delta^\perp(F) = \liminf \delta^\perp(C_\ell)$. We would like to study the "limit" value of $\alpha_q(\omega, C_\ell)$. Set

$$\alpha_F(\omega) = \limsup \frac{1}{n} \log_q A_{\lceil n\omega \rceil} ,$$

where $0 \le \omega \le 1$, the limit is taken over all $[n,k,d]_q$-codes $C \in F$, $n \longrightarrow \infty$, and $\lceil \cdot \rceil$ denotes the integer part. Since $A_j \le q^k$ for any j, we see that $\alpha_F(\omega) \le R$.

Here goes the principal result of this section. Let

$$\alpha_q^0(\omega) = R - 1 + H_q(\omega) ,$$

H_q being the q-ary entropy function. It is well known that, the spectrum of a random code asymptotically behaves as $\alpha_q^0(\omega)$. In particular, the following statement is valid.

Proposition 1. *For any* R *there exists a family* F *of linear codes such that* $R(F) = R$ *and*

$$\alpha_F(\omega) \le \begin{cases} -\infty & \text{for} \quad 0 \le \omega < \delta_{GV}(R) \\ \alpha_q^0(\omega) & \text{for} \quad \delta_{GV}(R) \le \omega \le 1 , \end{cases}$$

where $\delta_{GV}(R) = H_q^{-1}(1 - R)$, H_q^{-1} *being the inverse function to the* q*-ary entropy.* ∎

Let

$$\omega^* = (1 - \delta^\perp) H_q^{-1}\left(\frac{R - \delta^\perp + q\delta/(q-1)}{1 - \delta^\perp}\right) .$$

We shall prove

Theorem 1. *Let* F *be a family of linear* q*-ary codes with* $R(F) = R$, $\delta(F) \geq \delta$, *and* $\delta^{\perp}(F) \geq \delta^{\perp}$. *Then*

$$\alpha_F(\omega) \leq f_1(\omega) := \begin{cases} -\infty & \text{for } 0 \leq \omega < \delta \\ \alpha_q^0(\omega) + [1 - R - q\delta/(q-1)] & \text{for } \delta \leq \omega \leq \omega^* \\ \alpha_q^0(\omega) + (1-\delta^{\perp})(1-H_q(\omega/(1-\delta^{\perp}))) & \text{for } \omega^* \leq \omega \leq (q-1)(1-\delta^{\perp})/q \\ \alpha_q^0(\omega) & \text{for } (q-1)(1-\delta^{\perp})/q \leq \omega \leq (q-1)/q \\ R & \text{for } (q-1)/q \leq \omega \leq 1 \end{cases}$$

Proof: Since the function $f_1(\omega)$ is non-increasing in parameters δ and $\delta^{\perp}$, we can suppose that for the family $\{C_\ell\}$ the limits $\lim \delta(C_\ell)$ and $\lim \delta^{\perp}(C_\ell)$ do exist and equal δ and $\delta^{\perp}$, respectively.

For the first and the last segments the statement is clear and we assume $\delta \leq \omega \leq (q-1)/q$. Lemma 3 and the Stirling formula yield

$$\alpha_F(\omega) \leq H_2(\omega)\log_q 2 + R - \max_{0 \leq \nu \leq \delta^{\perp}} \{\nu + (1-\nu)(\log_q 2)H_2(\omega/(1-\nu))\}$$

(informally speaking, here ν "is" $\lim i/n$).

One easily checks that the function

$$f(\nu) = \nu + (1-\nu)(\log_q 2)H_2(\omega/(1-\nu))$$

attains its maximum for $\nu^* = 1 - \omega q/(q-1)$ and $f(\nu^*) = 1 - \omega\log_q(q-1)$.

If $(q-1)(1-\delta^{\perp})/q \leq \omega \leq (q-1)/q$ then $\nu^* \leq \delta^{\perp}$ and

$$\alpha_F(\omega) \leq H_2(\omega)\log_q 2 + R - f(\nu^*) =$$

$$= H_2(\omega)\log_q 2 - 1 + R + \omega\log_q(q-1) =$$

$$= H_q(\omega) + R - 1 = \alpha_q^0(\omega) .$$

If $\omega \le (q-1)(1-\delta^{\perp})/q$ then $\nu^{*} \ge \delta^{\perp}$ and the maximum on $[0,\delta^{\perp}]$ is attained at the end point $\nu = \delta^{\perp}$. Therefore

$$\alpha_F(\omega) \le R + H_2(\omega)\log_q 2 - \delta^{\perp} - (1-\delta^{\perp})(\log_q 2)H_2(\omega/(1-\delta^{\perp})) =$$

$$= \alpha_q^{o}(\omega) + (1-\delta^{\perp})(1 - H_q(\omega/(1-\delta^{\perp}))) .$$

For $\omega \le \omega^{*}$ this estimate can be ameliorated using the asymptotic Plotkin bound

$$\limsup K(n,d)/n \le 1 - q\delta/(q-1) .$$

Indeed, let i run over a family with $\lim \frac{i}{n} = \nu$, then we get

$$\limsup K(n-i,\, d)/n \le 1 - \nu - q\delta/(q-1) ,$$

and Lemma 4 yields

$$\alpha_F(\omega) \le H_2(\omega)\log_q 2 +$$

$$+ \min_{0\le\nu\le 1-\delta} (1 - \nu - q\delta/(q-1) - (1-\nu)(\log_q 2)H_2(\omega/(1-\nu))) =$$

$$= H_2(\omega)\log_q 2 + 1 - q\delta/(q-1) - \max f(\nu) =$$

$$= \alpha_q^{0}(\omega) + (1 - q\delta/(q-1) - R) ,$$

since $\nu^{*} = 1 - \omega q/(q-1) < 1 - \delta$ for $\delta \le \omega \le (q-1)/q$.

This bound is valid for any ω in this range, and one easily checks that for $\omega \le \omega^{*}$ it is better than the previous bound. ∎

Remark 1. One can also use other upper bounds for $K(n,d)$. The result becomes slightly better but the formulae are much more complicated and the calculations are very cumbersome.

Remark 2. Though it does not influence the error probability of decoding, it is interesting to bound $\alpha_F(\omega)$ for $\omega \ge (q-1)/q$, the estimate $\alpha_F(\omega) \le R$ being very rough. This can be done using somewhat different idea instead of Lemma 1 (cf. [4]).

Remark 3. The estimate of Theorem 1 is approximately valid for all linear codes (not only asymptotically). It is not difficult to estimate the error term. There are two points.

First we have calculated $f(\nu^*)$ at a non-integer point, second we have used the asymptotic Plotkin bound rather than its exact form. The deviation between integral and real maximum is very small and the error in the Plotkin bound also is not very important. We do not write out the exact answer which does not look very beautiful, let us just mention that the error term is $O(1/n^2)$.

2. Spectra of algebraic-geometric codes

For an algebraic-geometric code the Clifford theorem gives a better estimate for B_i when i is not too large, and thus yields a better estimate for the spectrum.

Let $C = (X,\mathcal{P},D)_L$ be an algebraic-geometric code. Here X is a smooth absolutely irreducible projective curve of genus g over $\mathbb{F}_q$, $\mathcal{P} \subseteq X(\mathbb{F}_q)$, $\mathcal{P} = \{P_1,\ldots,P_n\}$, and D is an $\mathbb{F}_q$-rational divisor of degree a. Let $g - 1 < a < n$. Then

$$k \geq k^* = a - g + 1 ,$$

$$d \geq d^* = n - a ,$$

$$d^\perp \geq d^{\perp *} = a - 2g + 2 .$$

The values k^*, d^*, and $d^{\perp *}$ are called the designed parameters of C. Let $R^* = k^*/n$, $\delta^* = d^*/n$, $\delta^{\perp *} = d^{\perp *}/n$.

To simplify the situation from now on we suppose that $a \geq 2g - 1$ so that $k = k^* = a - g + 1$ and $R^* = R$, and we do not distinguish between R and R^*.

Lemma 5. *Let* $C = (X,\mathcal{P},D)_L$. *If* $a - \ell \leq 2g$ *then*

$$B_\ell \leq \binom{n}{\ell}\left(q^{(n - d^* - \ell)/2 + 1} - 1\right) .$$

Proof: It is enough to show that

$$k(i_1,\ldots,i_\ell) \leq (n - d^* - \ell)/2 + 1 = (a - \ell)/2 + 1 .$$

Indeed,

$$C(i_1,\dots,i_\ell) = (X, \mathcal{P} - \{P_{i_1},\dots,P_{i_\ell}\}, D' = D - \sum_{j=1}^{\ell} P_{i_j})_L .$$

If D' is non-special, i.e. $\ell(D') = a - \ell - g + 1$, then $\ell(D') \le (a-\ell)/2 + 1$ since $a - \ell \le 2g$. Otherwise D' is special and the Clifford theorem shows that $\ell(D') \le (a-\ell)/2 + 1$. ∎

Now we proceed exactly as in the proof of Theorem 1 using Lemma 5 instead of Lemma 4 and the Plotkin bound. We get

Theorem 2. *Let* F *be a family of algebraic-geometric codes with* $R(F) = R$, $\delta^*(F) \ge \delta^*$, *and* $\delta^{\perp *}(F) \ge \delta^{\perp *}$. *Then*

$$\alpha_F(\omega) \le f_2(\omega) := \begin{cases} -\infty \quad \text{for } 0 \le \omega < \delta^* \\ \alpha_q^0(\omega) + (1 - R - q\delta^*/(q-1)) \quad \text{for } \delta^* \le \omega \le \omega^{**} \\ \alpha_q^0(\omega) + (1 - R - \delta^*/2 - \omega\log_q(\sqrt{q}+1)) \\ \qquad \text{for } \omega^{**} \le \omega \le (1-\delta^{\perp *})(\sqrt{q}-1)/\sqrt{q} \\ \alpha_q^0(\omega) + (1-\delta^{\perp *})(1-H_q(\omega/(1-\delta^{\perp *}))) \quad \text{for} \\ \qquad (1-\delta^{\perp *})(\sqrt{q}-1)/\sqrt{q} \le \omega \le (q-1)(1-\delta^{\perp *})/q \\ \alpha_q^0(\omega) \quad \text{for } (q-1)(1-\delta^{\perp *})/q \le \omega \le (q-1)/q \\ R \quad \text{for } (q-1)/q \le \omega \le 1 \end{cases}$$

where $\omega^{**} = \delta^* (1/(q-1) + 1/2)/\log_q(\sqrt{q}+1)$.

Proof: From Lemma 1 and Lemma 5 we see that

$$A_j \le \min_{d^{\perp *}-2 \le i \le n-j} \binom{n}{i} q^{(n-d^*-i)/2+1} \Big/ \binom{n-j}{i} =$$

$$= \binom{n}{j} q^{(n-d^*)/2+1} \min_{d^{\perp *}-2 \le i \le n-j} \frac{q^{-i/2}}{\binom{n-i}{j}}$$

(the condition $i \ge d^{\perp *} - 2$ being equivalent to $a - i \le 2g$). Thus we get

$$\alpha_F(\omega) \le H_2(\omega)\log_q 2 + (1-\delta^*)/2 - \max_{\delta^{\perp *} \le \nu \le 1-\omega} \{\nu/2 + (1-\nu)(\log_q 2)H_2(\tfrac{\omega}{1-\nu})\}.$$

The maximum of

$$g(\omega) = \nu/2 + (1-\nu)(\log_q 2) H_2\left(\frac{\omega}{1-\nu}\right)$$

is attained for $\nu^* = 1 - \omega\sqrt{q}/(\sqrt{q} - 1)$ and $g(\nu^*) = 1/2 - \omega \log_q(\sqrt{q} - 1)$.

Therefore if $\delta^{\perp *} \le \nu^* \le 1 - \omega$, i.e. if $\omega \le (1 - \delta^{\perp *})\sqrt{q}/(\sqrt{q} - 1)$ we have

$$\alpha_F(\omega) \le H_2(\omega)\log_q 2 + \omega \log_q(\sqrt{q} - 1) - \delta^*/2 =$$

$$= H_q(\omega) - \delta^*/2 - \omega \log_q(\sqrt{q} + 1) =$$

$$= \alpha_q^0(\omega) + 1 - R - \delta^*/2 - \omega \log_q(\sqrt{q} + 1) .$$

The rest follows from Theorem 1 (of course, one can put δ^* and $\delta^{\perp *}$ in place of δ and $\delta^{\perp}$ since our function is non-increasing in these parameters). ■

Remark 4. It is clear that Theorem 2 remains valid if we sustitute δ and $\delta^{\perp}$ for δ^* and $\delta^{\perp *}$. We prefer the above statement of the theorem since the parameters δ^* , $\delta^{\perp *}$ can be directly calculated in terms of algebraic-geometric parameters g , n and a . It is also useful to note that the "Clifford" part of Theorem 2 (i.e. that on the segment $[\omega^{**}, (1 - \delta^{\perp *})(\sqrt{q} - 1)/\sqrt{q}]$ can be written as

$$\alpha_q^0(\omega) + (1 - \delta^{\perp *})/2 - \omega \log_q(\sqrt{q} + 1) .$$

Let us also remark that $f_2(\omega)$ is differentiable everywhere except ω^{**} and δ^* (and at δ^* it is not even continuous).

Remark 5. Using slightly different ideas one can obtain similar bound for $\alpha_F(\omega)$ which is non-trivial for $\delta > \frac{q-1}{q}$ (see [4]).

The bound of Theorem 2 is valid for any family of codes on algebraic curves provided the limit $R(F)$ does exist. Using technique developed in [5] one can obtain a similar (but a better) bound for a "random code" on algebraic curves. We give here the answer only in the most important case of asymptotically maximal curves.

Let $q \geq 49$ be an even power of a prime, let $\{X\}$ be a family of asymptotically maximal curves, i.e. curves with

$$\lim_{g \longrightarrow \infty} \frac{N(X)}{g} = \sqrt{q} - 1$$

(the maximal possible ratio of the number of $\mathbb{F}_q$-points to the genus).

Let $\delta_1, \delta_2, \delta_3, \delta_4 \in [0, (q-1)/q]$ be defined by the following conditions:

i. $0 < \delta_1 < \delta_2 < \delta_3 < \delta_4 < (q-1)/q$;

ii. δ_1 and δ_4 are the roots of the equation

$$H_q(\delta) + \frac{q}{q-1}(1-\delta) = 1 + \gamma \ ;$$

iii. δ_2 and δ_3 are the roots of the equation

$$H_q(\delta) + (1-\delta)\log_q(q-1) = 1 + \gamma \ ;$$

where $\gamma = \gamma_q = (\sqrt{q} - 1)^{-1}$.

Then it is known (see [5], or [2] Theorem 3.4.11) that for any R there exists a choice of divisors D such that the family $F = \{C = (X, \mathcal{P}, D)_L\}$ has the following relation between R and δ (and in some precise sense almost any choice is such).

If $0 \leq \delta \leq \delta_1$ or $\delta_4 \leq \delta \leq (q-1)/q$ then $R = 1 - H_q(\delta)$. If $\delta_2 \leq \delta \leq \delta_3$ then $R = 1 - \gamma - \delta$. If $\delta_1 \leq \delta \leq \delta_2$ or $\delta_3 \leq \delta \leq \delta_4$ then

$$F_V(R,\delta) := (R+\gamma)\left(1 - H_q\left(\frac{1-\omega}{R+\gamma}\right)\right) + H_q(\delta) - 1 - \gamma = 0 \ .$$

For a given R such that $1 - H_q(\delta_1) \geq R \geq 1 - \gamma - \delta_2$ or $1 - \gamma - \delta_3 \geq R \geq 1 - H_q(\delta_4)$ let $\delta_V(R)$ be the (unique) solution of the last equation. Let $\delta_{GV}(R) = H_q^{-1}(1-R)$, where H_q^{-1} is the inverse function to the q-ary entropy. The following statement is proved in [4], the technique being the same as in [5].

Theorem 3. *Let* $q \geq 49$ *be an even power of a prime, let* $\{X\}$ *be a family of asymptotically maximal curves. Then for any* R *there exists a choice of divisors* D *such that the family* $F = \{(X, \mathcal{P}, D)_L\}$ *has the above relation between* R *and* δ *and*

If $R \in [0, 1 - H_q(\delta_4)] \cup [1 - H_q(\delta_1), 1]$ *then*

$$\alpha_F(\omega) \leq \begin{cases} -\infty & \text{for} \quad 0 \leq \omega < \delta_{GV}(R) \\ \alpha_q^0(\omega) & \text{for} \quad \delta_{GV}(R) \leq \omega \leq 1 \ . \end{cases}$$

If $R \in [1 - H_q(\delta_4), 1 - \delta_3 - \gamma] \cup [1 - \delta_2 - \gamma, 1 - H_q(\delta_1)]$ *then*

$$\alpha_q(\omega) \leq \begin{cases} -\infty & \text{for} \quad 0 \leq \omega < \delta_V(R) \\ \alpha_q^0(\omega) - (R+\gamma)\left(1 - H_q\left(\dfrac{1-\omega}{R+\gamma}\right)\right) & \text{for} \quad \delta_V(R) \leq \omega \leq \omega_0 \\ \alpha_q^0(\omega) & \text{for} \quad \omega_0 \leq \omega \leq 1 \ , \end{cases}$$

where $\omega_0 = 1 - (R+\gamma)(q-1)/q$.

If $R \in [1 - \delta_3 - \gamma \ , 1 - \delta_2 - \gamma]$ *then*

$$\alpha_q(\omega) \leq \begin{cases} -\infty & \text{for} \quad 0 \leq \omega < 1 - R - \gamma \\ \alpha_q^0(\omega) & \text{for} \quad 1 - R - \gamma \leq \omega \leq 1 \ . \end{cases}$$ ■

3. Error probability of decoding

Let C be a linear $[n, k, d]_q$-code. This code can be used to transmit information over a noisy q-ary symmetric channel. This means that we transmit code words $x \in C$, and the error probability per symbol equals some p ; moreover the probability that a given symbol is distorted to another given symbol equals $p/(q-1)$. The distorted word z is then transformed (decoded) into a code word y nearest to z (so called *maximum likelihood* or minimum distance decoding). Let $P_C(p)$ be the probability of wrong decoding $(y \neq x)$.

Let F be a family of codes C. Let

$$E_F(p) = \limsup_{C \in F} \frac{1}{n} \log_q P_C(p) ,$$

and let

$$E(p,R) = \lim\inf E_F(p)$$

over all families of codes whose rate equals R.

Let $e_q(n,w.p)$ be the probability that for $x = 0 \in \mathbb{F}_q^n$ the received word z is nearer to a given y of weight w than to x (note that $e_q(n,w.p)$ does not depend on y but only on w). The following result bounding $P_C(p)$ in terms of the spectrum is well known.

Lemma 6. $P_C(p) \le \sum_{i=1}^{n} A_i e_q(n,i,p)$.

Proof: If there is an error of decoding, y should be a code word of some weight i. The probability that y differs from 0 is at most the sum of probabilities that y equals some given non-zero code word. ∎

Let us calculate $e_q(n,i,p)$ and its asymptotical behaviour. Let

$$e_q(\omega,p) = \lim \frac{1}{n} \log_q e_q(n,\lceil \omega n \rceil,p) .$$

Lemma 7. a) $e_q(n,i,p) =$

$$= \sum \binom{n-i}{n_1} (q-1)^{n_1} \binom{i}{n_2} \binom{i-n_2}{n_3} (q-2)^{n_3} (1-p)^{n-n_1-n_2-n_3} \left(\frac{p}{q-1}\right)^{n_1+n_2+n_3}$$

where the sum is taken over all triples (n_1, n_2, n_3) *of non-negative integers subject to the conditions* $n_1 \le n - i$, $n_2 + n_3 \le i$, $2n_2 + n_3 \ge i$.

b) $e_q(\omega,p) = \omega \log_q \beta_q(p)$,

where

$$\beta_q(p) = p(q-2)/(q-1) + 2\sqrt{p(1-p)/(q-1)} .$$

Proof: Let $x = 0 \in \mathbb{F}_q^n$, $y \in \mathbb{F}_q^n$, $wt(y) = i$. Suppose that the first $n - i$ positions are zeroes. Let $z \in \mathbb{F}_q^n$ have $n_1 \le n - i$ non-zero entries in the first $n - i$ positions and

$n_2 + n_3 \le i$ non-zero entries in other positions. Besides we suppose that in n_2 positions the coordinates of z are equal to those of y. If n_1, n_2, and n_3 are fixed then the probability of receiving z as an output of a q-ary symmetric channel when the zero word was transmitted equals

$$\binom{n-i}{n_1}(q-1)^{n_1}\binom{i}{n_2}\binom{i-n_2}{n_3}(q-2)^{n_3}(1-p)^{n-n_1-n_2-n_3}\left(\frac{p}{q-1}\right)^{n_1+n_2+n_3}.$$

The distance between x and z equals $n_1 + n_2 + n_3$, and between y and z it equals $i - n_2 + n_1$. Thus $e_q(n,i,p) =$

$$\sum_{\substack{n_1 \le n-i \\ n_2+n_3 \le i \\ 2n_2+n_3 \ge i}} \binom{n-i}{n_1}(q-1)^{n_1}\binom{i}{n_2}\binom{i-n_2}{n_3}(q-2)^{n_3}(1-p)^{n-n_1-n_2-n_3}\left(\frac{p}{q-1}\right)^{n_1+n_2+n_3}$$

and the first statement is proved.

To prove the second statement we need to logarithmize the previous expression, and note that $\frac{1}{n}\log_q n$ is asymptotically small. So the asymptotic value of the sum equals the maximum of its summands. It is not difficult to see that this maximum is as stated in b). ∎

Remark 6. It is clear that $\beta_q(p)$ for $0 \le p \le (q-1)/q$ is an increasing concave function tangent to the vertical axis at $p = 0$ and to the horizontal line $\beta = 1$ at $p = (q-1)/q$.

We are ready to estimate the error probability for a given family F.

Lemma 8.

$$E_F(p) \le \max_{\delta \le \omega \le 1} \{\alpha_F(\omega) + \omega \log_q \beta_q(p)\}.$$

Proof: We logarithmize the expression of Lemma 6 and note that asymptotically the sum equals the maximal summand. Then we use part b of Lemma 7. ∎

The following statement is obtained by a direct calculation combining Lemma 8 and Proposition 1.

Theorem 4. *Let*

$$R_0(p) = 1 - H_q\left(\frac{(q-1)\beta_q(p)}{1+(q-1)\beta_q(p)}\right) .$$

Then

$$E(p,R) \le E_G(p,R) = \begin{cases} \delta_{GV}(R)\log_q\beta_q(p) & \text{for} \quad 0 \le R \le R_0(p) \\ (R+1) + \log_q(1+(q-1)\beta_q(p)) & \\ & \text{for} \quad R_0(p) \le R \end{cases} .$$ ■

Remark 7. This bound is called the Gallager bound without expurgation. The Gallager bound with expurgation is

$$E(p,R) \le E'_G(p,R) =$$

$$= \min_{0\le\rho\le1} (\rho(R-1)+(\rho+1)\log_q((p/(q-1))^{1/(1+\rho)}(q-1)+(1-p)^{1/(1+\rho)}))$$

which is better than $E_G(p,R)$ starting from some $R_* \ge R_0(p)$. Moreover, it is known that for $R_* \le R \le C = 1 - H_q(p)$ we have the complete answer

$$E(p,R) = E'_G(p,R) .$$

Because of that further on we are interested only in $R < R_*$, for which $E_G(p,R)$ is the best previously known bound for $E(p,R)$.

Let us combine Lemma 8 with Theorem 3. We get

Theorem 5. *Let* $q \ge 49$ *be an even power of a prime. Then*

a) *For* $R \in [0, 1 - H_q(\delta_4)] \cup [1 - H_q(\delta_1), 1]$ *we have*

$$E(p,R) \le E_G(p,R) .$$

b) *For* $R \in [1 - H_q(\delta_4), 1 - \delta_3 - \gamma] \cup [1 - \delta_2 - \gamma, 1 - H_q(\delta_1)]$ *we have*

$$E(p,R) \le E_V(p,R) = \begin{cases} \delta_{GV}(R)\log_q\beta_q(p) & \text{for } 0 \le \beta_q(p) < \beta_1(R) \\ F_V(R,\omega(p)) + \omega(p)\log_q\beta_q(p) & \\ \qquad \text{for } \beta_1(R) \le \beta_q(p) < \dfrac{1-R-\gamma}{(q-1)(R+\gamma)} & \\ E_G(p,R) \quad \text{for } \dfrac{1-R-\gamma}{(q-1)(R+\gamma)} \le \beta_q(p) & \end{cases}$$

where

$$\beta_1(R) = \frac{(\delta_V(R) + R + \gamma - 1)\delta_V(R)}{(1 - \delta_V(R))^2},$$

and $\omega(p)$ *is the unique positive solution of the equation* $(\omega + R + \gamma - 1)\omega = \beta_q(p)(1 - \omega)^2$.

c) *For* $R \in [1 - \delta_3 - \gamma , 1 - \delta_2 - \gamma]$ *we have*

$$E(p,R) \le E_{AG}(p,R) = \begin{cases} \alpha_q^0(1-R-\gamma)+(1-R-\gamma)\log_q\beta_q(p) & \\ \qquad \text{for } 0 < \beta_q(p) < \dfrac{1-R-\gamma}{(q-1)(R+\gamma)} & \\ E_G(p,R) \quad \text{for } \dfrac{1-R-\gamma}{(q-1)(R+\gamma)} \le \beta_q(p) & \end{cases}.$$

Sketch of proof: We apply Lemma 8 to the family F with $R(F) = R$. The direct maximization using formulae of Theorem 3 yields the result. ■

Remark 7. Since the bound of Theorem 3 is not worse than that of Proposition 1, the bound of Theorem 5 is never worse than that of Theorem 4 and often ameliorates it. For example, let $R \in (1 - \delta_2 - \gamma, 1 - H_q(\delta_1))$, and let

$$\beta_q(p) \le \min\left\{\frac{\delta_1}{(q-1)(1-\delta_1)} , \frac{(\delta_2 + R + \gamma - 1)\delta_2}{(1-\delta_3)^2}\right\}$$

Then

$$E_G(p,R) - E_V(p,R) = (\delta_{GV}(R) - \delta_V(R))\log_q\beta_q(p) > 0 .$$

Note that in this case $R < R_0(p) < R_*$.

Similar examples exist for $E_{AG}(p,R)$ for $R \in (1 - \delta_3 - \gamma , 1 - \delta_2 - \gamma)$.

Remark 8. Theorem 5 estimates the error probability of decoding for "almost any" family of algebraic geometric codes on asymptotically maximal curves. It is however difficult to produce an example of such a family. On the other hand, Theorem 2 bounds the spectra of any given family of algebraic geometric codes. One can substitute this result into Lemma 8, thus obtaining a bound for the error probability for the given family. For some explicit families (say, with $R = 1 - \delta - \gamma$ on the segment where this line ameliorates the Gilbert-Varshamov bound) this bound is sometimes also better than the Gallager bound.

References

1. G.L.Katsman, M.A.Tsfasman. *Spectra of algebraic-geometric codes.* Probl. Info. Trans., 23, 262-275 (1987)

2. M.A.Tsfasman, S.G.Vladuţ. *Algebraic-Geometric Codes.* Kluwer Acad. Publ., Dordrecht/Boston/London, 1991

3. R.G.Gallager. *Information Theory and Reliable Communications.* J.Wiley and sons, NY/London/Sidney/Toronto, 1968.

4. S.G.Vladuţ. *Two remarks on spectra of algebraic-geometric codes.* In preparation.

5. S.G.Vladuţ. *An exhaustion bound for algebraic-geometric modular codes.* Probl. Info. Trans., 23, 22-34 (1987).

On the True Minimum Distance of Hermitian Codes

Kyeongcheol Yang and P. Vijay Kumar

Communication Sciences Institute, EE-Systems,
University of Southern California
Los Angeles, CA 90089-2565, U.S.A.

Abstract

A class of geometric Goppa codes based on Hermitian curves was introduced by Stichtenoth [3]. These codes are parametrized by an integer m that governs both dimension and minimum distance of the code. In that paper, the exact minimum distance is given in the range that $0 \leq m \leq q^3 - q^2$ or $m \equiv 0 \ (mod\ q)$ with $m < q^3$. In this paper we determine the exact minimum distance of these codes for any m with $m \geq q^3 - q^2$. Taken together the two results give the exact minimum distance of Hermitian codes for all values of the parameter m.

I. Introduction

In [1], van Lint and Springer considered a class of codes defined by Goppa's algebraic-geometric construction over Hermitian curves and remarked that these codes are usually better than the corresponding Reed-Solomon codes with the same rate. Tiersma [2] studied these codes in more detail and provided a clear description of their dual codes. By working with an isomorphic curve having only one point at infinity, Stichtenoth [3] generalized and simplified the results of Tiersma. Codes of length $n = q^3$ and any dimension $0 \leq k \leq q^3$ over $GF(q^2)$ were considered by him. In particular, the exact minimum distance of these codes in the range that $0 \leq m \leq q^3 - q^2$ or $m \equiv 0 \ (mod\ q)$ with $m < q^3$, (where m is a parameter that governs both dimension and minimum distance of the code) was determined in his paper.

In this paper, we evaluate the exact minimum distance of those codes defined in [3] for any m with $m \geq q^3 - q^2$. Combined with Stichtenoth's results, these results give the exact minimum distance of Hermitian codes for all values of the parameter m.

This paper is organized as follows. In section II, we will review some properties of Hermitian curves based on Stichtenoth's results and give our results. In section III, we will give the proof of the main theorem. Partial results on the minimum distance obtained independently by other researchers are given at the end of this section.

II. Background and Results

We adopt almost all the notation in [3]. Let K be a finite field $K = GF(q^2)$ (q a power of some prime) and $F = K(x, y)$ be the function field of the Hermitian curve $y^q + y = x^{q+1}$. The genus g of the function field F/K is $g = (q^2 - q)/2$ and F/K has exactly $1 + q^3$ places of degree one. Let Q be the common pole of x and y, and $P_{\alpha,\beta}$ be a common zero of $x - \alpha$ and $y - \beta$ for any $\alpha \in K$ and any $\beta \in K$ such that $\beta^q + \beta = \alpha^{q+1}$. Then the divisor of $y - \beta$ is

$$(y-\beta) = \begin{cases} (q+1)P_{0\beta} - (q+1)Q, & \text{if } \beta^q + \beta = 0 \\ \sum_{\substack{\alpha\in K,\\ \alpha^{q+1}=\beta^q+\beta}} P_{\alpha,\beta} - (q+1)Q, & \text{if } \beta^q + \beta \neq 0. \end{cases} \quad (1)$$

Proposition 1 ([3], Proposition 1) *For each integer $m \geq 0$, $L(mQ)$ has a basis $B(m)$ such that*

$$B(m) \triangleq \{x^i y^j \mid 0 \leq i,\ 0 \leq j \leq q-1,\ iq + j(q+1) \leq m\}. \quad (2)$$

Let $D = \sum_{\substack{\alpha,\beta\in K\\ \alpha^{q+1}=\beta^q+\beta}} P_{\alpha,\beta}$ and $C_m = C(mQ, D)$. Let $d(C_m)$ denote the minimum distance of the code C_m. Note that C_m is a linear code of blocklength $n = q^3$ and if $m_1 \leq m_2$ then $d(C_{m_1}) \geq d(C_{m_2})$.

Proposition 2 ([3], Theorem 1) *For any $m \in Z$ the codes C_m and $C_{q^3+q^2-q-2-m}$ are dual to each other.*

An integer l is a gap number of Q if there is no function $f \in F$ such that $f \in L(lQ) \setminus L((l-1)Q)$. From Proposition 1, we know that $\{aq + b \mid 0 \leq a < b \leq q-1\}$ is the set of all gap numbers of Q. Now define

$$\bar{m} \triangleq max\{l \mid l = iq + j(q+1) \leq m,\ 0 \leq i,\ 0 \leq j \leq q-1\}. \quad (3)$$

If $m \geq 2g = q^2 - q$ then $\bar{m} = m$ by the Riemann-Roch theorem. Note that m is a gap number of Q if and only if $m \neq \bar{m}$, *i.e.*, $m > \bar{m}$.

Proposition 3 ([3], Theorem 4) *For any integer $m \geq 0$*

$$d(C_m) \geq n - \bar{m}. \quad (4)$$

Proposition 4 ([3], Theorem 5) *Assume $\bar{m} = iq + j(q+1) \leq q^3 - 1$ with $0 \leq i$, $0 \leq j \leq q-1$. If either $j = 0$ (i.e., $\bar{m} \equiv 0$ mod q) or $m \leq q^3 - q^2$ then*

$$d(C_m) = q^3 - \bar{m}. \quad (5)$$

Theorem 1 (Main Theorem) *The code C_m is a linear $[q^3, k, d(C_m)]$ code as shown in the Table 1.*

Remark: The dimension of the code C_m is completely given by Stichtenoth ([3], Theorem 2). By Proposition 4, $d(C_m)$ for any m with $m \leq q^3 - q^2$ is determined. We will prove $d(C_m)$ in the range of $m \geq q^3 - q^2$ in the next section.

Corollary 1
(a) The numbers of the form $q^2 - aq - b$ with $1 \le b \le a \le q-2$ are not achieved by $d(C_m)$.
(b) Given $d(C_m) = q^2 - aq$ with $0 \le a \le q-2$, the best code is parametrized by $m = q^3 - q^2 + aq + a$.
(c) Given $d(C_m) = q-a$ with $0 \le a \le q-1$, the best code is parametrized by $m = q^3+aq+a$.
(d) Except for (b) and (c), the given minimum distance uniquely determines the dimension of the code C_m.

Proof: It follows immediately from the above theorem. □

	m	Dimension k	Min. Dist. $d(C_m)$	Remarks
1)	$m < 0$	0	.	Trivial
2)	$0 \le m < q^2 - q$; $\tilde{m} = aq + b$, $0 \le b \le a \le q-1$	$a(a+1)/2$ $+b+1$	$n - \tilde{m}$	Stichtenoth
3)	$q^2 - q \le m < q^3 - q^2$	$m - g + 1$	$n - m$	Stichtenoth
4)	$q^3 - q^2 \le m < q^3$; $m = q^3 - q^2 + aq + b$, $0 \le a < b \le q-1$	$m - g + 1$	$n - m$	Theorem 2
5)	$q^3 - q^2 \le m < q^3$; $m = q^3 - q^2 + aq + b$, $0 \le b \le a \le q-1$	$m - g + 1$	$n - m + b$	Corollary 2
6)	$q^3 \le m \le q^3 + q^2 - q - 2$; $m_\perp = q^3 + q^2 - q - 2 - m$, $\tilde{m}_\perp = aq + b$, $0 \le b \le a \le q-1$	$q^3 - a(a+1)/2$ $-b-1$	$a+2$ if $b = a$ $a+1$ if $b < a$	Theorem 6
7)	$m > q^3 + q^2 - q - 2$	q^3	1	Trivial

Table 1: Parameters of the code $C_m = C(mQ, D)$.

III. Proof of the Main Theorem

A. The Minimum Distance of the Code C_m, in the Range that $q^3 - q^2 \le m < q^3$

Note that in this range $m = \tilde{m}$ and m can be expressed as $m = q^3 - q^2 + aq + b$, where $0 \le a, b \le q-1$. The gap sequence will play an important role in determining $d(C_m)$.

Lemma 1 *Assume $0 \le m \le n = q^3$ and let $\delta = n - m$. Then there exists a function $f \in L(mQ)$ having exactly m distinct zeroes in $supp(D)$ iff there exists $h \in L(\delta Q)$ having exactly δ distinct zeroes in $supp(D)$.*

Proof: Let $u = x^{q^2} - x$. The proof then follows from noting that $(u) = D - q^3 Q$ and from considering the function u/f for a given f satisfying the condition in the lemma. □

Theorem 2 *Assume* $m = q^3 - q^2 + aq + b$, *where* $0 \leq a, b \leq q-1$. *If either* $b = 0$ *or* $a < b$, *then*

$$d(C_m) = n - m = q^2 - aq - b. \tag{6}$$

Proof: If $b = 0$ then $d(C_m) = q^2 - aq$ from Proposition 4. Thus we can assume that $a < b$. It suffices to show that there exists a function $f \in L(mQ)$ such that f has exactly m distinct zeroes in $supp(D)$. Note that

$$\delta = n - m = q^2 - aq - b = (q - (a+1))q + (q - b).$$

Since $q - (a+1) \geq q - b$, δ is not a gap number of Q, so there exists a function $h \in L(\delta Q)$ having exactly δ distinct zeroes in $supp(D)$ by Proposition 4. An application of Lemma 1 completes the proof. □

Theorem 3 *Assume* $m = q^3 - q^2 + aq + a$ *where* $1 \leq a \leq q-1$. *Let* $f \in L(mQ)$ *be a nonzero function. Then* f *has at most* $m - a$ *distinct zeroes in* $supp(D)$.

Proof: Let $\delta_r = n - m + r$, where $0 \leq r \leq a-1$. Then we have

$$\begin{aligned} \delta_r &= q^2 - aq - a + r \\ &= (q - a - 1)q + (q - a + r). \end{aligned}$$

Note that $0 \leq q - a - 1 < q - a + r \leq q - 1$. So δ_r is a gap number of Q for any r with $0 \leq r \leq a-1$. Thus by Lemma 1 there does not exist a function $f \in L((m-r)Q)$, $m - a + 1 \leq m - r \leq m$, such that f has exactly $m - r$ distinct zeroes in $supp(D)$.

Now assume that there is a function $f \in L(mQ)$ such that f has exactly $m - r$, $m - a + 1 \leq m - r \leq m$, distinct zeroes in $supp(D)$. Then it must be that for some m' with $m - r \leq m' \leq m$

$$(f) = E + S - m'Q$$

where E is an effective divisor such that $E \leq D$, $deg(E) = m - r$, and S is an effective divisor such that $supp(S) \cap supp(D) \subset supp(E)$, $Q \notin supp(S)$, and $s \triangleq deg(S) = m' - (m - r)$. Note that $0 \leq s \leq r \leq q - 2$. Thus

$$\left(\frac{u}{f}\right) = \left(\frac{x^{q^2} - x}{f}\right) = D - E - S - (q^3 - m')Q.$$

Consider the constant field extension [5] F' which is the composite of F and the algebraic closure $\bar{K}$ of $K = GF(q^2)$. It is easy to show that F' has the same genus and the same gap sequence at Q. Our aim is to show that even in the larger field F', such a function u/f with divisor as given above does not exist. In F', all places (except Q) are of degree one and correspond to points (α, β) with coordinates in $\bar{K}$. Given a place of degree one (a point) $P_{\alpha,\beta}$ contained in the support of S, we consider the line $y - \beta$ passing through $P_{\alpha,\beta}$. Since $(y - \beta) = P_{\alpha,\beta} + J - (q+1)Q$ where $\beta^q + \beta = \alpha^{q+1}$, $J \geq 0$ and $Q \notin supp(J)$, we get $P_{\alpha,\beta} \equiv -J + (q+1)Q$. Replacing each place of degree one in the support of S with an equivalent divisor in this way, we get

$$S \equiv -R + s(q+1)Q$$

where $R \geq 0$ and $Q \notin supp(R)$. Thus

$$(\frac{u}{f}) \equiv D - E + R - (q^3 - m' + s(q+1))Q \tag{7}$$

where $D - E + R \geq 0$ and $Q \notin supp(D-E) \bigcup supp(R)$. Let $m' = m - j$ with $0 \leq j \leq r$, so $s = r - j$. Then

$$\begin{aligned} q^3 - m' + s(q+1) &= q^3 - m + j + s(q+1) \\ &= q^2 - aq - a + j + s(q+1) \\ &= (q - a + s - 1)q + (q - a + s + j). \end{aligned}$$

Note that

$$0 \leq q - a + s - 1 < q - a + s + j = q - a + r \leq q - 1.$$

Thus $q^3 - m' + s(q+1)$ is a gap number of Q for any m' with $m - r \leq m' \leq m$. This is a contradiction to our equivalence of (u/f) in (7). Therefore, there is no function $f \in L(mQ)$ such that f has exactly $m - r$ distinct zeroes in $supp(D)$ for any $m - r$ with $m - a + 1 \leq m - r \leq m$. □

Theorem 4 *If $m = q^3 - q^2 + aq + a$ where $1 \leq a \leq q - 1$, then*

$$d(C_m) = n - m + a = q^2 - aq. \tag{8}$$

Proof: Note that $d(C_m) \leq d(C_{q^3-q^2+aq}) = q^2 - aq$ by Proposition 4 since $m > q^3 - q^2 + aq$. By Theorem 3, there is no function $f \in L(mQ)$ having exactly $m - r$ distinct zeroes in $supp(D)$ for each $r = 0, 1, ..., a-1$. Thus $d(C_m) \geq n - m + a = q^2 - aq$. □

Corollary 2 *If $m = q^3 - q^2 + aq + b$ where $0 \leq b \leq a \leq q - 1$, then*

$$d(C_m) = n - m + b = q^2 - aq. \tag{9}$$

Proof: Since $m \geq q^3 - q^2 + aq$, $d(C_m) \leq d(C_{q^3-q^2+aq}) = q^2 - aq$ by Proposition 4. Since $m \leq q^3 - q^2 + aq + a$, $d(C_m) \geq d(C_{q^3-q^2+aq+a}) = q^2 - aq$ from the above theorem. □

B. The Minimum Distance of the Code C_m, in the Range that $m \geq q^3$

Note that $L(mQ - D)$ is no longer $\{0\}$ in this range. Also, the lower bound for $d(C_m)$ in Proposition 3 is not useful any more. Recall that $C_m^\perp = C_{q^3+q^2-q-2-m}$ for any integer $m \in Z$ from Proposition 2. To simplify notation, we define $m_\perp \triangleq q^3 + q^2 - q - 2 - m$. Thus $C_m^\perp = C_{m_\perp}$.

If $m > q^3 + q^2 - q - 2$, then $m_\perp < 0$, so $L(m_\perp Q) = \{0\}$ and $C_{m_\perp} = \{\underline{0}\}$. As a result, $C_m = K^n$ and $d(C_m) = 1$.

We can therefore restrict our attention to $q^3 \leq m \leq q^3 + q^2 - q - 2$. Then $0 \leq m_\perp \leq q^2 - q - 2 = 2g - 2$. Applying Proposition 1, we get that for this range of $m_\perp$

$$\tilde{m_\perp} = aq + b \tag{10}$$

(where $\tilde{m_\perp}$ denotes the $\sim$ operator applied to $m_\perp$) where $0 \le b \le a \le q-2$. Obviously, $C_{m_\perp} = C_{\tilde{m_\perp}}$.

Consider a generator matrix $\mathbf{H}$ for $C_{\tilde{m_\perp}}$ (*i.e.*, a parity check matrix for C_m) obtained by using a basis $B(\tilde{m_\perp})$. Note that each row in $\mathbf{H}$ corresponds to a function in the basis. We examine $\mathbf{H}$ and get an upper bound on $d(C_m)$.

Theorem 5 (Upper Bound) *If* $\tilde{m_\perp} = aq + b$ *where* $0 \le b \le a \le q-2$, *then*

$$d(C_m) \le \begin{cases} a+2 & \text{if } b = a \\ a+1 & \text{if } b < a. \end{cases} \tag{11}$$

Proof: We prove it by finding linearly dependent columns of $\mathbf{H}$ over K. For fixed $\alpha \in K$, we find a subset $\{P_i \mid P_i = P_{\alpha,\beta_i},\ i = 1, 2, ..., q\}$ of $supp(D)$. Note that $\beta_i \ne \beta_j$ for $i \ne j$ and $a + 2 \le q$.

(i) *case* $b = a$: A basis of $L(\tilde{m_\perp}Q)$ is $B(aq+a) = \{1,\ x,\ y, ...,\ x^a,\ x^{a-1}y,\ x^{a-2}y^2, ...,\ y^a\}$. Consider a submatrix $\mathbf{H_1}$ of $\mathbf{H}$ with columns corresponding to $P_1, P_2, ..., P_{a+2}$. By using the Gaussian elimination method, we can make $\mathbf{H_1}$ as follows.

$$\mathbf{H_1} = \begin{bmatrix} 1 & 1 & 1 & \cdots & 1 \\ \beta_1 & \beta_2 & \beta_3 & \cdots & \beta_{a+2} \\ {\beta_1}^2 & {\beta_2}^2 & {\beta_3}^2 & \cdots & {\beta_{a+2}}^2 \\ \vdots & \vdots & \vdots & & \vdots \\ {\beta_1}^a & {\beta_2}^a & {\beta_3}^a & \cdots & {\beta_{a+2}}^a \\ 0 & 0 & 0 & \cdots & 0 \\ \vdots & \vdots & \vdots & & \vdots \\ 0 & 0 & 0 & \cdots & 0 \end{bmatrix}. \tag{12}$$

Thus $rank(\mathbf{H_1}) = a+1$ and $\mathbf{H_1}$ has $a+2$ columns, so columns of $\mathbf{H_1}$ are linearly dependent. Therefore, $d(C_m) \le a+2$.

(ii) *case* $b < a$: Note that $\tilde{m_\perp} < aq + a$. Since y^a has $a(q+1)$ poles at Q, it can not be contained in $L(\tilde{m_\perp}Q)$. Thus the row corresponding to y^a is deleted from the above $\mathbf{H_1}$. Call as $\mathbf{H_2}$ the matrix obtained by deleting the row corresponding to y^a from $\mathbf{H_1}$. Then $rank(\mathbf{H_2}) = a$, so any $a+1$ columns of $\mathbf{H_2}$ are linearly dependent over K. Therefore, $d(C_m) \le a+1$. □

This theorem gives us an upper bound on $d(C_m)$ in the range that $q^3 \le m \le q^3 + q^2 - q - 2$. Our next goal is to find a lower bound and thus the true minimum distance.

At first, we focus on the case that $\tilde{m_\perp} = aq + a$ where $0 \le a \le q-2$. Consider a submatrix $\mathbf{A}$ of $\mathbf{H}$ obtained by choosing $a+1$ distinct columns from $\mathbf{H}$ arbitrarily. Since each column of $\mathbf{H}$ corresponds to a place $P_{\alpha,\beta}$ of degree one, we can reorder columns of $\mathbf{A}$ according to α. That is,

$$\begin{matrix} P_{\alpha_1,\beta_{1,1}}, & P_{\alpha_1,\beta_{1,2}}, & \cdots, & P_{\alpha_1,\beta_{1,b_1}} \\ P_{\alpha_2,\beta_{2,1}}, & P_{\alpha_2,\beta_{2,2}}, & \cdots, & P_{\alpha_2,\beta_{2,b_2}} \\ \vdots & & \vdots & \vdots \\ P_{\alpha_r,\beta_{r,1}}, & P_{\alpha_r,\beta_{r,2}}, & \cdots, & P_{\alpha_r,\beta_{r,b_r}} \end{matrix} \tag{13}$$

where α_i's are pairwise distinct and $b_1 + b_2 + ... + b_r = a + 1$ with $b_1 \geq b_2 \geq ... \geq b_r \geq 1$. It is easy to check that

$$x^{i-1}y^{j_i} \in B(\tilde{m_\perp}), \quad 0 \leq j_i \leq b_i - 1; \quad 1 \leq i \leq r. \tag{14}$$

We rewrite these basis elements in the form.

$$\begin{array}{ccccc} 1, & y, & y^2, & \cdots & y^{b_1-1} \\ x, & xy, & xy^2, & \cdots & xy^{b_2-1} \\ x^2, & x^2y, & x^2y^2, & \cdots & x^2y^{b_3-1} \\ \vdots & \vdots & \vdots & & \vdots \\ x^{r-1}, & x^{r-1}y, & x^{r-1}y^2, & \cdots & x^{r-1}y^{b_r-1} \end{array} \tag{15}$$

Then we can extract an $(a+1) \times (a+1)$ submatrix **B** of **A** as follows: i) Each row corresponds to a function in (15) in the given order. ii) Each column corresponds to a place of degree one in (13) in the given order. iii) Each entry of **B** is obtained by evaluation. That is,

$$\mathbf{B} = [\mathbf{B_{i,j}}], \quad i, j = 1, 2, ..., r \tag{16}$$

where $\mathbf{B_{i,j}}$ is a $(b_i \times b_j)$ matrix whose $(k,l)th$ entry is $\alpha_j{}^{i-1}\beta_{j,l}{}^{k-1}$, *i.e.*,

$$\mathbf{B_{i,j}} = \alpha_j{}^{i-1}\mathbf{D_{i,j}} \tag{17}$$

with

$$\mathbf{D_{i,j}} := \begin{bmatrix} 1 & 1 & 1 & \cdots & 1 \\ \beta_{j,1} & \beta_{j,2} & \beta_{j,3} & \cdots & \beta_{j,b_j} \\ \beta_{j,1}{}^2 & \beta_{j,2}{}^2 & \beta_{j,3}{}^2 & \cdots & \beta_{j,b_j}{}^2 \\ \vdots & \vdots & \vdots & & \vdots \\ \beta_{j,1}{}^{b_i-1} & \beta_{j,2}{}^{b_i-1} & \beta_{j,3}{}^{b_i-1} & \cdots & \beta_{j,b_j}{}^{b_i-1} \end{bmatrix}. \tag{18}$$

Using the Gaussian elimination method and the induction method, we get the following lemma.

Lemma 2 *Assuming the above notations,*

$$det(\mathbf{B}) = (\prod_{i=1}^{r} det(\mathbf{D_{i,i}})) \cdot (\prod_{j=2}^{r} \tau_j{}^{b_j}) \tag{19}$$

where

$$\tau_j \triangleq \prod_{i=1}^{j-1}(\alpha_j - \alpha_i), \quad j = 2, 3, ..., r. \tag{20}$$

Lemma 3 *Assume* $\tilde{m_\perp} = aq + a$ *where* $0 \leq a \leq q - 2$. *Then any* $a+1$ *columns of* **H** *are linearly independent over* K.

Proof: Consider any $a+1$ distinct columns of **H** and reorder those columns according to α of $P_{\alpha,\beta}$. Then we can construct matrices **A** and **B**, *etc.*, as before. Since α_i's are pairwise distinct for $i = 1, 2, ..., r$, we have $\tau_j \neq 0$ for any $j = 2, 3, ..., r$. Since $\beta_{i,j}$'s are pairwise distinct for $j = 1, 2, ..., b_i$ and a given i, we get $det(\mathbf{D_{i,i}}) \neq 0$. Thus $det(\mathbf{B}) \neq 0$ by Lemma 2. This means that $a + 1 = rank(\mathbf{B}) \leq rank(\mathbf{A}) \leq a + 1$, so $rank(\mathbf{A}) = a + 1$. Therefore, the columns of **A** are linearly independent over K. □

Theorem 6 *Assume $\tilde{m_\perp} = aq + b$ where $0 \le b \le a \le q-2$. Then*

$$d(C_m) = \begin{cases} a+2 & \text{if } b = a \\ a+1 & \text{if } b < a. \end{cases} \tag{21}$$

Proof: If $b = a$ then any $a+1$ columns of $\mathbf{H}$ are linearly independent over K by Lemma 3, so $d(C_m) \ge a+2$. But, we have $d(C_m) \le a+2$ by Theorem 5. Therefore, we get $d(C_m) = a+2$.

If $b < a$ then let $\tilde{m'_\perp} = (a-1)q + (a-1)$. Since $\tilde{m'_\perp} \le \tilde{m_\perp}$, we have $C_m \subseteq C_{m'}$, so $d(C_m) \ge d(C_{m'}) = (a-1)+2 = a+1$. But, $d(C_m) \le a+1$ by Theorem 5. □

Note that this theorem tells us the true minimum distance of C_m for any m with $q^3 \le m \le q^3 + q^2 - q - 2$. As said before, if $m > q^3 + q^2 - q - 2$ then $d(C_m) = 1$. Finally, we have:

Corollary 3 *If $q^3 - q \le m \le q^3$ then*

$$d(C_m) = q. \tag{22}$$

Proof: If $m = q^3$, then $m_\perp = q^3+q^2-q-2-m = q^2-q-2$, so $\tilde{m_\perp} = (q-2)q+(q-2) = m_\perp$. Thus, $d(C_{q^3}) = (q-2)+2 = q$ from Theorem 6. Recall that $d(C_{q^3-q}) = q$ by Proposition 4. Since $q^3 - q \le m \le q^3$, $C_{q^3-q} \subseteq C_m \subseteq C_{q^3}$, so $q = d(C_{q^3}) \le d(C_m) \le d(C_{q^3-q}) = q$. □

Remark: The referee has brought to our attention that the exact minimum distance of C_m for m in the range $n - q \le m \le n-1$ has also been determined by Xing [8]. It is also shown in [8] that if $m = qr + t$, $0 \le t \le q-1$, and $2g - 1 \le m \le n-1$, then

$$d(C_m) \le n - m + t.$$

Pellikaan [9] also shows that the minimum distance of the codes C_{n+a}, $0 \le a < q$, satisfies

$$d(C_{n+a}) \ge q - a.$$

The exact minimum distance in cases $m = n, n+1$ is also given here [9]. Some special cases of m are also considered by Garcia and Lax in [10], where the true value of $d(C_m)$ for $q^3 - q^2 < m < q^3$ and $m \equiv 3 \pmod{q}$ is given.

References

[1] J. H. van Lint and T. A. Springer, "Generalized Reed-Solomon Codes from algebraic Geometry," *IEEE Trans. Inform. Theory*, vol. IT-33, pp. 305-309, May 1987.

[2] H. J. Tiersma, "Remarks on Codes from Hermitian Curves," *IEEE Trans. Inform. Theory*, vol. IT-33, pp. 605-609, July 1987.

[3] H. Stichtenoth, "A Note on Hermitian Codes," *IEEE Trans. Inform. Theory*, vol. IT-34, pp. 1345-1348, Sept. 1988.

4] H. Stichtenoth, "Self-Dual Goppa Codes," *J. Pure and Appl. Math.*, vol. 55, pp. 199-211, 1988.

[5] H. Stichtenoth, *Algebraic Function Fields and Codes*, in preparation.

[6] J. H. van Lint, "Algebraic Geometric Codes," *meetkunde 6/21/1988 DRAFT.*

[7] F. J. MacWilliams and N. J. A. Sloane, *The Theory of Error Correcting Codes*, Amsterdam: North-Holland, 1977.

[8] C. P. Xing, "A Note on the Minimum Distance of Hermitian Codes," submitted to the *IEEE Trans. Inform. Theory.*

[9] R. Pellikaan, "On the Gonality of Curves, Abundant Codes and Decoding," presented at the 3rd Conference on *Algebraic Geometry and Coding Theory*, C.I.R.M., Marseilles, June 17-21, 1991.

.0] A. Garcia and R. F. Lax, "Goppa Codes and Weierstrass Points," presented at the 3rd Conference on *Algebraic Geometry and Coding Theory*, C.I.R.M., Marseilles, June 17-21, 1991.

Sphere Packings Centered at S-units of Algebraic Tori

Boris È. KUNYAVSKII
Saratov Institute for Mechanization in Agriculture
60 Sovetskaya , SARATOV 410740, USSR

0. Introduction

This paper is inspired by a series of articles [L/Ts], [R/Ts], [Ts], introducing several new explicit constructions of dense sphere packings in $\mathbb{R}^n$ which have good asymptotic properties while n tends to infinity. We present here a generalization of these constructions which seems quite natural. The main idea is to consider the group of S-units of an algebraic torus defined over a global field and to embed the torsion-free component of this finitely generated group into $\mathbb{R}^n$ with the help of the logarithmic map (in the case of a number ground field), or the divisorial map (in the case of a function field). (For the trivial torus $T=\mathbb{G}_m$ it is nothing but a construction of [R/Ts]). The resulting lattice in $\mathbb{R}^n$ generates a sphere packing whose parameters can be estimated with the help of the technique developed in [Sh1-Sh3], [V2] while investigating arithmetical properties of algebraic tori. The key points are the (generalized) Dirichlet unit theorem valid for S-units of algebraic tori and some class number relations.

Section 1 contains most necessary definitions concerning sphere packings and tori. The main construction is described in Section 2. In Section 3 we estimate the parameters of the obtained lattices. Section 4 deals with several examples where we succeed with the calculation of the numerical values of all the constants. In Section 5 we present a generalization of the main construction to the case of a congruence lattice associated to a divisor. Some final remarks and open questions are collected in Section 6.

The long-term collaboration with M.A.Tsfasman was the source of my interest in sphere packings, and I thank him for many stimulating discussions. I am also grateful to V.E.Voskresenskii and A.A.Klyachko for the constant interest in my work and valuable remarks.

This work has been partly done during my stay at the University Paris-7.

I especially thank J.-J. Sansuc for his hospitality. I also thank I.H.E.S. where this work was finished.

1. Definitions

See [C/S] for a detailed exposition of the problems concerning dense sphere packings, and [O2], [V1] for those concerning algebraic tori.

1.1. Packings

Let L_n be a lattice in $\mathbb{R}^n$. Let us consider a set P of non-overlapping open spheres centered at the points of L_n, and let V denote a sphere of the radius r. We call $\Delta(L_n) := \limsup_{r\to\infty} \frac{\mathrm{vol}(P\cap V)}{\mathrm{vol}(V)}$ *the density* of the sphere packing associated to L_n, and $\lambda(L_n) := (-\log_2 \Delta(L_n))/n$ - *the density exponent.*

Let $\mathcal{L}$ be a family of lattices $L_n \subset \mathbb{R}^n$ with n tending to infinity. Let us define $\lambda(\mathcal{L}) := \liminf_{n\to\infty} \lambda(L_n)$. We say that $\mathcal{L}$ is *asymptotically good* if $\lambda(\mathcal{L}) < \infty$.

1.2. Problem

Give an explicit construction of an asymptotically good family $\mathcal{L}$ and calculate (or, at least, estimate) $\lambda(\mathcal{L})$.

1.3. Bounds for the density

We recall the best known upper and lower bounds for λ given by Minkowski [Mi]: $\lambda(\mathcal{L}) \le 1$ for some $\mathcal{L}$ (which we cannot construct in an effective way), and by Kabatiansky - Levenshtein [K/L]: $\lambda(\mathcal{L}) \ge 0.599$ for any $\mathcal{L}$.

See [L/Ts], [R/Ts], [Ts] for a series of explicit constructions of asymptotically good families. Note that the obtained density exponents are worse than the Minkowski bound.

1.4. Algebraic tori

Let K be a field, and let T be an algebraic K-torus split by a field L, i.e. $T\times_K L$ is isomorphic to the product of n copies of the multiplicative group $G_{m,L}$. The minimal field among such L's is called *the splitting field* of T, and let G be *the splitting group* (the Galois group of L over K). Let $\hat{T}$=Hom($T\times_K L$,

$\mathbb{G}_{m,L}$) be *the character module* of T. It is a torsion-free $\mathbb{Z}$-module of a finite $\mathbb{Z}$-rank (equal to the K-dimension of T); we can consider $\hat{T}$ as a G-module.

Let $(\hat{T})_K$ denote the submodule of $\hat{T}$ consisting of the characters defined over K (i.e. $(\hat{T})_K=\hat{T}^G$ is the submodule of G-invariant elements). We call T *anisotropic* if $(\hat{T})_K=(0)$. Denote by $T(K)$ the group of K-points of T.

From now on let K denote a *global field*, i.e. a *number field* (a finite extension of $\mathbb{Q}$ of degree m), or a *function field* ($K=\mathbb{F}_q(X)$ with X denoting a smooth projective curve over a finite field of q elements). Let S be a finite set of places of K containing S_∞ (the set of all archimedean places). For each $v\in S$ let $T_v:=T(K_v)$ denote the group of K_v-points of T, and let $(\hat{T})_v$ be the submodule of $\hat{T}$ consisting of the characters defined over K_v. (K_v denotes, as usual, the completion of K at v).

2. Construction of packings

In this section we present a natural generalization of a construction introduced by Rosenbloom and Tsfasman [R/Ts].

2.1. S-units of an algebraic torus

Let $T_{\mathbb{A}}(S):=\prod_{v\in S}T_v \times \prod_{v\notin S}T_v^c$, T_v^c denoting the maximal compact subgroup of T_v. In other terms, $T_v^c=\{x\in T_v:\ \|\chi(x)\|_v=1$ for all $\chi\in(\hat{T})_v\}$, $\|\ \ \|_v$ denoting the normalized absolute value.

<u>Definition</u> [V1], [Sh1]. We call $T_K(S):=T(K)\cap T_{\mathbb{A}}(S)$ <u>the group of S-units of T</u>.

<u>Remark</u>. In the case of the trivial torus $T=\mathbb{G}_m$ we obtain the usual group of S-units of K.

Set $T_{\mathbb{A}}^c:=\prod_v T_v^c$, $W(T):=T(K)\cap T_{\mathbb{A}}^c$. $W(T)$ is a finite group coinciding with the group of the roots of 1 in K in the case $T=\mathbb{G}_m$.

For each v set $r_v:=\mathrm{rank}_{\mathbb{Z}}(\hat{T})_v$, $r(S):=\sum_{v\in S} r_v$. Suppose that $r_v>0$ for all $v\in S$, and fix a $\mathbb{Z}$-basis $\{\chi_{v,1},\ldots,\chi_{v,r_v}\}$ of $(\hat{T})_v$.

2.2. Number field case

Let $l_S : T_{\mathbb{A}}(S) \longrightarrow \mathbb{R}^{r(S)}$ be a map defined as follows:

$$l_S(\{t_v\}) := (\ln\|\chi_{v,i}(t_v)\|_v)_{\substack{v\in S\\ 1\le i\le r_v}}$$

(here and further on ln denoting log_e).

In fact, for $v \notin S$ we have $\|\chi_{v,1}(t_v)\|_v=1$ because $t_v \in T_v^c$, and so the corresponding components vanish.

Definition. Set $N_S := l_S(T_K(S))$.

The following result generalizes Dirichlet's unit theorem.

Theorem. [V1], [Sh1], [K]

N_S is a lattice of $\mathbb{Z}$-rank $r(S)-r$, where $r:=\text{rank}(\hat{T})_K$.

This lattice is the main object of our consideration.

Remarks.

1. $\text{Ker}(l_S)=W(T)$.
2. For $S=S_\infty$ the above theorem was proved in [O1].

2.3. Function field case

In this section $K=\mathbb{F}_q(X)$ is the field of rational functions of a smooth projective curve X over a finite field $\mathbb{F}_q$.

Recall a construction from [R/Ts], [Ts]. Let $S \subseteq X(\mathbb{F}_q)$ be a finite set of places of K, and let $O_S^*=\{f \in K^* : \text{Supp}(f) \subseteq S\}$. We have a natural map $\varphi_S : O_S^* \longrightarrow \text{Div}_S(X) \simeq \mathbb{Z}^n$ (divisors supported in S), $\varphi_S(f):=\text{div}(f)$.

A generalization of this construction to the group of S-units of a K-torus T gives rise to a map $\psi_S : T_K(S) \longrightarrow \mathbb{Z}^n$. (We set $(\psi_S(t))_v=\text{ord}_v(\chi_{v,1}(t_v))$). We have $n=r(S)$ as in the previous section, and define F_S as the image of ψ_S. As above, we have $\text{rank}F_S=r(S)-r$, and $\text{Ker}\psi_S=W(T)$ (for $T=\mathbb{G}_m$ the latter group coincides with $\mathbb{F}_q^*$).

3. Estimation of parameters

3.1. The main asymptotic formula [L/Ts], [R/Ts], [Ts].

If $\mathcal{L}$ is a family of lattices L_n of growing rank, the Stirling formula yields $\lambda(\mathcal{L}) \approx -\log\sqrt{\pi e/2} + \log\sqrt{n} - \log d(L_n) + \frac{1}{n}\log(\det L_n)$,

where for brevity we denote $log=log_2$, $\det L$ is the determinant of L (the volume of a fundamental parallelotope of L), $d(L)$ is the minimum distance of L $(:=\inf_{x \in L\setminus\{0\}}|x|)$, and $\approx$ means an asymptotic equality.

The above formula shows that we have to obtain an upper bound for $\det L_n$ and a lower bound for $d(L_n)$.

From now on we assume T *anisotropic*. In this case we have $n=r(S)$.

3.2. Number field case

Denote by L the splitting field of T, and let p be the degree of L/K. Let $S=S_\infty \cup S_f$, and let S' be the set of places of L over S. Denote by $O^*_{S'}$ the group of S'-units in L, and let W denote the group of roots of 1 in L. Set $\mu := \min_{v\in S} r_v$.

Theorem.

(i) $\det N_S \le h(T)\, R(T) \prod_{v\in S_f} r_v \ln N(v)$,

$h(T)$ denoting the class number of T [O2], [V1], and $R(T)$ denoting the regulator of T [O2]; $N(v)$ is the cardinality of the residue field at v.

(ii) Assume that the following condition holds.

(H) For each $t\in T_S(K)\setminus W(T)$ and any collection $\{\chi_v \in (\hat{T})_v\}_{v\in S}$ there exists $f\in O^*_{S'}$ such that $\|f\|_w$ and $\|\chi_v(t)\|_v$ (w denotes a place over v) are as near as we wish.

Then $d(N_S) \ge \dfrac{\mu}{p\sqrt{n}} \inf_{f\in O^*_{S'}\setminus W} \sum_{w\in S'} |\ln\|f\|_w|$

In particular, if L is totally real, and $S=S_\infty$, we have

$\det N_S \le h(T)\, R(T)$,

$d(N_S) \ge \dfrac{\mu\, m}{\sqrt{n}} \ln \dfrac{1+\sqrt{5}}{2}$,

m denoting the degree of K.

Proof.

(i) We go along the lines of the proof of Lemma 4.1 (iii) in [Ts]. Recall that $h(T)=[T_A : T(K)T_A(S_\infty)]$, where T_A denotes the inductive limit of $T_A(S)$. If $T=\mathbb{G}_m$, we have $h(T)=h(K)$ - the usual class number of K.

The regulator can be defined as follows.

First suppose T to be defined over $\mathbb{Q}$. Set $r_\infty := \operatorname{rank}(\hat{T})_{\mathbb{R}}$ (we have an obvious equality $r_\infty = r(S_\infty) := \sum_{v\in S_\infty} r_v$); fix a basis $\{\chi_i\}_{1\le i\le r_\infty}$ of $(\hat{T})_{\mathbb{R}}$ such that $\chi_1,\dots,\chi_r$ form a basis of $(\hat{T})_{\mathbb{Q}}$, and a basis $\{\varepsilon_j\}_{r+1\le j\le r_\infty}$ of $T_{\mathbb{Q}}(S_\infty)$. A positive number $\pm\det(\ln|\chi_i(\varepsilon_j)|)_{r+1\le i,\, j\le r_\infty}$ is called the regulator of T over $\mathbb{Q}$ (it depends neither on $\{\chi_i\}$, nor on $\{\varepsilon_j\}$).

For a torus T defined over an arbitrary K let $T_1=R_{K/\mathbb{Q}}(T)$ be the $\mathbb{Q}$-torus obtained by Weil's restrictions of scalars [V1], and we define $R(T):=R(T_1)$. If $T=\mathbb{G}_m$, we have $R(T)=R(K)$ - the usual regulator of K.

If T is anisotropic, the corresponding lattice of S-units N_S is a full lattice in $\mathbb{R}^n$, so we can decompose $\det N_S$ into two factors: "archimedean" h_1 and "non-archimedean" h_2 (to obtain this decomposition we can order the coordinates in $\mathbb{R}^n$ in such a way that the first coordinates correspond to archimedean places). The first factor h_1 equals $R(T)$, the second one, being the determinant of a lattice whose index in $\bigoplus_{v\in S_f} \mathbb{Z}r_v \ln N(v)$ does not exceed $h(T)$, satisfies an inequality $h_2 \le h(T) \prod_{v\in S_f} r_v \ln N(v)$.

(ii) The idea is to reduce everything to the case of the trivial torus $T=\mathbb{G}_m$ using the assumption (H), and to exploit [Ts], Lemma 4.1 (i), (ii).

Let $t\in T_K(S)\setminus W(T)$. For $v\in S$ and $1\le i\le r_v$ set $y_{v,i}:=\ln\|\chi_{v,i}(t_v)\|_v$, where t_v denotes the image of t under the natural embedding $T(K)\longrightarrow T(K_v)$. Let $Y=\{y_{v,i}\}_{\substack{v\in S\\ 1\le i\le r_v}}$. All the elements of Y are non-zero. We have

$$d(N_S) = \inf_Y \sqrt{\sum_{v\in S}\ \sum_{i=1}^{r_v} y_{v,i}^2}\ .$$

Let $f\in O_{S'}^*\setminus W$. For $w\in S'$ set $x_w:=\ln\|f\|_w$. Let $X=\{x_w\}_{w\in S'}$.

For each $v\in S$ let $z_v = \min_{1\le i\le r_v} y_{v,i} = \ln\|c_v\|_v$, where c_v is the value of some of the $\chi_{v,i}$'s at t_v. Using the assumption (H) we can find $f\in O_{S'}^*$ such that f and c_v are as near (in v-adic metric) as we wish. We can say the same thing about x_v and z_v, $\ln$ being continuous.

Thus $d(N_S) \ge \inf_Y \sqrt{\sum_{v\in S} r_v z_v^2} \ge \inf_X \sqrt{\sum_{w\in S'} r_v x_w^2}$ (we take in the right-hand sum one w for each v). To complete the proof of the assertion (ii), it remains to use the inequality

$$\sqrt{\sum_{i=1}^N b_i\alpha_i^2} \ge \frac{1}{\sqrt{\sum_{i=1}^N b_i}} \sum_{i=1}^N b_i|\alpha_i|,$$

valid for any positive b_i's and any α_i's, and remark that over each place $v\in S$ there is at most p places $w\in S'$.

To prove the assertion concerning the case of a totally real field we act as in [R/Ts] and use Schinzel's inequality [Schi]:

$$\prod_{w\in S'} \|x\|_w > \left(\frac{1+\sqrt{5}}{2}\right)^{M/2}$$

valid for any $x\ne 1$ such that $\|x\|_w>1$ for all $w\in S'$, M denoting the degree of the extension $L/\mathbb{Q}$.

To complete the proof of the theorem, it remains to note that

$\sum_{\substack{w\in S' \\ \|f\|_w>1}} |ln\|f\|_w| = 2\sum_{w\in S'} |ln\|f\|_w|$ because of the product formula written for $f\in O^*_{S'}$. ∎

3.3. Function field case

In this section $\pi\colon Y \longrightarrow X$ denotes a Galois covering of degree p, $K=\mathbb{F}_q(X)$, $L=\mathbb{F}_q(Y)$, $S'\subseteq Y(\mathbb{F}_q)$, $S=\pi(S')$, $O^*_{S'}=\{f\in L^*:\ \mathrm{Supp} f\subseteq S'\}$, F_S is the corresponding lattice of S-units of T (see 2.3).

Arguing as in the previous section, i.e. using the assumption (H) and the results of [Ts], we obtain the following estimates.

<u>Theorem</u>.

(i) $\det F_S \le h(T)$

(ii) Under the assumption (H)

$$d(F_S) \ge \frac{\mu}{\sqrt{p}} \inf_{f\in O^*_{S'}\setminus\mathbb{F}^*_q} \sqrt{2\deg f} \ge \frac{\mu}{\sqrt{p}} \sqrt{\frac{2\,\#Y(\mathbb{F}_q)}{q+1}}\,.$$

<u>Remark</u>. Maybe the above lower bounds for d are valid without the assumption (H). Anyway this assumption is fulfilled in the examples below.

4. Particular cases

4.1. Tori of the norm type

In this section T denotes a *torus of the norm type* defined over a number field K of degree m, i.e. T is the kernel of the norm map $R_{L/K}\mathbb{G}_m \longrightarrow \mathbb{G}_m$ (which is the usual norm $L^* \longrightarrow K^*$ if we consider the K-points), where L/K is an arbitrary finite extension. We restrict ourselves to the case of *cyclic* extensions, where we can use the results of Voskresenskii and Shyr concerning some class number problems for algebraic tori. Note that tori of the norm type are anisotropic. Throughout this section we assume $S=S_\infty$.

4.1.1. Quadratic extensions

First, consider the simplest interesting case of *a quadratic* extension L/K. In this case we have $\dim T = 1$, the splitting group G is isomorphic to $\mathbb{Z}/2\mathbb{Z}$.

Let us make the estimates of the previous section more explicit. First of all we note that the rank of N_S equals $n:=\#\{v\in S\colon\ v$ splits in $L\}$ because $r_v=1$ for such v's, and $r_v=0$ otherwise [Sh2]. It is obvious that $\mu=1$. The assumption (H) is fulfilled. In fact, for each $v\in S$ splitting in L a generator χ of $\hat{T}$ is defined over K_v, and we can view it as a generator χ_v of $(\hat{T})_v$. For $t\in T_K(S)$ we can take $f=\chi(t)\in O_S^*$, with $\|f\|_w=\|\chi_v(t)\|_v$.

Furthermore, according to [V2], [Sh3], we have

$$h(T)R(T) = \frac{h(L)R(L)}{h(K)R(K)}.w.2^{1-u},$$

where $w:=[W(K):N_{L/K}(W_L)] = 1$ or 2, and $u:=\#\{v\in S\colon\ v$ is ramified in $L\}$.

Now we use some rough estimates in order to obtain the numerical values of constants and see that our method leads to asymptotically good packings. Assume K totally real so that $m=n$. According to Theorem 3.2, we have $d(N_S) \geq \sqrt{n}\ ln((1+\sqrt{5})/2)$. Further, as in [Ts], we can give an upper bound for $h(L)R(L)$: $h(L)R(L) \ll |D_L|/\pi^{2n}$, D_L denoting the discriminant of L. Letting n tend to infinity we obtain an asymptotic inequality for the denominator ([L], ch. XVI, §4): $log\ h(K)R(K)\ /\ log \sqrt{|D_K|} \gtrsim 1$, whence

$\lambda(N_S) \lesssim -log \sqrt{\pi^5 e/2} - log\ ln((1+\sqrt{5})/2) + (log(|D_L|/\sqrt{|D_K|})/n - (u-1)/n$.

Now, as in [Ts], let K run over an unramified tower of totally real fields beginning from K_0 (so $(log|D_K|)/n$ is constant), and let L/K lie inside the tower so that L/K is unramified and $u=0$. For such a family $\mathcal{L}$ we finally obtain:

$$\lambda(\mathcal{L}) \lesssim -log \sqrt{\pi^5 e/2} - log\ ln((1+\sqrt{5})/2) + \frac{3}{2}(log|D_{K_0}|)/n_0$$

(n_0 denoting the degree of K_0). If, as in [Ts], we use Martinet's tower beginning from $K_0=\mathbb{Q}(\sqrt{2},\sqrt{70035})$ with $n_0=4$, $|D_{K_0}|=2^8 3^2 5^2 7^2 23^2 29^2$ [Ma], we get $\lambda\lesssim 11.78$.

4.1.2. Cyclic extensions: K growing, L growing

We can consider an analogue of the above family taking an arbitrary cyclic extension L/K of degree p instead of a quadratic extension. In this case we have $r_v=p-1$ if v splits in L, and $r_v=0$ otherwise, so $\mu=p-1$, $n=m(p-1)$, and Theorem 3.2 gives for $d(N_S)$ the same lower bound as above. Using the results of [V2], [Sh2] we can obtain an analogous upper bound for $\det N_S$:

$$\det N_S \ll \frac{h(L)R(L)}{h(K)R(K)} p^{1-u},$$

and construct asymptotically good families (whose parameters are worse than those obtained for $p=2$).

4.1.3. Cyclic extensions: K fixed, L growing

Let us try to go in another direction. Namely, we can fix K (for example, $K=\mathbb{Q}$) and consider a family arising from the tori of the norm type corresponding to totally real cyclic extensions $L/\mathbb{Q}$ of degree p with p tending to infinity. In this case we have $S=S_\infty=\{\infty\}$, $n=\mathrm{rk}(\hat{T})_\infty=p-1$, $\mu=p-1=n$, $m=1$, and we obtain the same bounds for $d(N_S)$ and $\det N_S$ as above. Note that here $L/\mathbb{Q}$ is ramified, and we cannot guarantee that this way leads to an asymptotically good family.

4.2. Function field case

The constructions of the previous section have functional analogues. We assume q being prime to the degree of L/K in order to avoid wild ramification. Here we only consider a quadratic case.

Let X be a smooth projective curve over $\mathbb{F}_q$, and let $\pi:Y\longrightarrow X$ be a covering of degree 2. Denote by $K=\mathbb{F}_q(X)$ and $L=\mathbb{F}_q(Y)$ the corresponding fields of rational functions; as above, T will denote the K-torus of the norm type corresponding to the extension L/K. Suppose $S=X(\mathbb{F}_q)$, and let F_S be the lattice of S-units of T (see 2.3, 3.3). Let us estimate the parameters of F_S.

As in 4.1, $n=\sum_{v\in S} r_v = \#\{v\in S\colon v \text{ splits in } L\}$. As to $h(T)$, we have an analogue of the results of Voskresenskii and Shyr (which were formulated for tori over number fields):

$$h(T)=\frac{\#J_Y(\mathbb{F}_q)}{\#J_X(\mathbb{F}_q)}.w.2^{1-u},$$

where u is the number of ramification (scheme) points of π, $w=[\mathbb{F}_q^*:\mathbb{F}_q^{*2}]=2$, and J_X (resp. J_Y) is the Jacobian of X (resp. Y).

Arguing as in [Ts], consider a family of curves X of growing genus g_X such that $\frac{\# X(\mathbb{F}_q)}{g_X}\to\sqrt{q}-1$ (this is the maximal possible value of such a limit [V/D]). Suppose that π is unramified, so that $u=0$, $g_Y=2g_X-1$ (the Hurwitz formula), hence $\frac{\# Y(\mathbb{F}_q)}{g_Y}\to\sqrt{q}-1$. According to [R/Ts], we have an asymptotic formula valid for both X and Y:

$$\frac{1}{g}\log\#J(\mathbb{F}_q)\approx\log q+(\sqrt{q}-1)\log\frac{q}{q-1}.$$

As in 4.1.1, the assumption (H) is fulfilled. Since $\mu=1$, and $\#Y(\mathbb{F}_q)=2\#X(\mathbb{F}_q)$, π being unramified, we have the same lower bound as in [Ts]:

$$d(F_S)\geq\sqrt{\frac{2\ \#X(\mathbb{F}_q)}{q+1}}$$

which yields the same asymptotic inequality as in [R/Ts]:

$$\lambda(\mathcal{L}) \lesssim - \log \sqrt{\pi e} + \log \frac{\sqrt{q+1}}{q-1} + \frac{\sqrt{q}}{\sqrt{q}-1} \log q$$

For $q=9$ we obtain, as in [Ts], the best estimate: $\lambda \lesssim 1.87$.

5. Congruence sublattices

In this section we generalize our construction imitating [Ts], §6. We only consider the function field case where the parameters are better.

5.1. Construction and parameters

Let $D = \sum a_jP_j$ be a positive divisor on X such that Supp $D \cap S = \emptyset$, and denote by $a = \sum a_jm_j$ its degree, m_j denoting the degree of P_j. Denote by $\mathcal{P}$ the set of P_j's. Consider a subgroup

$$T_{K,D}(S) := \{t \in T_K(S) : \chi_{v,i}(t_v) \equiv 1 \pmod{D}\}_{\substack{v \in \mathcal{P} \\ 1 \le i \le r_v}} ,$$

and denote by FD_S its image in $\mathbb{R}^n$ under ψ_S (see 2.3). Let us estimate the parameters of FD_S .

Denote by D' the inverse image $\pi^{-1}(D)$, and let a' be the degree of D'.

Theorem.

(i) $\det FD_S \le \frac{h(T)}{w(T)} q^a \prod_j r_j(1-q^{-m_j})$,

where $w(T)=\#W(T)$.

(ii) Under the assumption (H),

$d(FD_S) \ge \mu \sqrt{2a'/p}$

Proof.

(i) As in [Ts], we estimate the index of FD_S in F_S and use Theorem 3.3(i). Consider an embedding $T_K(S) \longrightarrow \prod T_K(P_j)$, where $T_K(P_j) = \{ t \in T_{P_j} : \chi(t) \in \hat{O}^*_{P_j}$ for all $\chi \in \hat{T}_{P_j} \}$, $\hat{O}^*_{P_j}$ denoting the group of units in the completion of the local ring at P_j . Let

$\hat{O}_{P_j,a_j} = \{ x \in \hat{O}^*_{P_j} : x \equiv 1 \pmod{P_j^{a_j}} \}$, and let

$T_K(P_j,a_j) = \{ t \in T_K(P_j) : \chi(t) \in \hat{O}_{P_j,a_j}$ for all $\chi \in \hat{T}_{P_j} \}$. We have

$T_{K,D}(S) = T_K(S) \cap \prod T_K(P_j,a_j)$,

and $[T_K(S):T_{K,D}(S)] \le [\prod T_K(P_j):T_K(P_j,a_j)] \le \prod r_j((q^{m_j}-1)q^{m_j(a_j-1)})$.

Since $\mathrm{Ker}\psi_S = W(T)$, we have $[T_K(S):T_{K,D}(S)] = w(T)\ [F_S:FD_S]$, whence the required inequality.

(ii) As in Theorem 3.3 (ii) , we have

$$d(FD_S) \geq \frac{\mu}{\sqrt{p}} \inf_{f\in O^*_{S'D'}\setminus \mathbb{F}^*_q} \sqrt{2\deg f}\ ,$$

where $O^*_{S'D'} := \{f\in O^*_{S'}:\ f\equiv 1(\mathrm{mod} D')\}$. (In fact, using the weak approximation theorem, we may add a finite number of restrictions in order to obtain $f\in O^*_{S'D'}$.) To conclude, remark that $\deg f = \deg(f-1) \geq \deg D' = a'$. ■

5.2. Particular case

Now, as in 4.2, consider the torus of the norm type corresponding to a quadratic extension L/K , where $K=\mathbb{F}_q(X)$, $L=\mathbb{F}_q(Y)$, π: $Y \longrightarrow X$ is a covering of degree 2. Here we have $r_j=1$, $W(T)=(0)$, $w(T)=1$, $\mu=1$, $h(T) = \frac{\#J_Y(\mathbb{F}_q)}{\#J_X(\mathbb{F}_q)}.w.2^{1-u}$, where u is the number of ramification points of π.

Suppose $S=X(\mathbb{F}_q)$, and take X and Y as in 4.2. We have $a'=pa$, π being unramified. As in [Ts], let us choose D in such a way that $\frac{\deg D}{\#X(\mathbb{F}_q)} \longrightarrow \frac{1}{2\ \ln q}$. Then we obtain the same asymptotic inequality as in [Ts]:

$$\lambda(\mathcal{L}) \lesssim -\ \log \sqrt{\pi/2} + \frac{1}{2}\log(\ln q) + \frac{\sqrt{q}}{\sqrt{q}-1}\ \log q\ -\ \log(q-1).$$

For $q=47^2$ we obtain the same constant as in [Ts]: $\lambda \lesssim 1.389$.

6. Remarks and questions

6.1. It seems quite natural to interpret the quotient $\frac{\#J_Y(\mathbb{F}_q)}{\#J_X(\mathbb{F}_q)}$ as the number of $\mathbb{F}_q$-points on the Prymian of π (this remark is due to A. A. Klyachko). It is interesting whether we can ameliorate the estimates for $\det F_S$ exploiting geometric properties of Prymians, for instance, their distinction from Jacobians.

6.2. One can try to consider some tori T other than those of the norm type. The problem consists in estimating $h(T)R(T)$. There exists an analogue of the analytic class number formula [Sh2]:

$$h(T)R(T)=\tau(T)\rho(T)w(T)\sqrt{D(T)}\ ,$$

where $\tau(T)$ is the Tamagawa number [O2], [V1]; $w(T)=\#W(T)$, $\rho(T)$ is the residue of the L-function $L(T,s)$ at $s=1$ [V1], [Sh2], and $D(T)$ is the "quasi-discriminant" (roughly speaking, $\sqrt{D(T)}$ is the factor distinguishing the two measures on $T_{\mathbb{A}}$: the Tamagawa measure and the product of natural local measures - the archimedean measures dt_v/t_v on $(\mathbb{R}^*_+)^{r_v}$ and the canonical discrete measures on $\mathbb{Z}^{r_v}$; see [Sh2] for more details). Is it possible to use the above relation to obtain better estimates?

6.3. Another way to strengthen the estimates for λ is to ameliorate the lower bounds for d. The estimates given here do not reflect the arithmetic properties of T. It may be possible to obtain better estimates studying a function of the "height" type like the following one.

Let $X=\{\chi_{v,i}\}_{v\in S,\ 1\leq i\leq r_v}$ be a collection of bases of $(\hat{T})_v$ (recall that $r_v:=\mathrm{rank}(\hat{T})_v$, and $W(T):=T(K)\cap T^c_{\mathbb{A}}$). Let us define

$$ht_X(T):=\inf_{t\in T_S(K)\setminus W(T)} \sum_v \sum_{i=1}^{r_v} |ln\|\chi_{v,i}(t_v)\|_v|$$

How can one obtain lower bounds for $ht_X(T)$? Is it possible to use here a general theory of heights on commutative group schemes [D]?

6.4. One can consider other lattices corresponding to an algebraic torus T. For example, let O be the ring of integers in K, and let $\mathcal{T}$ denote the Néron model of T over O. Denote by $\mathcal{T}^\circ$ its connected component, and let O_1 be the ring of S-integers in K. We can study the lattice $\mathcal{T}^\circ(O_1)$ which has a finite index in the lattice of S-units of T [Schn]. In the case $S=\{v\in K \text{ dividing } p\}$, p being a prime integer, there exists a non-degenerate pairing between $\mathcal{T}^\circ(O_1)$ and $\oplus_{v\in S}(\hat{T})_v$, whose determinant enters in the expression for the value of L-function of T at $s=0$; see [Schn] for more details.

6.5. In general, if L/K is a Galois extension with the group G, and M is a G-module of arithmetical interest related to L, we can consider a group $\mathrm{Hom}_G(\hat{T},M)$, $\hat{T}$ being the character module of a K-torus T split by L; if $M=O^*_L$ is the group of units of L, we obtain the group of units of T [K].

6.6. The authors of [B/M] introduce another characteristic of the density of a lattice. If we denote $\gamma(L_n) := 4(\Delta(L_n)/V_n)^{2/n}$ (V_n denoting the volume of the unit sphere in $\mathbb{R}^n$), set $\gamma'(L_n) := \sqrt{\gamma(L_n)\gamma(L^\circ_n)}$, L°_n denoting the dual lattice. It is obvious that $\gamma'(L_n) = \sqrt{d(L_n)d(L^\circ_n)}$. We can introduce

$\lambda'(L_n) := -\frac{1}{n} \log \frac{\gamma'(L_n)^{n/2} V_n}{2^n}$ and pose the problem: to construct families $\mathcal{L}$ with $\lambda'(\mathcal{L}) := \liminf_{n\to\infty} \lambda'(L_n) < \infty$. We have an obvious asymptotic equality $\lambda'(\mathcal{L}_n) \approx -\log\sqrt{\pi e/2} + \log\sqrt{n} - (\log d(L_n) + \log d(L_n^{\circ}))/2$. Is it possible to choose among the families of lattices of S-units of tori one with $\lambda'(\mathcal{L}) < \infty$? Surely it is, if there exists such a family with $L_n \simeq L_n^{\circ}$.

References

[B/M] A.-M.Bergé, J.Martinet, Sur un problème de dualité lié aux sphères en géometrie des nombres, *J.Number Theory*, 1989, v.32, N 1, 14-42.

[C/S] J.H.Conway, N.J.A.Sloane, Sphere packings, lattices and groups, *Springer*, N.Y., 1988.

[D] P.Deligne, Preuve des conjectures de Tate et de Shafarevitch (d'après G.Faltings), *Astérisque*, 1985, v.121-122, 25-41.

[K/L] G.A.Kabatiansky, V.I.Levenshtein, Bounds for packing on a sphere and in space, *Problemy Peredachi Informatsii*, 1978, v.14, N 1,3-25; Engl. transl.: *Problems Inform. Transmission*, 1978, v.14, Nx1, 1-17.

[K] H.K.Kim, Unit theorems on algebraic tori, *Nagoya Math. J.*, 1988, v.112, 117-124.

[L] S.Lang, Algebraic number theory, *Addison - Wesley*, Reading, 1970.

[L/Ts] S.N.Litsyn, M.A.Tsfasman, Constructive high-dimensional sphere packings, *Duke Math.J.*, 1987, v.54, N 1, 147-161.

[Ma] J.Martinet, Tours de corps de classes et estimation de discriminants, *Invent. Math.*, 1978, v.44, N 1, 65-73.

[Mi] H.Minkowski, Diskontinuitätsbereich für arithmetische Aequivallenz, *J. Reine Angew. Math.*, 1905, v.129, 220-274.

[O1] T.Ono, On some arithmetic properties of linear algebraic groups, *Ann. of Math.*, 1959, v.70, N 2, 266-290.

[O2] T.Ono, Arithmetic of algebraic tori, *Ann. of Math.*, 1961, v.74, N 1, 101-139.

[R/Ts] M.Ju.Rosenbloom, M.A.Tsfasman, Multiplicative lattices in global fields, *Invent. Math.*, 1990, v.101, N 3, 687-696.

[Schi] A.Schinzel, On the product of the conjugates outside the unit circle of an algebraic number, *Acta Arith.*, 1973, v.24, 385-399; 1975, v.26, 329-331.

[Schn] P.Schneider, Iwasawa L-functions of varieties over algebraic number fields. A first approach, *Invent. Math.*, 1983, v.71, N 2, 251-293.

[Sh1] J.-M.Shyr, A generalization of Dirichlet's unit theorem, *J. Number Theory*, 1977, v.9, N 2, 213-217.

[Sh2] J.-M.Shyr, On some class number relations of algebraic tori, *Mich. Math. J.*, 1977, v.24, N 3, 365-377.

[Sh3] J.-M.Shyr, Class numbers of binary quadratic forms over algebraic number fieds, *J. Reine Angew. Math.*, 1979, v.307-308, 353-364.

[Ts] M.A.Tsfasman, Global fields, codes and sphere packings, Preprint, 1989.

[V/D] S.G.Vlăduţ, V.G.Drinfeld, Number of points on an algebraic curve, *Funktsional. Anal. i Prilozhen.*, 1983, v.17, N 1, 68-69; Engl.transl.: *Functional Anal. Appl.*, 1983, v.17, N 1, 53-54.

[V1] V.E.Voskresenskii, Algebraic tori, *Nauka*, Moscow, 1977 (Russian).

[V2] V.E.Voskresenskii, The arithmetic of algebraic groups and homogeneous spaces, *Studies in Number Theory. Algebraic Number Theory*. Saratov Univ. Press, 1987, 7-38; Engl. transl: *Selecta Math. Soviet.*, 1990, v.9, N 1, 22-48.

A FUNCTION FIELD RELATED TO THE REE GROUP

Jens Peter Pedersen[†]

Abstract

We construct an algebraic function field over a finite field of characteristic 3, which has the Ree group as automorphism group. In this way, we obtain an explicit construction of the Ree group. We also prove, that the function field has as many rational places as possible, and that the number for certain extensions of the ground field reaches the Hasse-Weil bound.

Keywords: algebraic function fields, Ree groups, Hasse-Weil bound.

I. Introduction

Lef F be an algebraic function field of one variable over an algebraically closed field K. (see e.g. Chevalley [3]). F has characteristic p and genus $g \geq 2$. Let $G = \mathrm{Aut}(F/K)$ denote the group of automorphisms of F fixing K. For $p = 0$, Hurwitz [8] proved that G is finite and satiesfies $|G| \leq 84(g-1)$. For $p > 0$, the finiteness of G was proved by Schmid [15], who also noted that the bound above didn't apply in this case.

Stichtenoth [19] proved that $|G|$ is bounded by $|G| \leq 16g^4$, except in the case where the function field is Hermitian. Henn [7] improved the bound to $|G| < 8g^3$ by excluding three more cases, and in a footnote he claims that the latter bound can be strengthened to $|G| < 3(2g)^{5/2}$ by excluding another two cases. However, in [6] Hansen presented an irreducible algebraic curve (and thereby a function field) for which $|G| \geq 3(2g)^{5/2}$ without being on the list of excluded cases. This curve is a Deligne-Lusztig variety, [4], [11], defined for $p = 3$, $g = g_R = \frac{3}{2}q_0(q-1)(q+q_0+1)$, where $q = 3^{2s+1}$, $q_0 = 3^s$, $s \in \mathbf{N}$, and $G = {}^2G_2(q)$, a Ree group of order $q^3(q-1)(q^3+1)$, [14]. This curve has q^3+1 $\mathbf{F}_q$-rational points.

Our aim is to construct explicitly a function field F_R over $\mathbf{F}_q$ with the parameters above. This is done in Section II, where we also write down all the automorphisms of $F_R/\overline{\mathbf{F}}_q$ and prove that they form a Ree group. Furthermore, we briefly discuss how F_R can be seen as the function field of a curve. It should be noted, that even though we obtain a function field with the desired parameters and automorphism group, it still remains to be shown, that it is unique and/or corresponds to a Deligne-Lusztig variety.

In [6], Hansen extracted from [4] a family of Deligne-Lusztig varieties in the following way. A connected reductive algebraic group gives rise to a Deligne-Lusztig variety $\mathcal{X}$

[†] The author is with Mathematical Institute, Building 303, The Technical University of Denmark, DK-2800 Lyngby, Denmark.

This research was done while the author visited The Department of Discrete Mathematics, Eindhoven Technical University, The Netherlands. The visit was supported by a grant from The Danish Research Academy.

with a group G of automorphisms. (see e.g. Gorenstein [5] and Carter [2] for a description of algebraic groups and the associated notation). The variety $\mathcal{X}$ is an irreducible algebraic curve and G is simple, if and only if

(i) G is the projective special unitary group ${}^2A_2(q^2)$, q is a prime power. In this case $\mathcal{X}$ has a function field with genus $g_H = \frac{1}{2}q(q-1)$. Over the algebraic closure $\overline{\mathbf{F}}_{q^2}$, the function field is unique up to isomorphism to the Hermitian function field $F_H = \overline{\mathbf{F}}_{q^2}(x,y)$ defined by $y^q + y = x^{q+1}$, see Segre [16] and Stichtenoth [20].

(ii) G is the Suzuki group ${}^2B_2(q)$, $q = 2^{2s+1}$, $s \in \mathbf{N}$. In this case $\mathcal{X}$ has a function field with genus $g_S = q_0(q-1)$, $q_0 = 2^s$. Over the algebraic closure $\overline{\mathbf{F}}_q$, the function field is unique up to isomorphism to the function field $F_S = \overline{\mathbf{F}}_q(x,y)$ defined by $y^q + y = x^{q_0}(x^q + x)$, see Henn [7].

(iii) G is the Ree group ${}^2G_2(q)$, $q = 3^{2s+1}$, $s \in \mathbf{N}$. In this case $\mathcal{X}$ has a function field defined over $\mathbf{F}_q$ with genus $g_R = \frac{3}{2}q_0(q-1)(q+q_0+1)$, $q_0 = 3^s$.

It is well-known, that $F_H/\mathbf{F}_{q^2}$ has the maximal number of $\mathbf{F}_{q^2}$-rational places for a function field of genus g_H w.r.t. the Hasse-Weil bound. Serre (cf. [17], [18]) proved that the function fields of case (ii) and (iii) have the maximal number of $\mathbf{F}_q$-rational places for function fields of their genus. In Section III, we exhibit the Zeta functions of F_S and F_R, and we obtain that the number of rational places in these functions fields in fact satisfies the Hasse-Weil bound for certain extensions of their ground field. So even though the connection between F_R and case (iii) is not established, we can conclude, that all three function fields corresponding to irreducible Deligne-Lusztig curves with automorphism groups which are simple, reaches the Hasse-Weil bound for suitable extensions of their ground fields.

II. The function field F_R and its automorphisms

In [12] it was shown, that the equation $y^q - y = x^{q_0}(x^q - x)$ yields a function field $\mathbf{F}_q(x,y)$ with $[\mathbf{F}_q(x,y) : \mathbf{F}_q(x)] = q$ for any prime power q and $q_0^2 \mid q$, $0 < q_0 < \sqrt{q}$. The construction of these function fields were inspired by the construction of F_S, and motivates the following definition of F_R.

Definition 1: *Let $q = 3^{2s+1}, s \in \mathbf{N}$, and $q_0 = 3^s$. Define the function field $F_R = \mathbf{F}_q(x, y_1, y_2)$ by*

$$y_1^q - y_1 = x^{q_0}(x^q - x) \tag{1}$$
$$y_2^q - y_2 = x^{q_0}(y_1^q - y_1) \tag{2}$$

From now on we always consider F_R as an extension of $\mathbf{F}_q(x)$. Let Q_∞ be the place at infinity of $\mathbf{F}_q(x)$ and let P_∞ be a place of F_R lying above Q_∞. We denote the discrete valuation at P_∞ by ν_∞ and the ramification index of P_∞ w.r.t. Q_∞ by e_∞. In Appendix A, we have constructed some functions of F_R and computed their valuations at P_∞. These functions will be used throughout the paper.

Theorem 1:
(i) $e_\infty = q^2$ *and* P_∞ *is the only place of* F_R *lying above* Q_∞.
(ii) F_R *has* q^3+1 $\mathbf{F}_q$*-rational places.*
(iii) F_R *has genus* $g_R = \frac{3}{2}q_0(q-1)(q+q_0+1)$.

Proof: (i) From Appendix A we have $\nu_\infty(w_8) = -(1+\frac{1}{q_0}+\frac{2}{q}+\frac{1}{q_0q}+\frac{1}{q^2})e_\infty$, implying $e_\infty = eq^2, e \in \mathbf{N}$. On the other hand $e_\infty \leq [F_R : \mathbf{F}_q(x)] \leq q^2$, and we conclude $e_\infty = q^2$. Consequently, P_∞ is the only place of F_R lying above Q_∞.
(ii) From (i) we have, that P_∞ is a $\mathbf{F}_q$-rational place. Let $Q_\beta, \beta \in \mathbf{F}_q$, be a $\mathbf{F}_q$-rational place of $\mathbf{F}_q(x)$ outside infinity. Then there are q^2 $\mathbf{F}_q$-rational places $P_{\beta\gamma\delta}, \gamma, \delta \in \mathbf{F}_q$, of F_R lying above Q_β. ($P_{\beta\gamma\delta}$ denotes the place corresponding to the solution $x=\beta, y_1 = \gamma, y_2 = \delta$ of (1) and (2)). Hence, F_R has q^3+1 $\mathbf{F}_q$-rational places.
(iii) P_∞ is the only place of F_R which ramifies w.r.t. $\mathbf{F}_q(x)$, and we have

$$2g_R - 2 = -2q^2 + \nu_\infty\left(\frac{dx^{-1}}{du}\right), \tag{3}$$

where $u \in F_R$ is a prime element of P_∞. From Appendix A we get, that $u = w_6/w_8$ is a prime element. It is straightforward to compute that

$$\frac{dx^{-1}}{du} = \frac{w_8^2}{x^2(w_6 w_7^{3q_0} - w_4^{3q_0} w_8)}$$

and

$$\nu_\infty\left(\frac{dx^{-1}}{du}\right) = 3q_0q^2 + 3q^2 - q - 3q_0 - 2. \tag{4}$$

Substituting (4) into (3) proves (iii). ◇

F_R is the function field of a curve $\mathcal{X}_R$. The obvious thing to do is to try to embed $\mathcal{X}_R$ in the projective space $\mathbf{P}_q^3$. Let

$$f_1 = Z^{q_0}(Y_1^q - Y_1Z^{q-1}) - X^{q_0}(X^q - XZ^{q-1}),$$
$$f_2 = Z^{q_0}(Y_2^q - Y_2Z^{q-1}) - X^{q_0}(Y_1^q - Y_1Z^{q-1}).$$

f_1 and f_2 gives us F_R, if we divide them by Z^{q+q_0} and put $x = X/Z$, $y_1 = Y_1/Z$ and $y_2 = Y_2/Z$. However, f_1 and f_2 define a reducible curve $\tilde{\mathcal{X}}_R$. It is easy to see, that $\tilde{\mathcal{X}}_R = \mathcal{X}_R \cup L$, where L is a line defined by $h_1 = X^{q_0}$ and $h_2 = Z^{q_0}$. In order to get the equations for $\mathcal{X}_R \subset \mathbf{P}_q^3$, we have to apply linkage theory. (see e.g. Peskine and Szpiro [13]). We get that $\mathcal{X}_R$ is a non-complete intersection defined by f_1, f_2 and $f_3 = \det\begin{pmatrix} a_{11} & a_{12} \\ a_{21} & a_{22}\end{pmatrix}$, where

$$f_1 = a_{11}h_1 + a_{12}h_2 = -(X^q - XZ^{q-1})h_1 + (Y_1^q - Y_1Z^{q-1})h_2,$$
$$f_2 = a_{21}h_1 + a_{22}h_2 = -(Y_1^q - Y_1Z^{q-1})h_1 + (Y_2^q - Y_2Z^{q-1})h_2.$$

Hence,

$$f_3 = (Y_1^q - Y_1Z^{q-1})^2 - (X^q - XZ^{q-1})(Y_2^q - Y_2Z^{q-1}),$$

nd we observe, that the place P_∞ corresponds to the solution $X = Z = Y_1 = 0$ and $_2 = 1$ of $f_1 = f_2 = f_3 = 0$. Equivalently, $\mathcal{X}_R$ can be seen as the (2×2)-minors of the natrix

$$\begin{pmatrix} X^{q_0} & -(Y_1^q - Y_1 Z^{q-1}) & -(Y_2^q - Y_2 Z^{q-1}) \\ Z^{q_0} & -(X^q - X Z^{q-1}) & -(Y_1^q - Y_1 Z^{q-1}) \end{pmatrix}.$$

t should be noted, that this model of $\mathcal{X}_R$ is singular. This can be seen in the following vay. The curve is non-singular outside infinity (Z=0). At infinity, the point P_∞ is ingular, since the ideal (f_1, f_2, f_3) corresponding to the curve, is contained in m_∞^2, vhere m_∞ is the maximal ideal of the local ring of P_∞.

Naturally, $\mathcal{X}_R$ has a plane model. It turns out to be easy to determine an equation or a plane model. Consider the function $w_2 = xy_1^{3q_0} - y_2^{3q_0}$, which satisfy $w_2^q - w_2 = {}_1^{3q_0}(x^q - x)$. (see Appendix A). By using this equation and (1), we obtain

$$f(x, w_2) = w_2^{q^2} - [1 + (x^q - x)^{q-1}]w_2^q + (x^q - x)^{q-1} w_2 - x^q(x^q - x)^{q+3q_0} = 0.$$

Define the function field $F_R' = \mathbf{F}_q(x, w_2)$ by $f(x, w_2) = 0$. Then F_R is isomorphic to F_R' nder the isomorphism

$$\mu : \begin{cases} x & \mapsto x \\ y_1 & \mapsto \left(\frac{w_2^q - w_2}{x^q - x}\right)^{q_0} - x^{q_0}(x^q - x) \\ y_2 & \mapsto x^{q_0}\left(\frac{w_2^q - w_2}{x^q - x}\right)^{q_0} - x^{2q_0}(x^q - x) - w_2^{q_0} \end{cases}$$

Hence, the curve defined by $Z^{q(q+3q_0+1)} f(X/Z, W_2/Z)$ is a plane model of $\mathcal{X}_R$.

Now, we will determine the automorphism group $G_R = \mathrm{Aut}(F_R/\overline{\mathbf{F}}_q)$. It turns out hat $G_R = \mathrm{Aut}(F_R/\mathbf{F}_q)$, i.e. all the automorphisms are $\mathbf{F}_q$-rational. G_R acts as a ermutation group on the set of places $\mathbf{P}_{F_R}$ in $F_R/\overline{\mathbf{F}}_q$. Let $\mathbf{P}_{F_R}(\mathbf{F}_q) \subset \mathbf{P}_{F_R}$ denote the et of places which are $\mathbf{F}_q$-rational in $F_R/\mathbf{F}_q$.

Theorem 2: *G_R is the Ree group ${}^2G_2(q)$, $|G_R| = q^3(q-1)(q^3+1)$.*

In order to prove Theorem 2, we need the following notation and lemmas. Let T be nonempty subset of $\mathbf{P}_{F_R}$. We denote by $G_R(T)$ the stabilizer of T, i.e. the subgroup f G_R fixing T.

Lemma 3: *$G_R(P_\infty) = \{\psi_{\alpha\beta\gamma\delta} \mid \alpha \in \mathbf{F}_q^*,\ \beta, \gamma, \delta \in \mathbf{F}_q\}$, where*

$$\psi_{\alpha\beta\gamma\delta} : \begin{cases} x & \mapsto \alpha x + \beta \\ y_1 & \mapsto \alpha^{q_0+1} y_1 + \alpha\beta^{q_0} x + \gamma \\ y_2 & \mapsto \alpha^{2q_0+1} y_2 - \alpha^{q_0+1}\beta^{q_0} y_1 + \alpha\beta^{2q_0} x + \delta \end{cases}$$

Proof: Let $\psi \in G_R(P_\infty)$. Then ψ maps $L(kP_\infty)$ bijectively to $L(kP_\infty)$. We are nterested in the case where $k = q(q + 2q_0)$. Obviously, $1, x, y_1, y_2 \in L(q(q+2q_0)P_\infty)$, ince these functions have poles only at P_∞ and $0 = -\nu_\infty(1) < -\nu_\infty(x) < -\nu_\infty(y_1) < -\nu_\infty(y_2) = q(q + 2q_0)$. (Appendix A and Theorem 1 (i)). Furthermore, by a result of ewittes [10, Th. 1.b] we get that $z \in L(kP_\infty) \setminus \mathbf{F}_q \Rightarrow -\nu_\infty(z) \geq q^2$. Hence, a basis for

$L(q(q+2q_0)P_\infty)$ is $1, x, z_{1,1}, \ldots, z_{1,m_1}, y_1, z_{2,1}, \ldots, z_{2,m_2}, y_2$, where the elements have strictly increasing pole order at P_∞. Now, we have that ψ is of the form

$$\psi : \begin{cases} x \mapsto \alpha x + \beta, & \alpha \in \overline{\mathbf{F}}_q^*, \beta \in \overline{\mathbf{F}}_q \\ y_1 \mapsto \lambda_1 y_1 + z_1 + \lambda_2 x + \gamma, & \lambda_1 \in \overline{\mathbf{F}}_q^*, \lambda_2, \gamma \in \overline{\mathbf{F}}_q \\ y_2 \mapsto \lambda_3 y_2 + z_3 + \lambda_4 y_1 + z_2 + \lambda_5 x + \delta, & \lambda_3 \in \overline{\mathbf{F}}_q^*, \lambda_4, \lambda_5, \delta \in \overline{\mathbf{F}}_q \end{cases},$$

where z_1, z_2 are linear combinations of $z_{1,1}, \ldots, z_{1,m_1}$, and z_3 of $z_{2,1}, \ldots, z_{2,m_2}$. From (1) and (2) we have that ψ has to satisfy

$$\psi(y_1)^q - \psi(y_1) = \psi(x)^{q_0}(\psi(x)^q - \psi(x)),$$
$$\psi(y_2)^q - \psi(y_2) = \psi(x)^{q_0}(\psi(y_1)^q - \psi(y_1)),$$

and from these equations, it is easy to conclude that ψ must be as in the Lemma. ◇

Lemma 4: *Let $S_{\lambda_0\lambda_1\lambda_2} = \{\psi_{\alpha\beta\gamma\delta}(P_{\lambda_0\lambda_1\lambda_2}) \mid \psi_{\alpha\beta\gamma\delta} \in G_R(P_\infty)\}$ be the $G_R(P_\infty)$-orbit of $P_{\lambda_0\lambda_1\lambda_2} \in \mathbf{P}_{F_R} \setminus \{P_\infty\}$. Then $S_{\lambda_0\lambda_1\lambda_2} = \mathbf{P}_{F_R}(\mathbf{F}_q) \setminus \{P_\infty\}$ if $P_{\lambda_0\lambda_1\lambda_2} \in \mathbf{P}_{F_R}(\mathbf{F}_q)$, and $|S_{\lambda_0\lambda_1\lambda_2}| = q^3(q-1)$ if $P_{\lambda_0\lambda_1\lambda_2} \notin \mathbf{P}_{F_R}(\mathbf{F}_q)$.*

Proof: From Lemma 3, it is easy to see, that $G_R(P_\infty)$ acts transitively on $\mathbf{P}_{F_R}(\mathbf{F}_q) \setminus \{P_\infty\}$, proving the first statement of the Lemma.
Suppose $P_{\lambda_0\lambda_1\lambda_2} \notin \mathbf{P}_{F_R}(\mathbf{F}_q)$, implying that $\lambda_0, \lambda_1, \lambda_2 \in \overline{\mathbf{F}}_q \setminus \mathbf{F}_q$. If $|S_{\lambda_0\lambda_1\lambda_2}| < q^3(q-1)$, then there exits distinct automorphisms $\psi_{\alpha_1\beta_1\gamma_1\delta_1}, \psi_{\alpha_2\beta_2\gamma_2\delta_2} \in G_R(P_\infty)$ s.t. $\psi_{\alpha_1\beta_1\gamma_1\delta_1}(P_{\lambda_0\lambda_1\lambda_2}) = \psi_{\alpha_2\beta_2\gamma_2\delta_2}(P_{\lambda_0\lambda_1\lambda_2})$. It is easy to see, that this is impossible. ◇

Proof of Theorem 2: First we determine the group $\tilde{G}_R \subset G_R$ acting on $\mathbf{P}_{F_R}(\mathbf{F}_q)$. Clearly, $G_R(P_\infty) \subset \tilde{G}_R$. Now, consider the mapping

$$\phi : \begin{cases} x & \mapsto w_6/w_8 \\ y_1 & \mapsto w_{10}/w_8 \\ y_2 & \mapsto w_9/w_8 \end{cases},$$

where the functions involved are defined in Appendix A. It can be checked, that

$$\phi(y_1)^q - \phi(y_1) = \phi(x)^{q_0}(\phi(x)^q - \phi(x)),$$
$$\phi(y_2)^q - \phi(y_2) = \phi(x)^{q_0}(\phi(y_1)^q - \phi(y_1)),$$

implying that ϕ is an automorphism. $\phi^{-1} = \phi$ and ϕ acts on $\mathbf{P}_{F_R}(\mathbf{F}_q)$ mapping P_∞ to P_{000}. Hence, $\tilde{G}_R$ acts transitively on $\mathbf{P}_{F_R}(\mathbf{F}_q)$, and we have that $|\tilde{G}_R| = q^3(q-1)(q^3+1)$. Suppose $\tilde{G}_R$ is not the full automorphism group. Then $\tilde{G}_R$ is a proper subgroup of G_R and there exists an automorphism mapping P_∞ to a place outside $\mathbf{P}_{F_R}(\mathbf{F}_q)$. Therefore, G_R must act transitively on a finite set of places

$$S = \mathbf{P}_{F_R}(\mathbf{F}_q) \cup S_1 \cup \cdots \cup S_m, \qquad m > 0,$$

where $S_1, \ldots, S_m$ are distinct $G_R(P_\infty)$-orbits outside $\mathbf{P}_{F_R}(\mathbf{F}_q)$. Hence, $|G_R| = |S| \cdot |G_R(P_\infty)|$, and by Lemma 3 and 4 we get

$$|G_R| = [mq^3(q-1) + q^3 + 1]q^3(q-1).$$

On the other hand, $|\tilde{G}_R|$ has to divide $|G_R|$, implying that $\frac{1}{2}(q^3+1)$ divides m. Therefore,

$$|G_R| \geq [\tfrac{1}{2}(q^3+1)q^3(q-1)+q^3+1]q^3(q-1) > 16g_R^4,$$

contradicting the bound of Stichtenoth [19], and we conclude $G_R = \tilde{G}_R$.
Now, we have to prove that G_R is a Ree group. First of all, we note that G_R acts 2-transitively on $\mathbf{P}_{F_R}(\mathbf{F}_q)$. Secondly, we observe, that the stabilizer of P_∞ and P_{000}, $G_R(P_\infty, P_{000}) = \{\psi_{\alpha 000} \mid \alpha \in \mathbf{F}_q^*\}$, has a unique nontrivial element ψ_{-1000} fixing more than two places. Then, by a result of Kantor, O'Nan and Seitz [9, Th. 2.3], we have that G_R is of Ree type. Since the only groups of Ree type are ${}^2G_2(q)$, $q = 3^{2s+1}$, we are finished. (Bombieri [5]). ◇

III. The number of places in F_R

Let F be a function field of genus g defined over a finite field $\mathbf{F}_q$, and let N_m denote the number of places of degree one in $F/\mathbf{F}_{q^m}$. The number N_m is bounded by the Hasse-Weil bound

$$N_m \leq q^m + 1 + 2g\sqrt{q^m}.$$

If equality holds in the Hasse-Weil bound, $F/\mathbf{F}_{q^m}$ is said to be a *maximal* function field.

Theorem 5:
(i) *F_R has the maximal number of $\mathbf{F}_q$-rational places possible for a function field of genus g_R.*
(ii) *The Zeta function of F_R is*

$$Z(F_R, \mathbf{F}_q)(t) = \frac{(1+3q_0t+qt^2)^{q_0(q^2-1)}(1+qt^2)^{q_0(q-1)(q+3q_0+1)/2}}{(1-t)(1-qt)}.$$

(iii) *The number of places of degree one in $F_R/\mathbf{F}_{q^m}$ is*

$$N_m = q^m + 1 - q_0\sqrt{q}(q-1)[(q+3q_0+1)\cos m\pi/2 + 2(q+1)\cos 5m\pi/6].$$

In particularly, $F_R/\mathbf{F}_{q^m}$ is a maximal function field if and only if $m \equiv 6 \mod 12$.

Proof: (i) follows from [17], [18]. From [11], we get that

$$Z(F_R, \mathbf{F}_q)(t) = \frac{\prod_{j=1}^{g_R}(1-\alpha_j t)(1-\bar{\alpha}_j t)}{(1-t)(1-qt)},$$

where

$$\alpha_j \in \left\{\pm\sqrt{q}e^{i\pi/2},\ \sqrt{q}e^{\pm i5\pi/6}\right\}, \qquad j = 1, \dots, g_R.$$

Let a be the number of j's for which $\alpha_j = \pm\sqrt{q}e^{i5\pi/6}$ and b the number of j's for which $\alpha_j = \pm\sqrt{q}e^{i\pi/2}$. Then we have

$$Z(F_R, \mathbf{F}_q)(t) = \frac{(1+3q_0t+qt^2)^a(1+qt^2)^b}{(1-t)(1-qt)}. \tag{5}$$

It is well-known, that $N_m = q^m + 1 - \sum_{j=1}^{g_R}(\alpha_j^m + \bar{\alpha}_j^m)$, and we obtain

$$N_m = q^m + 1 - 2\sqrt{q^m}(a\cos m5\pi/6 + b\cos m\pi/2). \tag{6}$$

Combining Theorem 1 (ii) and (6) yields $N_1 = q^3 + 1 = q + 1 + 3q_0 a$, i.e. $a = q_0(q^2 - 1)$, and since $a + b = g_R$ we get $b = \frac{1}{2}q_0(q-1)(q + 3q_0 + 1)$. Substituting a and b into (5) and (6) proves (ii) and (iii), respectively. ◇

Computations, similar to those in the proof of Theorem 7, can be carried out for the function field $F_S = \mathbf{F}_q(x, y)$, $q = 2^{2s+1}$, $q_0 = 2^s$, $y^q - y = x^{q_0}(x^q - x)$, associated with the Suzuki group. In this case we get

$$Z(F_S, \mathbf{F}_q)(t) = \frac{(1 + 2q_0 t + qt^2)^{q_0(q-1)}}{(1-t)(1-qt)},$$

and that $F_S/\mathbf{F}_{q^m}$ is a maximal function field if and only if $m \equiv 4 \mod 8$.

Acknowledgement

The author wishes to thank Johan P. Hansen, The University of Aarhus, Denmark, for initiating the work at this subject and for helpfull discussions throughout the process. He also wishes to thank Ruud Pellikaan and Iwan Duursma, Eindhoven Technical University, The Netherlands, for suggestions and discussions on the subject.

Appendix A: Some functions of F_R and their valuations at P_∞

By definition we have $\nu_\infty(x) = -e_\infty$. Furthermore,

$$y_1^q - y_1 = x^{q_0}(x^q - x) \quad \Longrightarrow \quad \nu_\infty(y_1) = -(1 + \tfrac{1}{3q_0})e_\infty, \tag{A1}$$

$$y_2^q - y_2 = x^{q_0}(y_1^q - y_1) \quad \Longrightarrow \quad \nu_\infty(y_2) = -(1 + \tfrac{2}{3q_0})e_\infty. \tag{A2}$$

From (A1) and (A2), we define

$$\begin{aligned} w_1 &= x^{3q_0+1} - y_1^{3q_0} \\ w_1^{q_0} &= x^{q_0+1} - y_1 \quad \Longrightarrow \quad \nu_\infty(w_1) = -(1 + \tfrac{1}{q_0})e_\infty \\ w_1^q - w_1 &= x^{3q_0}(x^q - x) \end{aligned} \tag{A3}$$

and

$$\begin{aligned} w_2 &= xy_1^{3q_0} - y_2^{3q_0} \\ w_2^{q_0} &= x^{q_0}y_1 - y_2 \quad \Longrightarrow \quad \nu_\infty(w_2) = -(1 + \tfrac{1}{q_0} + \tfrac{1}{q})e_\infty \\ w_2^q - w_2 &= y_1^{3q_0}(x^q - x). \end{aligned} \tag{A4}$$

From (A1), (A2) and (A3) we get

$$w_1^q - w_1 = x^{q_0}(y_2^q - y_2), \tag{A5}$$

and we define

$$\begin{aligned} w_3 &= x y_2^{3q_0} - w_1^{3q_0} \\ w_3^{q_0} &= x^{q_0} y_2 - w_1 \quad \Longrightarrow \quad \nu_\infty(w_3) = -(1 + \tfrac{1}{q_0} + \tfrac{2}{q})e_\infty \\ w_3^q - w_3 &= y_2^{3q_0}(x^q - x). \end{aligned} \tag{A6}$$

From (A3) and (A4) we get

$$y_1^{3q_0}(w_1^q - w_1) = x^{3q_0}(w_2^q - w_2), \tag{A7}$$

and we define

$$\begin{aligned} w_4 &= x w_2^{q_0} - y_1 w_1^{q_0} \\ w_4^{3q_0} &= x^{3q_0} w_2 - y_1^{3q_0} w_1 \quad \Longrightarrow \quad \nu_\infty(w_4) = -(1 + \tfrac{2}{3q_0} + \tfrac{1}{q})e_\infty \\ w_4^q - w_4 &= w_2^{q_0}(x^q - x) - w_1^{q_0}(y_1^q - y_1). \end{aligned} \tag{A8}$$

From (A3) and (A6) we get

$$y_2^{3q_0}(w_1^q - w_1) = x^{3q_0}(w_3^q - w_3), \tag{A9}$$

and we define

$$\begin{aligned} v &= x w_3^{q_0} - y_2 w_1^{q_0} \\ v^{3q_0} &= x^{3q_0} w_3 - y_2^{3q_0} w_1 \quad \Longrightarrow \quad \nu_\infty(v) = -(1 + \tfrac{1}{q_0} + \tfrac{1}{q})e_\infty \\ v^q - v &= w_3^{q_0}(x^q - x) - w_1^{q_0}(y_2^q - y_2). \end{aligned} \tag{A10}$$

From (A4) and (A6) we get

$$y_2^{3q_0}(w_2^q - w_2) = y_1^{3q_0}(w_3^q - w_3), \tag{A11}$$

and we define

$$\begin{aligned} w_5 &= y_1 w_3^{q_0} - y_2 w_2^{q_0} \\ w_5^{3q_0} &= y_1^{3q_0} w_3 - y_2^{3q_0} w_2 \quad \Longrightarrow \quad \nu_\infty(w_5) = -(1 + \tfrac{1}{q_0} + \tfrac{1}{q} + \tfrac{1}{3q_0 q})e_\infty \\ w_5^q - w_5 &= w_3^{q_0}(y_1^q - y_1) - w_2^{q_0}(y_2^q - y_2). \end{aligned} \tag{A12}$$

From (A4), (A8) and (A10) we get

$$v^q - v = w_2^q - w_2 - x^{q_0}(w_4^q - w_4), \tag{A13}$$

and we define

$$\begin{aligned} w_6 &= v^{3q_0} - w_2^{3q_0} + x w_4^{3q_0} \\ w_6^{q_0} &= v - w_2 + x^{q_0} w_4 \quad \Longrightarrow \quad \nu_\infty(w_6) = -(1 + \tfrac{1}{q_0} + \tfrac{2}{q} + \tfrac{1}{q_0 q})e_\infty \\ w_6^q - w_6 &= w_4^{3q_0}(x^q - x). \end{aligned} \tag{A14}$$

Since $w_4 = y_1^2 - xy_2$, combining (A4), (A6) and (A14) yields

$$w_6^q - w_6 = y_1^{3q_0}(w_2^q - w_2) - x^{3q_0}(w_3^q - w_3), \tag{A15}$$

and we define

$$\begin{aligned} w_7 &= y_1 w_2^{q_0} - x w_3^{q_0} - w_6^{3q_0} \\ w_7^{3q_0} &= y_1^{3q_0} w_2 - x^{3q_0} w_3 - w_6 \quad \Longrightarrow \quad \nu_\infty(w_7) = -(1 + \tfrac{2}{3q_0} + \tfrac{1}{q} + \tfrac{1}{3q_0 q})e_\infty \\ w_7^q - w_7 &= w_2^{q_0}(y_1^q - y_1) - w_3^{q_0}(x^x - x). \end{aligned} \tag{A16}$$

From (A12) and (A16) we get

$$w_5^q - w_5 = -x^{q_0}(w_7^q - w_7), \tag{A17}$$

and we define

$$\begin{aligned} w_8 &= w_5^{3q_0} + x w_7^{3q_0} \\ w_8^{q_0} &= w_5 + x^{q_0} w_7 \quad \Longrightarrow \quad \nu_\infty(w_8) = -(1 + \tfrac{1}{q_0} + \tfrac{2}{q} + \tfrac{1}{q_0 q} + \tfrac{1}{q^2})e_\infty \\ w_8^q - w_8 &= w_7^{3q_0}(x^q - x). \end{aligned} \tag{A18}$$

From (A4) and (A14) we get

$$y_1^{3q_0}(w_6^q - w_6) = w_4^{3q_0}(w_2^q - w_2), \tag{A19}$$

and we define

$$\begin{aligned} w_9 &= w_2^{q_0} w_4 - y_1 w_6^{q_0} \\ w_9^{3q_0} &= w_2 w_4^{3q_0} - y_1^{3q_0} w_6 \quad \Longrightarrow \quad \nu_\infty(w_9) = -(1 + \tfrac{1}{q_0} + \tfrac{2}{q} + \tfrac{1}{3q_0 q})e_\infty \\ w_9^q - w_9 &= w_2^{q_0}(w_4^q - w_4) - w_6^{q_0}(y_1^q - y_1). \end{aligned} \tag{A20}$$

From (A6) and (A14) we get

$$y_2^{3q_0}(w_6^q - w_6) = w_4^{3q_0}(w_3^q - w_3), \tag{A21}$$

and we define

$$\begin{aligned} w_{10} &= y_2 w_6^{q_0} - w_3^{q_0} w_4 \\ w_{10}^{3q_0} &= y_2^{3q_0} w_6 - w_3 w_4^{3q_0} \quad \Longrightarrow \quad \nu_\infty(w_{10}) = -(1 + \tfrac{1}{q_0} + \tfrac{2}{q} + \tfrac{1}{3q_0 q})e_\infty \\ w_{10}^q - w_{10} &= w_6^{q_0}(y_2^q - y_2) - w_3^{q_0}(w_4^q - w_4). \end{aligned} \tag{A22}$$

References

1. E. Bombieri, *Thompson's Problem* ($\sigma^2 = 3$), Inv. Math. **58**, p. 77–100 (1980).
2. R.W. Carter, "Finite Groups of Lie type", John Wiley & Sons Ltd., (1985).
3. C. Chevalley, "Introduction to the Theory of Algebraic Functions of One Variable", A.M.S., Providence, RI (1951).

4. P. Deligne and G. Lusztig, *Representations of Reductive Groups over Finite Fields*, Ann. Math. **103**, p. 103–161 (1976).
5. D. Gorenstein, "Finite Simple Groups", Plenum Press, New York (1982).
6. J.P. Hansen, *Deligne-Lusztig Varieties and Group Codes*, to appear in these proceedings.
7. H.-W. Henn, *Funktionenkörper mit großer Automorphismengruppe*, J. Reine Angew. Math. **172**, p. 96–115 (1978).
8. A. Hurwitz, *Über algebraische Gebilde mit eindeutigen Transformationen in sich*, Math. Ann. **41**, p. 403–442 (1893).
9. W.M. Kantor, M.E. O'Nan and G.M. Seitz, *2-Transitive Groups in which the Stabilizer of Two Points is Cyclic*, J. Alg. **21**, p. 17–50 (1972).
10. J. Lewittes, *Places of Degree One in Function Fields over Finite Fields*, J. Pure App. Alg. **69**, p. 177–183 (1990).
11. G. Lusztig, *Coxeter Orbits and Eigenspaces of Frobenius*, Invent. Math. **38**, p. 101–159 (1976).
12. J.P. Pedersen and A.B. Sørensen, *Codes from certain Algebraic Function Fields with many Rational Places*, Mat-Report 1990-11, Tech. Uni. Denmark.
13. C. Peskine and L. Szpiro, *Liaison des Variétés Algébriques I*, Inv. Math. **26**, p. 271–302 (1974).
14. R. Ree, *A Family of Simple Groups associated with the Simple Lie Algebra of Type* (G_2), Am. J. Math. **83**, p. 432–462 (1961).
15. H.L. Schmid, *Über die Automorphismen eines algebraischen Funktionenkörpers von Primzahlcharacteristik*, J. Reine Angew. Math. **179**, p. 5–15 (1938).
16. B. Segre, *Forme e Geometrie Hermitiane, con particolare riguardo al Caso Finito*, Ann. Mat. Pura Appl. **70**, p. 1–201.
17. J.-P. Serre, *Sur le Nombre des Points Rationnels d'une Courbe Algébrique sur un Corps Fini*, C. R. Acad. Sci. Paris Sér. I Math. **296**, p. 397–402 (1983).
18. J.-P. Serre, *Algèbre et Géométrie*, Ann. Collège de France 1983–1984, p. 79–84.
18. H. Stichtenoth, *Über die Automorphismengruppe eines algebraischen Funktionenkörpers von Primzahlcharacteristik, Teil I: Eine Abschätzung der Ordnung der Automorphismengruppe*, Archiv Math. **24**, p. 527–544 (1973).
20. H. Stichtenoth, *Über die Automorphismengruppe eines algebraischen Funktionenkörpers von Primzahlcharacteristik, Teil II: Ein spezieller Typ von Funktionenkörpern*, Archiv Math. **24**, p. 615–631 (1973).

On the gonality of curves, abundant codes and decoding

Ruud Pellikaan *

1 Introduction

Let $\mathcal{X}$ be a curve defined over the finite field $\mathbf{F}_q$ with q elements. The genus of $\mathcal{X}$ is denoted by $g(\mathcal{X})$, or more often by g. Let $P_1, \ldots, P_n$ be n distinct rational points on the curve $\mathcal{X}$. Let D be the divisor $P_1 + \ldots + P_n$. Let G be a divisor on $\mathcal{X}$ of degree m. The code $C_L(D,G)$ is defined as the image of $L(G)$ in $\mathbf{F}_q^n$, under the evaluation map $f \longmapsto (f(P_1), \ldots, f(P_n))$. Goppa [5] showed that the functional code $C_L(D,G)$ has dimension at least $m+1-g$ and minimum distance at least $n-m$ in case $m < n$. If moreover $m > 2g-2$, then the dimension is equal to $m+1-g$. We call $n-m$ the Goppa designed minimum distance of $C_L(D,G)$, and denote it by d_G. Tsfasman, Vlăduţ, Zink and Ihara showed that modular curves have many rational points with respect to the genus, if q is a square, that is to say $N \sim (\sqrt{q}-1)g$, see [21, 4.1.52]. In case $q \geq 49$ the Tsfasman-Vlăduţ-Zink bound R_{TVZ} gives an improvement of the Gilbert-Varshamov bound R_{GV}, see [21, 3.4.4]. In the corners, where the graphs of R_{TVZ} and R_{GV} meet, Vlăduţ made a slight improvement, moreover he showed that there are codes, comming from curves, with parameters lying on the maximum of the above mentioned bounds, see [21, 3.4.11]. Later Pellikaan, Shen and van Wee [14] proved that every linear code can be represented with a curve, but if one imposes the condition $m < n$, then long binary algebraic-geometric codes have information rate at most $\frac{1}{2}$. So the question of finding good codes can be restated in the question: Which divisors give good codes ?

There are two ways to improve the bounds Goppa gave. In the first place by taking divisors such that the dimension is bigger than $m+1-g$. These are so called special divisors. If the field of constants is algebraically closed and $g - k(g-m+k-1) \geq 0$, then there exists a divisor of degree m and dimension at least k , by Brill-Noether theory. But this is no longer true over a finite field. The second possibility, which we will pursue in this paper, is to try to improve the bound on the minimum distance.

In section 2 we show that the minimum distance is at least $\mathrm{t}(\mathcal{X}) - a$ for an abundant divisor of abundance a, that is for a divisor which is equivalent to $D+A$, where A is an effective divisor of degree a. The number $\mathrm{t}(\mathcal{X})$ is the minimal degree of a map from the curve to the projective line, also called the gonality of the curve. This was proved

*Department of Mathematics and Computing Science of the Eindhoven University of Technology, P.O. Box 513, MB 5600 Eindhoven, The Netherlands, email ruudp@win.tue.nl

y Goppa [5, section 10] in the case $a = 0$, he called such codes canonical. It is easily een that $t(\mathcal{X}) \geq N/(q+1)$, where N is the number of rational points of the curve $\mathcal{X}$. In ection 3 we show that abundant codes give asymptotically good codes for small relative ninimum distance, better than the TVZ-bound, but still worse than the GV-bound. We iscuss upper bounds of the gonality in section 4, using upper bounds for the parameters f codes. We will give an application to the error correcting capacity of some codes in ection 5. We use the book of Tsfasman and Vlăduţ [21] as a reference, but our notation s different. We denote the functional code $(\mathcal{X}, D, G)_L$ by $C_L(D, G)$, and the residue code $\mathcal{X}, D, G)_\Omega$ by $C_\Omega(D, G)$.

2 Abundant codes

n this section we show that divisors equivalent with $D + A$, so called abundant divisors, ive an improvement on the bound of Goppa on the minimum distance. As a preparation e define the gonality of a curve and proof some simple properties. At the end we give ome examples with the Hermitian curve and weighted Reed-Muller codes.

Definition 2.1 The *gonality* of a curve $\mathcal{X}$ over a field $\mathbf{F}$ is the smallest degree of a on-constant map, defined over the field $\mathbf{F}$, from $\mathcal{X}$ to the projective line. We denote the onality of $\mathcal{X}$ by $t(\mathcal{X})$, or shortly by t.

Lemma 2.2 *If E is a divisor of degree smaller than $t(\mathcal{X})$, then $l(E) \leq 1$.*

Proof If $l(E) > 1$, then there exists a non-constant rational function f such that $f) \geq -E$, so $(f)_\infty \leq E$. One can consider f as a non-constant map, defined over the ield of constants, from $\mathcal{X}$ to the projective line. The degree of this map is equal to $\deg((f)_\infty)$, which is at most $\deg(E)$, and smaller than $t(\mathcal{X})$, by assumption. This gives a ontradiction with the definition of the gonality of $\mathcal{X}$. □

Corollary 2.3 $t \leq g + 1$.

Proof Over a finite field there exists a divisor of degree $g + 1$, see [11, Theorem 3.2], nd such a divisor has dimension at least 2, by the Riemann-Roch Theorem. Now the laim follows from Lemma 2.2. □

Corollary 2.4 *If $\mathcal{X}$ has a rational point and $\deg(E) \geq t - 1$, then $l(E) \leq \deg(E) + 2 - t$.*

Proof Let P be a rational point of $\mathcal{X}$. Let $b = \deg(E) - t + 1$. Then $b \geq 0$, since $\deg(E) \geq t - 1$. Let $F = E - bP$. If $l(E) > \deg(E) + 2 - t$, then $l(F) \geq l(E) - b > 1$. But $\deg(F) = t - 1$, which contradicts Lemma 2.2. □

Remark 2.5 The gonality is one if and only if $\mathcal{X}$ is isomorphic with the projective line. The gonality of a curve is two if and only if the curve is elliptic or hyperelliptic.
For a plane curve of degree m with a rational point P one can project the curve with enter P to a line outside this point. In this way we get a map from the curve to the rojective line of degree at most $m - 1$, so the gonality is at most $m - 1$. In fact the onality is equal to $m - 1$, which follows from Namba [12] in characteristic zero, and

Homma [6] in general. Coppens [2] considered the gonality of plane curves with nodes. If the field of constants is algebraically closed, then $t(\mathcal{X}) \leq \lfloor (g+3)/2 \rfloor$, by Brill-Noether theory, and equality holds for a general curve. The above inequality is not true over a finite field, as the following example shows. Consider a smooth plane curve of degree four over a finite field without rational points. Such curves exist. Take for example the curve with equation $X^4+Y^4+Z^4=0$ over $\mathbf{F}_5$. The fourth power of an element in $\mathbf{F}_5$ is either zero or one, so $x^4+y^4+z^4$ is equal to 0,1,2 or 3 whenever 3,2,1 or 0, respectively, out of the three elements $x,y,z \in \mathbf{F}_5$ are zero. Thus this curve has no $\mathbf{F}_5$-rational points. Another example is the curve with equation $X^4+Y^4+Z^4+Y^2Z^2+X^2Z^2+X^2Y^2+X^2YZ+XY^2Z+XYZ^2$ over $\mathbf{F}_2$. If there exists an effective special divisor of degree m and dimension k on a curve of genus g, then there exists an effective special divisor of degree $2g-2-m$ and dimension $k-m-1+g$, by the Riemann-Roch Theorem. A smooth plane curve of degree 4 has genus 3. If such a curve has no rational points, then its gonality is 4. Otherwise there exists an effective divisor of degree 3 and dimension 2, so this divisor is special and by the above remark there is an effective special divisor of degree one, thus there is a rational point, which is a contradiction. Therefore the gonality of such curves is 4, which is greater than $\lfloor (g+3)/2 \rfloor$.
This example shows that the gonality can change if one extends the field of constants. The above example also shows that the upperbound $g+1$ in Corollary 2.3 can be obtained for curves of genus greater than 1. If a curve has gonality $g+1>2$, then it has no rational points, it even has no effective divisors of degree $g-2$. Otherwise, let A be an effective divisor of degree $g-2$, then there exists a canonical divisor K with support disjoint from A. So $l(K-A) \geq l(K) - \deg(A) = 2$, and $\deg(K-A) = g$. Thus $t \leq g$, by Lemma 2.2. Thus a curve of gonality $g+1>2$ has no effective divisors of degree $g-2$, and therefore it has no rational points over an extension of degree $g-2$. So $g < 2\log_q(2g)+1$, by the Hasse-Weil bound, thus $g \leq 10$ and $q \leq 31$.
In section 4 we will consider upper bounds for the gonality of curves over a finite field using coding theory.

Remark 2.6 Over a finite field one has the following lower bound, which is proved and used by Rosenbloon and Tsfasman [15, Lemma 1.1]. A similar reasoning can be found in a paper of Lewittes [9], where it is proved that $N \leq qa+1$, if a is non-gap of a rational point. A third place where this reasoning can be found is in a paper of Lachaud and Martin-Deschamps [8], where it is applied to all finite extensions of $\mathbf{F}_q$ to get a relation between the zeta function and the gonality of the curve.

Proposition 2.7 *Let $\mathcal{X}$ be a curve defined over $\mathbf{F}_q$. Let N be the number of rational points of $\mathcal{X}$. Then $t(\mathcal{X}) \geq N/(q+1)$.*

Proof Let f be a non-constant map, defined over $\mathbf{F}_q$, of degree $t = t(\mathcal{X})$. Then the rational points on $\mathcal{X}$ are mapped to the $q+1$ rational points on the projective line. The map has degree t, so there are at most t rational points on $\mathcal{X}$ above a rational point on the projective line. So the number of rational points N on $\mathcal{X}$ is at most $t(q+1)$. □

Definition 2.8 A divisor with support disjoint from the support of D and which is linear equivalent with $D+A$ for some effective divisor A is called *abundant*, and $\deg(A)$ is called the *abundance* of G. If G is an abundant divisor, then the code $C_L(D,G)$ is called *abundant*.

Remark 2.9 The name abundant refers to the fact that the degree of such divisors is at least n, whereas in Goppa's bound one assumes $m < n$. In [14, Proposition 11] it is proved that binary AG codes (i.e. binary codes of the form $C_L(D,G)$ such that $m < n$) of length longer than 13 have information rate smaller than 1/2.
Note that $C_L(D,G) = C_\Omega(D, K - A)$, for some canonical divisor K [18, Corollary 2.6]. So abundant codes of abundance zero are also of the form $C_\Omega(D,K)$, where K is a canonical divisor. These codes were called *canonical* by Goppa [5, section 10].
For every effective divisor A there exists a divisor G with support disjoint from the support of D, which is equivalent with $D + A$, by the independance of valuations [1]. But it is not true that for every divisor G of degree at least n there exists an effective divisor A such that G is equivalent to $D + A$.

Lemma 2.10 *Let G be equivalent with $D+A$, for some effective divisor A. If there exists a non-zero code word in $C_L(D,G)$ of weight d, then $l(P_{i_1} + \ldots + P_{i_d} + A) > 1$ for some $i_1 < \ldots < i_d$.*

Proof Let c be a non-zero codeword of weight d. After a permutation of $P_1, \ldots P_n$, we may assume that the first d coordinates of c are non-zero. So there exists an $f \in L(G)$ such that $f(P_i) = c_i$. Thus $(f) \geq -G$ and $f(P_i) = 0$ for all $i > d$. Therefore $(f) \geq -G + P_{d+1} + \ldots + P_n$, since G has disjoint support with D. Let $E = (f) + G - (P_{d+1} + \ldots + P_n)$. Then E is effective and

$$E \sim G - (P_{d+1} + \ldots + P_n) \sim D + A - (P_{d+1} + \ldots + P_n) \sim P_1 + \ldots + P_d + A$$

If $P_i \in \mathrm{supp}(E)$ for some $i \leq d$, then $c_i = f(P_i) = 0$ for $i \leq d$, which is a contradiction. Thus E and $P_1 + \ldots + P_d + A$ are equivalent but not equal. Thus $l(P_1 + \ldots + P_d + A) > 1$. □

Theorem 2.11 *Let $\mathcal{X}$ be a curve over $\mathbf{F}_q$ with N rational points and of gonality t. If G is an abundant divisor of abundance a and $a < t$, then $C_L(D,G)$ is an $[n,k,d]$ code such that $d \geq t - a$ and $k \geq n + a - g$. If moreover $n + a > 2g - 2$, then $k = n + a - g$.*

Proof The special case $a = 0$ is treated by Goppa [5, section 10]. The statement about the minimum distance follows directly from Lemma 2.10 and 2.2. The code $C_L(D,G)$ is the image under the evaluation map of the vector space $L(G)$ and $l(G) \geq n + a + 1 - g$, by the Riemann-Roch Theorem, and equality holds in case $n + a > 2g - 2$. The kernel of this map is equal to $L(G - D)$, which is isomorphic to $L(A)$, since G is equivalent with $D + A$. The dimension of $L(A)$ is 1, by lemma 2.2, since $a < t$. Thus the dimension of the code is at least $n + a - g$, and equality holds in case $n + a > 2g - 2$. □

Remark 2.12 If G is equivalent with $D - A$, for some effective divisor A, then the minimum distance of $C_L(D,G)$ is at least t, since $C_L(D,G)$ is contained in $C_L(D, G + A)$ and the last one is abundant of abundance zero.

Definition 2.13 We call t the *abundant* designed minimum distance, and denote it by d_A.

Example 2.14 A specific case of an abundant code is studied in [14, Proposition 15]. There it is shown that the binary [7,4,3] Hamming code is canonical, see also [11, 5.7.1]. The curve used is a smooth plane curve of degree 4, hence of genus 3, which goes through all the 7 rational points of the projective plane. If we take for D the sum of these 7 points, and for G a divisor equivalent with D, and disjoint support with D, we get a $[7,4,\geq 3]$ code, by Theorem 2.11, which must be the [7,4,3] Hamming code.

Example 2.15 The Hermitian plane curve with equation $X^{r+1}+Y^{r+1}+Z^{r+1}=0$ over $\mathbf{F}_q$, where $q=r^2$, has r^3+1 rational points and genus $r(r-1)/2$. So the gonality is at least $(r^3+1)/(r^2+1)$, by Proposition 2.7, and at most r, since the curve is a plane curve of degree $r+1$, see Remark 2.5. Thus the gonality is equal to r. Fix any of the rational points and call it P_∞, and let D be the sum of the remaining r^3 rational points. Let $n=r^3$ and $G_m=mP_\infty$. The code C_m, where $C_m=C_L(D,G_m)$ is extensively studied. If $m<n$, then the Goppa designed minimum distance d_G is equal to $n-m$. If $m<n$ and m is a multiple of r or $m\leq n-q$, then the true minimum distance of C_m is equal to d_G, see Stichtenoth [19, Theorem 4]. Yang and Kumar [24] and Xing [23] proved that C_m has minimum distance at most $n-m+t$, in case $m<n$ and $m=rs+t$ and $0\leq t\leq r-1$. Furthermore the divisor r^3P_∞ is equivalent with D, see [20] or [19]. Thus if $m\geq n$, then G_m is an abundant divisor of abundance $m-n$. Therefore C_{n+a} is an $[r^3,r^3+a-r(r-1)/2,\geq r-a]$ code for all $0\leq a<r$, by Theorem 2.11. Every line over $\mathbf{F}_q$ either intersects the curve in exactly $r+1$ different rational points or is tangent to a rational point with multiplicity $r+1$, i.e. the rational points of a Hermitian curve form a $2-(r^3+1,r+1,1)$ design. If l is the tangent line to the curve at P_∞ and m is another line through P_∞ containing the points $P_1,\ldots P_r,P_\infty$, then the rational function m/l has divisor $P_1+\ldots+P_r-rP_\infty$, so $l(P_1+\ldots+P_r)>1$. Thus the minimum distance of C_n is exactly r, by Lemma 2.10. Similarly there exists a rational function with divisor $P_1+\ldots+P_{r-1}+P_\infty-rP_r$, thus the minimum distance of C_{n+1} is exactly $r-1$. We also get as a result that the codes C_m have at least minimum distance r, for all $n-r\leq m<n$, by Remark 2.13. Yang and Kumar [24] and Xing [23] proved that in fact equality holds.

Definition 2.16 Let $\mathbf{w}=(w_1,\ldots,w_l)$ be an l-tuple of positive integers such that $w_1\leq\ldots\leq w_l$. Let

$$V(\mathbf{w},m,l,q)=\{f\in\mathbf{F}_q[X_1,\ldots,X_l]|\deg_{\mathbf{w}}(f)\leq m\},$$

where $\deg_{\mathbf{w}}X_1^{e_1}\ldots X_l^{e_l}=\sum_i w_ie_i$, and $deg_{\mathbf{w}}(f)$ is the largest weighted degree of a monomial in f with a non-zero coefficient. Let $P_1,\ldots,P_n$ be the q^l rational points of the affine space over $\mathbf{F}_q$ of dimension l. Define $WARM(\mathbf{w},m,l,q)$, the *weighted affine Reed-Muller* code of weight w, order m and lenth q^l over q, as follows

$$WARM(\mathbf{w},m,l,q)=\{(f(P_1),\ldots,f(P_n))|f\in V(\mathbf{w},m,l,q)\}.$$

Remark 2.17 The definition of the weighted affine Reed-Muller code is due to Sørensen [17], who proved that the minimum distance is equal to $(q-c)q^{l-b-1}$, where $b=\max\{i|m\geq(q-1)(w_1+\ldots+w_i)\}$, and $c=\max\{s|m\geq(q-1)(w_1+\ldots w_b)+sw_{b+1}\}$. This code appears on the curve $\mathcal{X}(l,q)$ in [14]. This curve has q^l+1 rational points over $\mathbf{F}_q$, one of them is called $\tilde{P}_\infty$ and $P_1,\ldots,P_n$ are the remaining rational points. Let $D=P_1+\ldots+P_n$. Then $WARM(\mathbf{w},m,l,q)=C_L(D,m\tilde{P}_\infty)$, see [17, Example 4.1].

Proposition 2.18 *Let $w_i = q^{l-i}(q+1)^{i-1}$, for $1 \leq i \leq l$, and $\mathbf{w} = (w_1, \ldots, w_l)$. If $m \geq q^l$, then $WARM(\mathbf{w}, m, l, q)$ is an abundant code.*

Proof One has

$$\prod_{\alpha}(z_1 - \alpha) = D - n\tilde{P}_\infty,$$

where α runs over all the elements of $\mathbf{F}_q$. Hence G is equivalent with $D + (m-n)\tilde{P}_\infty$. Thus G is abundant, since $m = \nu(q+1)^{l-1} \geq q^l = n$, and therefore $WARM(\mathbf{w}, m, l, q)$ is abundant, by Remark 2.18. □

Remark 2.19 The curve $\tilde{\mathcal{X}}(l,q)$ has $q^l + 1$ rational points over $\mathbf{F}_q$. Thus the gonality of $\mathcal{X}(l,q)$ is at least $(q^l+1)/(q+1)$. The value q^{l-1} is a non-gap of the point $\tilde{P}_\infty$, so $\leq q^{l-1}$. If $0 \leq b \leq q-1$ and $m = q^l + bq^{l-2}$, then the abundant designed distance of $C_L(D, m\tilde{P}_\infty)$ is $d_A = t - bq^{l-2}$, so $(q^l+1)/(q+1) - bq^{l-2} \leq d_A \leq q^{l-1} - bq^{l-2}$, by our results. Sørensen [17] proved that the true minimim distance is $(q-b)q^{l-2}$, see Remark 2.18. So we are left with the following question: Is the gonality of the curve $\tilde{\mathcal{X}}(l,q)$ equal to q^{l-1} ?

3 Abundant codes are asymptotically good

In this section we consider the parameters of asymptotically good abundant codes. As usual we denote the information rate k/n by R, and the relative minimum distance d/n by δ.

Definition 3.1 Let $A(q)$ be the limes superior of all quotients N/g, taken over all curves $\mathcal{X}$ over $\mathbf{F}_q$ with N rational points and genus g. Let $\gamma_q = 1/A(q)$.

Definition 3.2 Define the function R_A as follows.

$$R_A(\delta) = \begin{cases} 1 + \frac{1}{q+1} - \gamma_q - \delta & \text{for } 0 \leq \delta \leq \frac{1}{q+1} \\ 1 - (q+1)\gamma_q\delta & \text{for } \frac{1}{q+1} \leq \delta \leq \frac{\sqrt{q}-1}{q+1} \end{cases}$$

We call the graph of R_A the abundant- or A-bound.

Proposition 3.3 *If q is a square of a prime power, then for every $0 \leq \delta \leq (\sqrt{q}-1)/(q+1)$ there exists a sequence of abundant codes of increasing length such that the relative minimum distance is at least δ and the information rate is at least $R_A(\delta)$.*

Proof By Theorem 2.11 and Proposition 2.7, there exist abundant codes such that $k \geq n + a + N/(q+1) - g$ and $d \geq N(q+1) - a$, for all n and a such that $n \leq N$ and $0 \leq a < N/(q+1)$, if there exists a curve with N rational points. It follows from the work of Tsfasman, Vlăduţ, Zink and Ihara, see [21], that $\gamma_q = 1/(\sqrt{q}-1)$. If we take $n = N$ and $0 \leq a < N/(q+1)$, then we get the result for δ in the first interval; and if $a = 0$ and $0 < n < N$, then for δ in the second interval. □

Remark 3.4 In Figures 1, 2, 3, and 4, respectively, we give the graphs of these functions in case $q = 4, 9, 16$ and 25, respectively, on scale. The A-bound is always below the

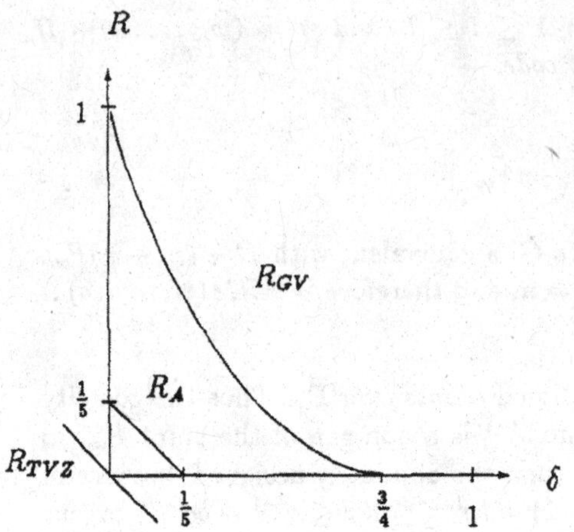

Figure 1: Bounds for $q = 4$, on scale

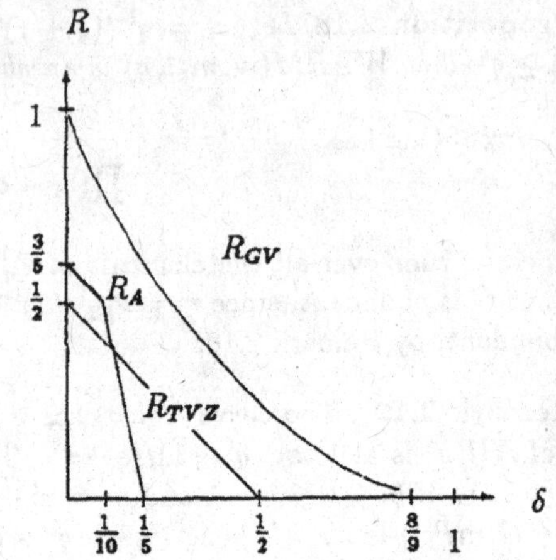

Figure 2: Bounds for $q = 9$, on scale

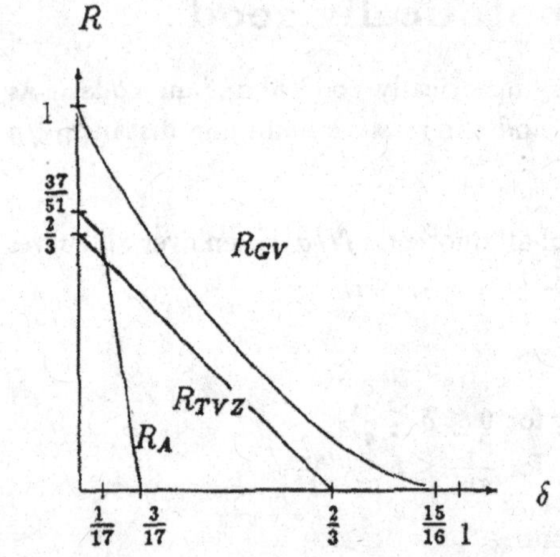

Figure 3: Bounds for $q = 16$, on scale

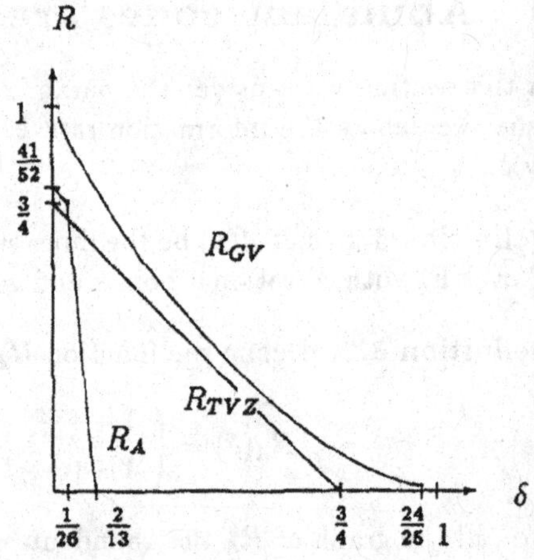

Figure 4: Bounds for $q = 25$, on scale

GV-bound, and for $0 \leq \delta < q - \sqrt{q} + 2$, the A-bound is above the TVZ-bound.
It is not difficult to show that the codes obtained by Katsman and Tsfasman [21, 3.4.23], using subfield subcodes of algebraic-geometric codes, are always better than the A-bound. Concatenation of abundant codes over $\mathbf{F}_4$ with a [3,2,2] binary code, gives binary codes above the line $R + \delta = 2/15$, also this bound is below the KT-bound
Codes on the TVZ- and KT-bound, and also abundant codes have a polynomial construction. It is not known whether codes on the GV-bound have a polynomial construction.

4 Applications of bounds on codes to bounds on the gonality

Remark 4.1 We used that the gonality of a curve is always at least $N/(q+1)$, but maybe there exists a sequence of curves with many rational points and high gonality. Let $\mathcal{X}_{2^m}$ be the modular curve over $\mathbf{F}_p$, p a prime, as defined in [21, 4.1.50]. Then this curve has at least $2^{m-3}(p-1)$ rational points over $\mathbf{F}_q$, where $q = p^2$, and genus $2^{m-3} + 1 - \delta_m$, where δ_m is equal to $3.2^{m/2-2}$ in case m is even, and $2^{(m-1)/2}$ in case m is odd, see [21, 4.1.59]. So the gonality of the curve is at least $2^{m-3}(\sqrt{q}-1)/(q+1)$, by Proposition 2.7. The divisor $2^{m-4}\infty$ has dimension 2 and degree 2^{m-4}, see [21, 4.1.60]. So the gonality of the curve is at most 2^{m-4}, which is asymptotically of the size $g/2$, see Remark 2.5.

Question Is there a sequence of curves over $\mathbf{F}_q$ with an increasing number of rational points such that $\mathcal{X}(\mathbf{F}_q) \sim (\sqrt{q}-1)g$ and $t \sim g/2$?

A positive answer would give a considerable improvement of the asymptotic parameters of abundant codes. So we are looking for curves with high gonality. We already remarked in 2.5 that $t(\mathcal{X}) \leq \lfloor (g+3)/2 \rfloor$, for a curve $\mathcal{X}$ over an algebraically closed field. In the sequel we investigate upper bounds on the gonality over finite fields, by applying bounds on codes.

Proposition 4.2 *If there exists a curve over* $\mathbf{F}_q$ *of genus* g *and gonality* t *with* N *rational points, then*

$$g \geq \log_q\Big(\sum_{i=0}^{e} \binom{N}{i}(q-1)^i\Big),$$

where $e = \lfloor (t-1)/2 \rfloor$.

Proof This result is due to Goppa [5, section 10]. There exists an $[N,k,d]$ code such that $k \geq N - g$ and $d \geq t$. The Hamming bound gives

$$q^k \leq A(N,d) \leq q^N / V_q(N,e),$$

see [21, 1.1.41]. Hence $V_q(N,e) \leq q^{N-k} \leq q^g$. Taking the q-logarithm gives the desired result. □

Proposition 4.3 *Let $\theta = (q-1)/q$. Let n and N be positive integers such that $n \leq N$. If there exists a curve over $\mathbf{F}_q$ of genus g, gonality t and N rational points, then*

$$n + log_q(1 - \theta N/t) \leq g \quad if \quad t > \theta n$$

$$t - \theta log_q(t) < \theta g + 1 \quad if \quad t \leq \theta N$$

Proof The proof is similar to the application of the Hamming bound on abundant codes, but now with the Plotkin bound on abundant codes, see [10, 5.2.4 and 5.2.5]. □

Remark 4.4 We give another application of the Plotkin bound, now on the existence of special divisors. Brill-Noether says that if $g - k(g - m + k - 1 \geq 0$, then on a curve of genus g there exists a divisor of degree m and dimension at least k. Take the particular case $m = g - 1$ and $k^2 \geq g$, then there exists a divisor of degree g-1 and dimension at least $\sqrt{g}$. If we apply this to Goppa's construction on this curve, then we get a code of length equal to the number of rational points on the curve, the dimension is at least $\sqrt{g}$, and the minimum distance is at least $n - g + 1$. If we use a sequence of modular curves, then we get a sequence of codes with information rate $R \geq \sqrt{\gamma_q}$, and relative minimum distance $\delta \geq 1 - \gamma_q$, see [21]. In particular for $q = 9$ we get $\gamma_9 = 1/2, R \geq \sqrt{2}/2$, and $\delta \geq 1/2$. But the Plokin bound gives the upper bound $R \leq 1 - \delta q/(q-1)$ for such a sequence, so in the above case, we get $R \leq 7/16$, which is a contradiction. This is one example, but for many m and q one gets a lower bound for R conflicting with an upper bound, like the Plotkin bound.

Remark 4.5 We are interested in asymptotically good abundant codes, and the asymptotic Plotkin bound implies that curves of gonality t and N rational points such that $t > N(q-1)/q$ do not give asymptotically good codes. We therefore give the following definition.

Definition 4.6 Let λ_q be the limes superior of the set of quotients t/g, taken over all curves over $\mathbf{F}_q$ of genus g, gonality t and at least $tq/(q-1)$ rational points.

Proposition 4.7

$$\lambda_q \leq (q-1)/q$$

$$(\sqrt{q} - 1)/(q+1) \leq \lambda_q \quad if\ q\ is\ a\ square$$

Proof The first inequality follows immediately from Proposition 4.3, since $\theta \log_q(t)/g \leq \theta \log_q(g+1)/g$, by Corollary 2.3, and the last expression tends to zero if g tends to infinity. The second inequality follows from Proposition 2.7 and the existence of curves with $N \sim (\sqrt{q} - 1)g$ rational points whenever q is a square, see [21]. □

Remark 4.8 The asymptotic Hamming bound gives $H_q(\delta/2) \leq \gamma$, if δ and γ are the asymptotic relative distance and relative genus of a sequence of curves. But one verifies in a straightforward way that $H_q(\gamma_q/2) \leq \gamma_q$, if q is a square, so this gives not an improvement of the bound $\lambda_q \leq 1$.

5 An application to decoding

Remark 5.1 The dual of the functional code $C_L(D,G)$ is the residue code $C_\Omega(D,G)$, which is the image of the residue map from $\Omega(G-D)$ to $\mathbf{F}_q^n$, see [21], and is an [n,k,d] code such that $k \geq n-m+g-1$ and $d \geq m-2g+2$, if $m > 2g-2$. If morover $m < n$, then $k = n-m+g-1$. We call $m-2g+2$ the Goppa designed minimum distance of $C_\Omega(D,G)$, and denote it by d_G. It is known that every functional code $C_L(D,G)$ is also a residue code, i.e. of the form $C_\Omega(D,G')$, where $G' = K+D-G$, for some canonical divisor K, see [18, Corollary 2.6]. Now we discuss the decoding of $C_\Omega(D,G)$. Let F be a divisor such that F has disjoint support with D and furthermore

1. $\dim(L(F)) > e$
2. $d(C_\Omega(D,G-F)) > e$
3. $d(C_L(D,G)) + d(C_\Omega(D,G)) > n$

Then one can decode e errors with F, see [7], [16] and [21, 3.3]. There always exists such a divisor F if $e = \lfloor (d_G-1-g)/2 \rfloor$. By applying the above idea several times one can decode $\lfloor (d_G-1)/2 \rfloor$ errors, see [13] and[22], but an efficient algorithm to find these divisors is known only in case of (hyper)elliptic curves, see [16] and [3]. Ehrhard [4] showed that one can decode $\lfloor (d_G-1-g+t)/2 \rfloor$ errors with $t+1$ divisors, given explicitly.
Now we want to apply section 2 to decoding. If the minimum distance of $C_\Omega(D,G-F)$ is more than one expects, i.e. $\deg(G-F)-2g+2$, then maybe one can decode more than $e = \lfloor (d_G-1-g)/2 \rfloor$ errors with a single F; and indeed this is the case. Before we do that we need a slight change of the conditions 2 and 3. Remark that condition 2 is equivalent with

4. $C_\Omega(Q,G-F) = 0$ for all Q such that $0 \leq Q \leq D$ and $\deg(Q) \leq e$

Furthermore $C_\Omega(Q,G-F) = 0$ if and only if $\Omega(G-F) = \Omega(G-F-Q)$. Consider the following condition

5. $\Omega(G-F-Q) = 0$ for all Q such that $0 \leq Q \leq D$ and $\deg(Q) \leq e$

Duursma [3] showed that conditions 1 and 5 imply that one can decode e errors with F.

Lemma 5.2 *Let $\mathcal{X}$ be a curve of gonality t. Let $P_1, \ldots, P_n, P_\infty$ be $n+1$ distinct rational points on $\mathcal{X}$. Let $D = P_1 + \ldots + P_n$. Let F_0 be a divisor of dimension at least $e+2$. Let A be an effective divisor of degree $t-e-1$ and disjoint support with D andP_∞. Let K be a canonical divisor with disjoint support with D and such that P_∞ is not in the negative part of K. Let $G = F_0 + K - A$ and $F = F_0 - P_\infty$. Then one can decode e errors of $C_\Omega(D,G)$ with F.*

Proof The dimension of $L(F)$ is at least $e+1$, since the dimension of $L(F_0)$ is at least $e+1$ by assumption. Furthermore $G-F_0 = K-A$, so $C_\Omega(D,G-F_0)$ is an abundant code of abundance $t-e-1$, by Remark 2.10, and therefore its minimum distance is at least $e+1$. So $\Omega(G-F_0) = \Omega(G-F_0-Q)$ for all Q such that $0 \leq Q \leq D$ and $\deg(Q) \leq e$, by Remark 5.1. Now $\Omega(G-F_0) = \Omega(K-A)$, which is equal to $\Omega(K)$, since A is effective and of degree smaller than the gonality t. Thus $\Omega(G - F_0)$ is one dimensional and generated by a differential ω with support K. So $\Omega(G - F_0 - Q) = \Omega(K - A - Q)$, is generated by ω, and the valuation of ω and $K - A$ at P_∞ are the same, since P_∞ is not in the negative part of K and not in the support of $A + Q$. Hence $\Omega(G - F - Q) = 0$, for all Q such that $0 \leq Q \leq D$ and $\deg(Q) \leq e$. Thus conditions 1 and 5 are satisfied, and one can decode e errors with F. □

Proposition 5.3 *Let $\mathcal{X}$ be a curve of gonality t. Let $P_1, \ldots, P_n, P_\infty$ be $n+1$ distinct rational points on $\mathcal{X}$. Let $D = P_1 + \ldots + P_n$. Let G be a divisor of degree m. Let $e = \lfloor (d_G - 2 - g + t)/2 \rfloor$. If there is an effective divisor of degree $e - t - 1$ and disjoint support with D and P_∞, then there exists a divisor F such that one can decode e errors of the code $C_\Omega(D, G)$ with F.*

Proof Let A be an effective divisor of degree $t - e - 1$ and disjoint support with D and P_∞. Such a divisor exsits by assumption. Let K be a canonical divisor such that the support of K is disjoint from D, and P_∞ is not in the negative part of K. Such a divisor exists, by the independence of valuations. Let $F_0 = G - K + A$, and let $F = F_0 - P_\infty$. Then F_0 has degree $m - 2g + 1 + t - e$, which is at least $g + e + 1$, by the choice of e. Hence the dimension of $L(F_0)$ is at least $e + 2$, by the Riemann-Roch Theorem. Thus one can decode e errors of $C_\Omega(D, G)$ with F, by Lemma 5.2. □

Remark 5.4 If the number of rational points is at least $n + 2$, then the existence of the divisor A is assured. One can take $A = (t - e - 1)Q_\infty$, where Q_∞ is a rational point different from P_∞ and not contained in D. Now $0 \leq t - e - 1$, since $m \leq 3g - 2 + t$. Hence A is effective.

Example 5.5 The code $C_\Omega(D, mP_\infty)$ on the Hermitian curve over $\mathbf{F}_q$, where $q = r^2$, has Goppa designed minimum distance $m - r^2 + r + 2$. It follows from work of Tiersma [20], in case q is even, and Stichtenoth [19] for arbitrary q, that $C_\Omega(D, mP_\infty)$ is the dual of C_m which is equal to $C_{r^3+r^2-r-2-m}$. Suppose $m \leq r(3r-1)/2 - 2$. Let $e = \lfloor (m - r(3r-5)/2)/2 \rfloor$. Suppose $r - e - 1 > 2$. It follows from the zeta function of the Hermitian curve that there exist places of any degree not equal to 2. Let A be a place of degree $r - e - 1$. Let $K = (r^2 - r - 2)P_\infty$. Then K is a canonical divisor. If we take $F = (m - r^2 + r + 1)P_\infty + A$, then we can decode e errors of the code $C_\Omega(D, mP_\infty)$ with F, by Proposition 5.3.

Example 5.6 If not only the minimum distance of $C_\Omega(D, G - F)$ is greater than one expects, but also the dimension of $L(F)$ is greater than $\deg(F) + 1 - g$, then sometimes one can decode even more errors. Let F_0 be the divisor $i(r+1)P_\infty$, for $0 < i < r - 1$, on the Hermitian curve of degree $r + 1$. Then $L(F_0)$ has dimension $\binom{i+2}{2}$. Let $e = (i^2 + 3i - 2)/2$. Suppose $r - e - 1 > 2$. Let A be a place of degree $r - e - 1$. Let $K = (r^2 - r - 2)P_\infty$. Then K is a canonical divisor. Let $G = F_0 + K - A$ and $F = F_0 - P_\infty$. Then the code $C_\Omega(D, G)$ has Goppa designed minimum distance $i(i + 2r + 5)/2 - r$ and one can decode $e = (i^2 + 3i - 2)/2$ errors of this code with $(ir + i - 1)P_\infty$, by Lemma 5.2. In particular, if $i = 1$, then the designed minimum distance is 3 and one can decode 1 error with rP_∞.

Remark 5.7 The function $\alpha_q^{pol\ dec,lin}(\delta)$ of asymptotic good linear codes with a polynomial construction and polynomial contructable decoding algorithm, is bounded below by the Skorobogatov-Vlăduţ-bound $R_{SV}(\delta) = 1 - 2\gamma_q - \delta$, see [21, Theorem 3.4.34]. We have the following improvement.

Proposition 5.8

$$\alpha_q^{pol\ dec,lin}(\delta) \geq 1 - 2\gamma_q + (q+1)^{-1} - \delta \quad .$$

Proof This follows immediately from Proposition 5.2 and Remark 5.3, whenever $\delta \leq 2\lambda_q$. For arbitrary δ it is proved by Ehrhard [4], see also Remark 5.1.
□

References

[1] C. Chevalley, Introduction to the theory of algebraic functions in one variable, Math. Surveys VI, Providence, AMS 1951.

[2] M. Coppens, The gonality of general smooth curves with a prescribed plane nodal model, Math. Ann. **289** (1991), 89-95.

[3] I. Duursma, Algebraic decoding using special divisors, preprint april 1991.

[4] D. Ehrhard, Über das Dekodieren algebraisch-geometrischer Codes, Thesis, Düsseldorf University, 1991.

[5] V.D. Goppa, Algebraico-geometric codes, Izv. Akad. Nauk SSSR **46** (1982), = Math. USSR Izvestija **21** (1983), 75-91.

[6] M. Homma, Funny plane curves in char $p > 0$, Comm. in Algebra, **15** (1987), 1469-1501.

[7] J. Justesen, K.J. Larsen, H.E. Jensen, A. Havemose and T. Høholdt, Construction and decoding of a class of algebraic geometric codes, IEEE Trans. Inform. Theory IT-**35** (1989), 811-821.

[8] G. Lachaud and M. Martin-Deschamps, Nombre de points des jacobiens sur un corps fini, Acta Arithmetica **16** (1990), 329-340.

[9] J. Lewittes, Places of degree one in function fields over finite fields, J. Pure Appl. Algebra, **69** (1990), 177-183.

[10] J.H. van Lint, Introduction to coding theory, Grad. Texts Math.**86**, Springer-Verlag, Berlin, 1982.

[11] C.J. Moreno, Algebraic curves over finite fields, Cambridge Texts in Math. **97**, Cambridge University Press, Cambridge, 1991.

[12] M. Namba, Families of meromorphic functions on compact Riemann surfaces, Lect. Notes Math. **767**, Springer-Verlag, Berlin, 1979.

[13] R. Pellikaan, On a decoding algorithm for codes on maximal curves, IEEE Trans. Inform. Theory IT-**35** (1989), 1228-1232.

[14] R. Pellikaan, B.Z. Shen and G.J.M. van Wee, Which linear codes are algebraic-geometric ?, IEEE Trans. INform. Theory IT-**37** (1991), 583-602.

[15] M. Yu. Rosenbloom and M.A. Tsfasman, Multiplicative lattices in global fields, Invent. Math. **101** (1990), 687-696.

[16] A.N. Skorobogatov and S.G. Vlăduţ, On the decoding of algebraic-geometric codes, IEEE Trans. Inform. Theory IT-**36** (1990), 1051-1060.

[17] A.B. Sørensen, Weighted Reed-Muller codes, preprint june 1991.

[18] H. Stichtenoth, Selbst-dual Goppa codes, J. Pure Appl. Algebra **55** (1988), 199-211.

[19] H. Stichtenoth, A note on Hermitian codes over GF(q^2), IEEE Trans. Inform. Theory IT-**34** (1988), 1345-1348.

[20] H.J. Tiersma, Codes comming from Hermitian curves, IEEE Trans. Inform. Theory IT-**33** (1987), 605-609.

[21] M.A. Tsfasman, S.G. Vlăduţ, Algebraic-geometric codes, Mathematics and its Applications **58**, Kluwer Academic Publishers, Dordrecht, 1991.

[22] S.G. Vlăduţ, On the decoding of algebraic-geometric codes over GF(q) for q$\geq$ 16, IEEE Trans. Inform. Theory **36** (1990), 1461-1463.

[23] Xing C.-P., A note on the minimum distance of Hermitian codes, preprint, Un. Science and Techn. of China, Hefei, Anhui, june 1991.

[24] K. Yang and P.V. Kumar, On the true minimum distance of Hermitian codes, preprint, june 1991.

Curves with Many Points and Multiplication in Finite Fields

Igor E. SHPARLINSKI, Michael A. TSFASMAN, Serge G. VLADUT

I.Sh.: 2-V-41 Mosfilmovskaja, 119285 Moscow, U.S.S.R.
e-mail: shpar@plb.icsti.su
M.Ts.& S.Vl.: Institute of Information Transmission
19 Ermolovoi st., Moscow GSP-4, U.S.S.R.
e-mail: tsfasman@ippi.msk.su , vladut@ippi.msk.su

Introduction

The main purpose of this paper is to understand and develop a brilliant idea of David and Gregory Chudnovsky [1] who applied curves with many points over finite fields to construct fast bilinear multiplication algorithms in large extensions of a given finite field. One should admit that their paper [1] has some unclear arguments and not all of its results can be considered as really proved. In this paper we would like to achieve the following aims.

1. To put the problem of multiplication in finite fields clearly, especially in its asymptotic statement.

2. To point out which properties of error-correcting codes are essential to apply them to the multiplication problem in finite fields.

3. To explain in simple algebraic-geometric terms the main idea of the algorithm of David and Gregory Chudnovsky.

4. To state and to prove the results of David and Gregory Chudnovsky which we consider to be really proved.

5. To ameliorate their results using some additional arguments and to write down some corollaries.

6. To explain why we consider some results announced by David and Gregory Chudnovsky to be unproved, and to point out the gaps that should be filled in to get complete proofs.

In Section 1 we recall some basic definitions and results which connect algorithms of multiplication in finite fields with the notions of tensor rank, codes, and what we call supercodes. Section 2 contains a description of the algorithm of David and Gregory Chudnovsky [1]. In Section 3 we prove the main results which are improvements of those of [1]. The last section is devoted to the gaps to be filled in to get the results announced by David and Gregory Chudnovsky.

This paper was partly written when two of the authors (M.Ts. & S.Vl.) were visiting the Institut für Experimentelle Mathematik, University of Essen, and took its final form when they were visiting the University of Geneva; we would like to express our deep gratitude to the institute and both universities, and especially to G.Frey, H.Stichtenoth, and D.Coray for their warm hospitality. One of the authors (I.Sp.) would like to thank J. von zur Gathen who elucidated many points concerning multiplicative complexity.

The general reference for coding theory and algebraic geometry used below is [3].

1. Multiplicative complexity, tensor rank, codes, and supercodes

Multiplicative complexity of multiplication. Let K be a field and A a commutative K-algebra of dimension k. If $\{e_1, \ldots, e_m\}$ is a basis of A over K then for $x = \sum_{i=1}^{k} x_i e_i$, $y = \sum_{j=1}^{k} y_j e_j \in A$ their product $z = \sum_{h=1}^{k} z_h e_h$ equals

$$z = xy = \sum_{h=1}^{k} \left(\sum_{i,j=1}^{k} t_{ijh} x_i y_j \right) e_h , \qquad (1.1)$$

where

$$e_i e_j = \sum_{h=1}^{k} t_{ijh}\, e_h , \qquad (1.2)$$

$t_{ijh} \in K$ being some constants. The direct calculation of $z = (z_1, \ldots, z_k)$ using (1.1) requires k^2 non-scalar multiplications $x_i y_j$ plus some extra multiplications by constants (which do not depend on x and y) and some additions.

Suppose that we have found a set of linear forms a_ℓ and b_ℓ such that

$$z_h = \sum_{\ell=1}^{n} c_{\ell h} a_\ell(x) b_\ell(y) \ . \qquad (1.3)$$

Then the computation of $z = xy$ can be done as follows. First compute the values $a_\ell(x)$ and $b_\ell(y)$, this computation requires only additions and scalar multiplications (i.e. multiplications by constants). Then we have to make n non-scalar multiplications to compute $a_\ell(x) \cdot b_\ell(y)$, and some scalar multiplications and additions to compute z_h.

An expression (1.3) is called a *bilinear multiplication algorithm* $\mathfrak{A}$.

The number n of non-scalar multiplications in $\mathfrak{A}$ is called the *multiplicative complexity* $\mu(\mathfrak{A})$ of $\mathfrak{A}$. Since non-scalar multiplications are much more complicated than scalar multiplications and additions, the total complexity of $\mathfrak{A}$ is (roughly speaking) close to $\mu(\mathfrak{A})$.

The value

$$\mu_K(A) = \min_{\mathfrak{A}} \mu(\mathfrak{A})$$

$\mathfrak{A}$ running over all bilinear multiplication algorithms $\mathfrak{A}$ is called the *multiplicative complexity of multiplication.*

Tensor rank. Let us pass to a more invariant language. Multiplication in A is bilinear. Therefore it defines the map

$$A \otimes A \longrightarrow A$$

or, which is the same, the tensor

$$t_m \in A^\circ \otimes A^\circ \otimes A$$

where $A^\circ = \mathrm{Hom}_K(A, K)$. The product of x and y is the convolution of this tensor with $x \otimes y \in A \otimes A$, and if we fix a basis we obtain the formula (1.1), t_{ijh} being the components of t_m in that basis.

Any decomposition into a sum of rank one tensors (i.e. of decomposable tensors)

$$t_m = \sum_{\ell=1}^{n} a_\ell \otimes b_\ell \otimes c_\ell \qquad (1.4)$$

(where $a_\ell \in A^\circ$, $b_\ell \in A^\circ$, $c_\ell \in A$) yields (1.3) whenever we fix a basis, thus giving a bilinear multiplication algorithm of multiplicative complexity

n . It is also clear that any bilinear multiplication algorithm of multiplicative complexity n can be obtained from such a decomposition.

Thus we see that $\mu_K(A)$ equals the *tensor rank* $\mathrm{rk}(t_m)$, which is defined as the minimum possible number of summands in (1.4) .

Remark 1.1. The definitions we have given can be easily generalized to the case of any system of bilinear forms over K , i.e. to the case of any 3-tensor t instead of t_m .

Finite field case. From now on we suppose the ground field to be finite

$$K = \mathbb{F}_q$$

and A to be its unique field extension of degree k

$$A = \mathbb{F}_{q^k} .$$

We write

$$\mu_q(k) = \mu_K(A)$$

since it depends only on q and k .

From the definition one obtains

Lemma 1.2. $\mu_q(m) \le \mu_q(mn) \le \mu_q(n)\cdot\mu_{q^n}(m)$. ■

Asymptotic parameters. Our main concern in this paper is asymptotic behaviour of $\mu_q(k)$ for $k \longrightarrow \infty$ and q being fixed.

It is convenient to introduce the following asymptotic parameters

$$M_q = \limsup_{k\longrightarrow\infty} \mu_q(k)/k ;$$

$$m_q = \liminf_{k\longrightarrow\infty} \mu_q(k)/k .$$

It is not at all obvious that either of these values is finite since it means that the multiplicative complexity of multiplication is asymptotically linear in the degree of extension. Moreover, the paper [1] was the first to prove that $m_q < \infty$ for some values of q . In this paper we shall prove that both m_q and M_q are finite for any q .

Corollary 1.3. *For any positive integer* k

$$m_q \le m_{q^k} \cdot \mu_q(k)/k \ ,$$

$$M_q \le M_{q^k} \cdot \mu_q(k) \ .$$

Proof: The second inequality of Lemma 1.2 shows that

$$\mu_q(mk)/mk \le (\mu_q(k)/k) \cdot (\mu_{q^k}(m)/m)$$

which is enough for the lower limit. For the upper limit we need all values of m (not only those divisible by k), therefore we use

$$\mu_q(m)/m \le \mu_q(k) \cdot (\mu_{q^k}(m)/m) \ . \blacksquare$$

Connection with codes. We use the following notation: a linear error-correcting code C over $\mathbb{F}_q$ of length n, dimension k, and minimum distance d is called an $[n,k,d]_q$-code; the rate k/n of such a code is denoted by R, and its relative minimum distance d/n by δ.

One can construct a code using decomposition of t_m into a sum of rank one tensors. Indeed, if

$$t_m = \sum_{\ell=1}^{n} u_\ell \otimes v_\ell \otimes w_\ell$$

with $u_\ell, v_\ell \in \mathbb{F}^\circ_{q^k}$, $w_\ell \in \mathbb{F}_{q^k}$ then one defines an $\mathbb{F}_q$-linear map

$$\varphi : \mathbb{F}_{q^k} \longrightarrow \mathbb{F}_q^n \ ,$$

$$x \longmapsto \big(u_1(x), \dots, u_n(x)\big) \ .$$

Let $C = \operatorname{Im} \varphi$.

Proposition 1.4. C *is an* $[n,\ k,\ \ge k]_q$*-code.*

Proof: We have to check that φ is injective and that the weight of any codeword is at least k. For any $x \in \mathbb{F}^*_{q^k}$ let us consider the element

$$t_m(x) = \sum_{\ell=1}^{n} u_\ell(x) v_\ell \otimes w_\ell \in \mathrm{Hom}_{\mathbb{F}_q}(\mathbb{F}_{q^k}, \mathbb{F}_{q^k}) .$$

It is clear that $t_m(x)$ is the multiplication by x in $\mathbb{F}_{q^k}$

$$t_m(x) : y \longmapsto yx .$$

This gives the injectiveness of φ since $\mathbb{F}_{q^k}$ has no zero divisors. Since $t_m(x)$ is an $\mathbb{F}_q$-linear authomorphism of $\mathbb{F}_{q^k}$, its rank is k and thus the weight of $\varphi(x) = \big(u_1(x),\dots,u_n(x)\big)$ is at least k. Indeed, the image of $t_m(x)$ is spanned by those w_ℓ for which $u_\ell(x) \neq 0$. ■

Corollary 1.5. *Let* $n_q(k)$ *be the minimum length of a linear* $[n, k, k]$*-code. Then* $\mu_q(k) \geq n_q(k)$. ■

Let us recall ([3] 1.3.1) that there exists a continuous decreasing function $\alpha_q^{\mathrm{lin}}(\delta)$ on the segment $[0, 1 - 1/q]$ which is the "true" bound for the rate R of linear codes over $\mathbb{F}_q$ with relative minimum distance at least δ.

Corollary 1.6. *One has* $m_q \geq \delta_q^{-1}$, *where* δ_q *is the unique solution of the equation* $\alpha_q^{\mathrm{lin}}(\delta) = \delta$. ■

Any upper bound for $\alpha_q^{\mathrm{lin}}(\delta)$ gives an upper bound for δ_q and thus a lower bound for m_q.

Corollary 1.7. $m_2 \geq 3.52$.

Proof [1]: Apply the bound of "four" ([3] 1.3.2) for asymptotic parameters of binary codes. ■

Corollary 1.8. $m_q \geq 2\left(1 + \dfrac{1}{q-1}\right)$.

Proof [1]: Apply the asymptotic Plotkin bound ([3] 1.3.2). ■

Supercodes. We have seen that a decomposition of t_m into a sum of n summands of rank 1 produces an $[n, k, \geq k]_q$-code. One can ask if it is possible vice versa to construct such a decomposition using a linear $[n, k, k]_q$-code. Generally speaking this is impossible since decomposition (or a bilinear multiplication algorithm, which is the same) has to do something with the multiplicative structure which is not used in a code. To construct a bilinear multiplication algorithm we need the following notion.

Let $S \subseteq \mathbb{F}_{q^k} \oplus \mathbb{F}_q^n$ be an $\mathbb{F}_q$-linear subspace. We call S an $[n, k]_q$*-supercode* if the following conditions are satisfied:

1. The first projection

$$\pi_1 : \mathbb{F}_{q^k} \oplus \mathbb{F}_q^n \longrightarrow \mathbb{F}_{q^k}$$

restricted to S is surjective

$$\pi_1|_S : S \twoheadrightarrow \mathbb{F}_{q^k} .$$

2. Let $S^2 = \{s_1 s_2 \mid s_1, s_2 \in S\}$ where the multiplication is that in $\mathbb{F}_q$-algebra $\mathbb{F}_{q^k} \oplus \mathbb{F}_q^n$, and let $\langle S^2 \rangle$ be the subspace in $\mathbb{F}_{q^k} \oplus \mathbb{F}_q^n$ spanned by S^2. Then the second projection

$$\pi_2 : \mathbb{F}_{q^k} \oplus \mathbb{F}_q^n \longrightarrow \mathbb{F}_q^n$$

restricted to $\langle S^2 \rangle$ is injective

$$\pi_2|_{\langle S^2 \rangle} : \langle S^2 \rangle \hookrightarrow \mathbb{F}_q^n .$$

Note that if S is a supercode then $\pi_2|_S$ is also injective. Indeed, if $s \in S$, $s \neq 0$, and

$$\pi_2(s) = (u_1(s),\dots,u_n(s)) \in \mathbb{F}_q^n$$

then

$$\pi_2(s^2) = (u_1(s)^2,\dots,u_n(s)^2) \neq 0$$

since $\pi_2|_{S^2}$. is injective, and hence $\pi_2(S) \neq 0$.

We call a supercode S *exact* if $\pi_1|_S$ is an isomorphism, i.e. if $\dim S = k$. Any supercode contains an exact one.

Lemma 1.9. *Let* S *be an* $[n, k]_q$*-supercode, and* $C = \pi_2(S)$ *. Then* C *is an* $[n, \geq k, \geq k]_q$*-code.*

Proof: We have to prove only that the minimum distance of C is at least k . Indeed let the weight of $\pi_2(x)$ equal w . Since for any $y \in S$ the vector $\pi_2(xy) = \pi_2(x)\pi_2(y)$ has zeroes at all positions where $\pi_2(x)$ does, the cardinality of the set $\pi_2(xS)$ is at most q^w . Since $\pi_2|_{S^2}$ is injective, $|\pi_2(xS)| \geq |S| = q^k$ and thus $w \geq k$. ■

Note that if S is exact then C is an $[n, k, \geq k]_q$-code.

Supercodes and decompositions. We are going to show that the notion of an exact supercode is almost equivalent to that of a decomposition of t_m into a sum of n rank one tensors.

Let S be an exact supercode. Let V be a subspace of $\mathbb{F}_q^n$ such that $\pi_2(<S^2>) \oplus V = \mathbb{F}_q^n$ and let $W = <S^2> \oplus V'$ where $V' = \{(0,v) \mid v \in V\} \subseteq \mathbb{F}_{q^k} \oplus \mathbb{F}_q^n$. It is clear that $\pi_2|_W : W \xrightarrow{\sim} \mathbb{F}_q^n$ is an isomorphism and that $\pi_1(V') = 0$.

Since $\pi_1|_S$ and $\pi_2|_W$ are isomorphisms, we can consider inverse maps

$$\rho_1 : \mathbb{F}_{q^k} \xrightarrow{\sim} S$$

and

$$\rho_2 : \mathbb{F}_q^n \xrightarrow{\sim} W .$$

Let

$$\varphi = (\pi_2|_S)\circ\rho_1 : \mathbb{F}_{q^k} \longrightarrow \mathbb{F}_q^n$$

and

$$\psi = (\pi_1|_W)\circ\rho_2 : \mathbb{F}_q^n \longrightarrow \mathbb{F}_{q^k} .$$

Let $1_i = (0,\ldots,0,1,0,\ldots,0) \in \mathbb{F}_q^n$ and let $\ell_i \in (\mathbb{F}_q^n)^\circ$ be the linear form dual to 1_i , in the sense that $\ell_i(1_i) = 1$ and $\ell_i(1_j) = 1$ for $i \neq j$. Consider the tensor

$$T = \sum_{i=1}^{n} \varphi^\circ(\ell_i) \otimes \varphi^\circ(\ell_i) \otimes \psi(1_i) ,$$

where φ° denotes the dual map $\varphi^\circ : (\mathbb{F}_q^n)^\circ \longrightarrow (\mathbb{F}_{q^k})^\circ$. Note that T is well-defined, it does not depend on the choice of V since ψ is unique because of the condition $\pi_1(V') = 0$.

Proposition 1.10. $T = t_m$.

Proof: We have to show that $T(x\otimes y) = xy$. We have

$$T(x\otimes y) = \sum_{i=1}^{n} \varphi^\circ(\ell_i)(x)\cdot\varphi^\circ(\ell_i)(y)\cdot\psi(1_i) =$$

$$= \sum_{i=1}^{n} \ell_i(\varphi(x))\cdot\ell_i(\varphi(y))\cdot\psi(1_i) .$$

Let $s = \rho(x)$, $t = \rho(y)$. Then

$$\pi_2(s) = (\ell_1(\varphi(x)),\ldots,\ell_n(\varphi(x))) = \sum_{i=1}^{n} \ell_i(\varphi(x))\cdot 1_i ,$$

$$\pi_2(t) = (\ell_1(\varphi(y)),\ldots,\ell_n(\varphi(y))) = \sum_{i=1}^{n} \ell_i(\varphi(y))\cdot 1_i ,$$

and

$$\pi_2(st) = \pi_2(s)\cdot\pi_2(t) = \sum_{i=1}^{n} \ell_i(\varphi(x))\cdot\ell_i(\varphi(y))\cdot 1_i .$$

Thus

$$\rho_2(\pi_2(st)) = \sum_{i=1}^{n} \ell_i(\varphi(x))\cdot\ell_i(\varphi(y))\cdot\rho_2(1_i) .$$

Therefore

$$xy = \pi_1(st) = \sum_{i=1}^{n} \ell_i(\varphi(x))\cdot\ell_i(\varphi(y))\cdot\pi_1(\rho_2(1_i)) =$$

$$= \sum_{i=1}^{n} \ell_i(\varphi(x))\cdot\ell_i(\varphi(y))\cdot\psi(1_i)$$

and we are done. ■

Let $t_m = \sum_{i=1}^{n} u_i \otimes v_i \otimes w_i$. We call this decompositon *symmetric* if $u_i = v_i$. We denote the symmetric decomposition

$$t_m = \sum_{i=1}^{n} \varphi^{\circ}(\ell_i) \otimes \varphi^{\circ}(\ell_i) \otimes \psi(1_i)$$

by $\Sigma(S)$.

Consider an arbitrary symmetric decomposition $t_m = \sum_{i=1}^{n} u_i \otimes u_i \otimes w_i$ which we denote by Σ . Let $\varphi(x) = (u_1(x), \dots, u_n(x)) \in \mathbb{F}_q^n$. We shall study the subspace

$$S(\Sigma) = \{ \ (x;\varphi(x)) \ | \ x \in \mathbb{F}_{q^k} \ \} \subseteq \mathbb{F}_{q^k} \oplus \mathbb{F}_q^n \ .$$

Proposition 1.11. $S(\Sigma)$ is an exact supercode.

Proof: Of course, $\dim S(\Sigma) = k$ and $\pi_1|_{S(\Sigma)}$ is surjective. We have to check that $\pi_1|_{<S(\Sigma)^2>}$ is injective. Let

$$u = \sum_{j=1}^{r} a_j s_j t_j \in \ <S(\Sigma)^2> \ ,$$

$x_j = \pi_1(s_j)$, $y_j = \pi_1(t_j)$. Suppose that $\pi_2(u) = 0$, then

$$\sum_{j=1}^{r} a_j \sum_{i=1}^{n} u_i(x_j) \cdot u_i(y_j) \cdot 1_i = 0 \ .$$

Hence $\sum_{j=1}^{r} a_j \cdot u_i(x_j) \cdot u_i(y_j) = 0$ for any i . Therefore

$$\pi_1(u) = \sum_{j=1}^{r} a_j \cdot x_j \cdot y_j = \sum_{j=1}^{r} a_j \sum_{i=1}^{n} u_i(x_j) \cdot u_i(y_j) \cdot w_i =$$

$$= \sum_{i=1}^{n} \left(\sum_{j=1}^{r} a_j \cdot u_i(x_j) \cdot u_i(y_j) \right) \cdot w_i = 0 \ .$$

Thus $u = 0$ and we are done. ■

Let $\Sigma_1 = \sum_{i=1}^{n} u_i \otimes u_i \otimes w_i$ and $\Sigma_2 = \sum_{i=1}^{n} v_i \otimes v_i \otimes z_i$ be two symmetric decompositions of t_m . We call Σ_1 and Σ_2 *equivalent* if $u_i = v_i$ for every i.

Theorem 1.12. *The map* $S \longmapsto \Sigma(S)$ *gives a bijection between the set of exact supercodes and the set of equivalence classes of symmetric decompositions of* t_m *. Moreover,* $S(\Sigma(S)) = S$, $\Sigma(S(\Sigma(S))) = \Sigma(S)$.

Proof: First let us check that $S(\Sigma(S)) = S$. The decomposition $\Sigma(S)$ is written as

$$t_m = \sum_{i=1}^{n} \varphi^{\circ}(\ell_i) \otimes \varphi^{\circ}(\ell_i) \otimes \psi(1_i) .$$

Then

$$S(\Sigma(S)) = \{ (x; \varphi^{\circ}(\ell_1)(x), \dots, \varphi^{\circ}(\ell_n)(x)) \} =$$

$$= \{ (x; \ell_1(\varphi)(x), \dots, \ell_n(\varphi)(x)) \} = \{ (x; \varphi(x)) \} = S .$$

From the relation $S(\Sigma(S)) = S$ it follows that the map $\Sigma \longmapsto S(\Sigma)$ is onto and that the map $S \longmapsto \Sigma(S)$ is injective. Now it is sufficient to prove that any equivalence class of symmetric decomposition contains a unique element of the form $\Sigma(S)$. It is implied by the following statements.

(i) If Σ is equivalent to Σ' then $S(\Sigma) = S(\Sigma')$.

(ii) $\Sigma(S(\Sigma))$ is equivalent to Σ for any Σ .

(iii) $\Sigma(S(\Sigma(S))) = \Sigma(S)$.

Statement (i) is obvious since the definition of $S(\Sigma)$ involves only u_i ; (ii) is clear since if $\Sigma = \sum_{i=1}^{n} u_i \otimes u_i \otimes w_i$ then $\Sigma(S(\Sigma)) = \sum_{i=1}^{n} u_i \otimes u_i \otimes \psi(1_i)$ in the above notation; (iii) is also clear since $\Sigma(S)$ is of the form $\sum_{i=1}^{n} u_i \otimes u_i \otimes \psi(1_i)$. ■

Corollary 1.13. *Any supercode* $S \subseteq \mathbb{F}_{q^k} \oplus \mathbb{F}_q^n$ *yields a bilinear multiplication algorithm of multiplicative complexity* n .

Proof: Any supercode contains an exact sub-supercode. If we fix such a sub-supercode, we get a decomposition Σ which gives us a bilinear multiplication algorithm. ■

Remark 1.14. In fact, for Lemma 1.9 and Corollary 1.13 to hold, one can weaken the Condition 2 in the definition of supercode, requiring injectivity only on S^2 rather than on $\langle S^2 \rangle$.

The following question looks very interesting.

Problem 1.15. How can one characterize those $[n, \geq k, \geq k]_q$-codes which are projections of supercodes?

2. The algorithm of Chudnovsky & Chudnovsky

What David and Gregory Chudnovsky [1] discovered, is in our terms the fact that some algebraic-geometric codes can be lifted to supercodes and thus can be applied to construct multiplication algorithms.

Their algorithm uses the following data:

1. An absolutely irreducible smooth curve X over $\mathbb{F}_q$.
2. A point Q of X of degree k .
3. An $\mathbb{F}_q$-rational divisor D on X .
4. A set $\mathcal{P} = \{P_1,\dots,P_n\}$ of $\mathbb{F}_q$-rational points of X .

Let

$$(\mathcal{P} \cup \{Q\}) \cap \operatorname{Supp} D = \emptyset \qquad (2.1)$$

and let two condition be satisfied:

A. The evaluation map

$$\mathrm{ev}_Q : L(D) \longrightarrow \mathbb{F}_q(Q) = \mathbb{F}_{q^k}$$

$$f \longmapsto f(Q)$$

is surjective.

B. The evaluation map

$$\mathrm{ev}_{\mathcal{P}} : L(2D) \longrightarrow \bigoplus_{P\in\mathcal{P}} \mathbb{F}_q(P) = \mathbb{F}_q^n$$

$$f \longmapsto (f(P_1),\dots,f(P_n))$$

is injective.

Then we have

Proposition 2.1. $\mu_q(k) \le n$.

Proof: Let S be the image of the map

$$\mathrm{ev} = (\mathrm{ev}_Q, \mathrm{ev}_{\mathcal{P}}) \; : \; L(D) \longrightarrow \mathbb{F}_{q^k} \oplus \mathbb{F}_q^n \, .$$

This map is injective because of B. Since $L(D)^2 \subseteq L(2D)$ the conditions A and B show that $S = \mathrm{ev}(L(D))$ is a supercode and we are done by Corollary 1.13.

■

Remark 2.2. One can easily show that the corresponding multiplication algorithm can be derived as follows. Consider the commutative diagram

$$\begin{array}{ccccc}
\mathbb{F}_{q^k} \oplus \mathbb{F}_{q^k} & \xleftarrow{\mathrm{ev}_Q \oplus \mathrm{ev}_Q} & L(D) \oplus L(D) & \xrightarrow{\mathrm{ev}_{\mathcal{P}} \oplus \mathrm{ev}_{\mathcal{P}}} & \mathbb{F}_q^n \oplus \mathbb{F}_q^n \\
\downarrow & & \downarrow & & \downarrow \\
\mathbb{F}_{q^k} & \xleftarrow{\quad \mathrm{ev}_Q \quad} & L(2D) & \xrightarrow{\quad \mathrm{ev}_{\mathcal{P}} \quad} & \mathbb{F}_q^n
\end{array}$$

where the vertical darts denote multiplication maps. Since the diagram is commutative for any $x, y \in \mathbb{F}_{q^k}$ we have:

$$xy = \mathrm{ev}_Q\Big[\mathrm{ev}_{\mathcal{P}}^{-1}\Big(\mathrm{ev}_{\mathcal{P}}[\mathrm{ev}_Q^{-1}(x)] \cdot \mathrm{ev}_{\mathcal{P}}[\mathrm{ev}_Q^{-1}(y)]\Big)\Big] \, ,$$

where $\mathrm{ev}_Q^{-1}(z)$ for $z \in \mathbb{F}_{q^k}$ denotes an arbitrary function $f \in L(D)$ with $f(Q) = z$. In other words we represent $\mathbb{F}_{q^k}$ as the residue field of Q and to multiply elements we lift them to rational functions and multiply their values at $\mathbb{F}_q$-rational points of $\mathcal{P}$.

Remark 2.3. One can also dispense with the condition (2.1). Indeed, it is sufficient to consider adjusted evaluation maps ev'_Q and $\mathrm{ev}'_{\mathcal{P}}$ which are defined as follows: if $P \in X$, $f \in L(D)$ then

$$\mathrm{ev}'_{\mathcal{P}} = \begin{cases} f(P) \, , & \text{if } P \notin \mathrm{Supp}\, D \, , \\ \big(t_P^{\,a_P} \cdot f\big)(P), & \text{if } D = a_P P + D' \text{ with } P \notin \mathrm{Supp}\, D', \end{cases}$$

where t_P is an arbitrary (fixed) local parameter at P. Using ev'_Q, and $ev'_{\mathcal{P}}$ one can obtain Proposition 2.1 without (2.1).

3. Results

Let us now show that the idea of David and Gregory Chudnovsky in many cases gives good asymptotic estimates for multiplicative complexity.

Denote by $N(q,g)$ the maximum number of $\mathbb{F}_q$-rational points $N(X)$ on an absolutely irreducible smooth curve X over $\mathbb{F}_q$ of genus g and let

$$A(q) = \limsup_{g \longrightarrow \infty} N(q,g)/g .$$

Here and below we use the following convenient abuse of notation. For a family of curves $\{X_s\}$ of growing genus $g = g_s = g(X_s)$ we often omit the index s and write $\{X\}$, $g \longrightarrow \infty$, and so on.

It is known (cf. [3] 2.3.3) that

$$c\cdot\log_2 q \le A(q) \le \sqrt{q} - 1 , \qquad (3.1)$$

where $c > 0$ is some absolute constant (the lower bound is due to Serre, the upper one to Drinfeld and Vlăduţ). Moreover, for $q = p^{2m}$ being an even power of a prime

$$A(q) = \sqrt{q} - 1 .$$

The following theorem is a generalization and an improvement of the corresponding result from [1].

Theorem 3.1. *Let* $A(q) > 1$. *Then*

$$m_q \le 2\left(1 + \frac{1}{A(q) - 1}\right) .$$

Proof: Let $\{X_s\}$ be a family of absolutely irreducible smooth curves, let $g = g_s$ be the genus of $X = X_s$. Suppose that $g \longrightarrow \infty$ and

$$\lim_{g\longrightarrow\infty} |X(\mathbb{F}_q)|/g = A(q) ,$$

$X(\mathbb{F}_q)$ being the set of $\mathbb{F}_q$-rational points of X. Let us fix a small $\varepsilon > 0$. Let $n = |X(\mathbb{F}_q)|$. We set

$$a = \lceil (n + g(1 - \varepsilon))/2 \rceil ,$$

$$k = \lceil (n - g(1 + 3\varepsilon))/2 \rceil ,$$

$\lceil\cdot\rceil$ denoting the integer part.

Claim. If g is large enough then there exist a point $Q \in X$ of degree k and a divisor D of degree a such that Conditions A and B are satisfied for $\mathcal{P} = X(\mathbb{F}_q)$, Q and D.

The theorem follows from the claim since by Proposition 2.1 we see that $\mu_q(k) \le n$ and hence

$$m_q \le \lim_{g\to\infty} n/k = 2\cdot\lim_{g\to\infty} n/(n - g(1 + 3\varepsilon)) =$$

$$= 2\cdot A(q)/(A(q) - 1 - 3\varepsilon) .$$

Since $\varepsilon > 0$ is arbitrary, we are done.

Proof of the claim. First of all we have to prove that for a sufficiently large g there exists a point of degree k. Indeed for any k we have

$$N_k = \sum_{d|k} dB_d$$

where $N_k = |X(\mathbb{F}_{q^k})|$ and B_d is the number of points of degree d on X. Applying the Möbius inversion formula and the Weil bound we get

$$B_k \ge \frac{1}{k}\Big(N_k - \sum_{d\le k/2} N_d\Big) \ge$$

$$\ge \frac{1}{k}\Big(q^k - 2g\cdot q^{k/2} + 1 - \sum_{d\le k/2} (q^d + 2g\cdot q^{d/2} + 1)\Big)$$

and thus B_k is positive since k is proportional to g.

Let us fix an arbitrary point Q of degree k. We shall prove that for a sufficiently large g there exists a divisor D of degree a such that

$\alpha)\ \ell(K - D + Q) = 0$

and

$\beta)\ \ell(2D - \mathrm{P}) = 0$

where

$$\mathrm{P} = \sum_{P \in \mathcal{P}} P$$

and K is the canonical class of X.

Conditions α and β imply A and B. Indeed, from Condition α we have $\ell(K - D) = 0$ since the divisor Q is positive. By the Riemann-Roch theorem one has

$$\ell(D) - \ell(D - Q) = (\deg D - g + 1) - (\deg (D - Q) - g + 1) = \deg Q = k$$

and thus the evaluation map ev_Q is surjective since its kernel $L(D - Q)$ has the "correct" codimension k. Condition β implies that the kernel of $\mathrm{ev}_{\mathcal{P}}$ is trivial.

We shall use the approach of [4]. Consider a family $\{X\}$ of curves X over $\mathbb{F}_q$ of growing genus $g = g(X) \longrightarrow \infty$. The family is called *asymptotically exact* if for all m there exist limits

$$\beta_m = \lim_{g \longrightarrow \infty} \frac{B_m}{g},$$

where $B_m = B_m(X)$ is the number of points of degree m on X. A simple diagonal argument shows that each family of curves of growing genus contains an asymptotically exact subfamily. Passing to a subfamily with $\beta_1 = A(q)$ we can assume our family to be asymptotically exact. The following two statements follow from Corollary 2 and Theorem 6 of [4].

Lemma 3.2. *Let $\{X\}$ be an asymptotically exact family of curves. Let $h = h(X)$ be the number of $\mathbb{F}_q$-points on the Jacobian of X. Then*

$$\log_q h = g \cdot \left(1 + \sum_{m=1}^{\infty} \beta_m \cdot \log_q \frac{q^m}{q^m - 1}\right) + o(g). \blacksquare$$

Lemma 3.3. *Let D_b be the number of positive $\mathbb{F}_q$-divisors of degree $b = b(g)$ on X. If*

$$b \geq g \cdot \left(\sum_{m=1}^{\infty} \frac{m\,\beta_m}{q^m - 1} \right) - o(g)$$

then

$$\log_q D_b = b + g \cdot \sum_{m=1}^{\infty} \beta_m \cdot \log_q \frac{q^m}{q^m - 1} + o(g) \; . \blacksquare$$

Let D be a divisor of degree a such that either Condition α or Condition β fails. It means that either the divisor $K - D + Q$ is equivalent to an effective divisor or this is the case for $2D - \mathrm{P}$. Consider the degrees of $K - D + Q$ and $2D - \mathrm{P}$:

$$\deg\,(K - D + Q) = 2g - 2 - a + k \leq \lceil g \cdot (1 - \varepsilon) \rceil - 1 \; ,$$

$$\deg\,(2D - \mathrm{P}) = 2a - n \leq \lceil g \cdot (1 - \varepsilon) \rceil + 1 \; .$$

If D' lies in a linear equivalence class different from that of D, the corresponding effective divisors are different. Thus the number of linear equivalence classes of divisors of degree a for which either Condition α or Condition β fails is at most $D_{b'} + D_b \leq 2D_b$, where $b' = \lceil g \cdot (1 - \varepsilon) \rceil - 1$ and $b = \lceil g \cdot (1 - \varepsilon) \rceil + 1$. The total number of linear equivalence classes of degree b does not depend on b and equals h , which is given by Lemma 3.2. To apply Lemma 3.3 we need b to be sufficiently large. Indeed, for large enough g we have

$$\sum_{m=1}^{\infty} \frac{m\,\beta_m}{q^m - 1} \leq \frac{1}{\sqrt{q} + 1} \cdot \sum_{m=1}^{\infty} \frac{m\,\beta_m}{q^{m/2} - 1} < \frac{1}{2}$$

since $q \geq 2$ and $\sum_{m=1}^{\infty} \frac{m\,\beta_m}{q^{m/2} - 1} \leq 1$ (see Corollary 1 of [4]). Therefore $2D_b < h$ for g large enough. Hence there exist divisors D for which Conditions α and β are satisfied and we are done. $\blacksquare$

Since for $q = p^{2m}$ it is known that $A(q) = \sqrt{q} - 1$ we get

Corollary 3.4. *Let* $q = p^{2m} \geq 9$. *Then*

$$m_q \le 2\left(1 + \frac{1}{\sqrt{q} - 2}\right) . \blacksquare$$

In the case of arbitrary q we can try to use Serre's bound $A(q) \ge c\cdot\log_2 q$. For q large enough we get

Corollary 3.5. *There exists a positive constant* c *independent of* q *such that if* $q > 2^{1/c}$ *then*

$$m_q \le 2\left(1 + \frac{1}{c\cdot\log_2 q - 1}\right) . \blacksquare$$

Remark 3.6. One can also apply lower bounds for $A(q)$ due to Th.Zink and M.Perret (see [3] Theorem 2.3.25), which are valid for some special values of q . For example, if $q = p^{3m}$ then we get

$$m_q \le 2\left(1 + \frac{q^{1/3} + 2}{2q^{2/3} - q^{1/3} - 4}\right) .$$

Using Theorem 3.1 and Corollary 1.2 we can obtain some results for an arbitrary q (which are not covered by preceeding results when $q = 4$, or q is an odd power of a prime too small for Corollary 3.5).

Corollary 3.7. $m_2 \le 35/6$.

Proof: By Lemma 1.2 we have $\mu_2(6) \le \mu_2(2)\cdot\mu_4(3) = 15$ since $\mu_2(2) = 3$, $\mu_4(3) = 5$ (see [1]). By Corollary 1.3 we get

$$m_2 \le m_{64}\cdot\mu_2(6)/6 \le \frac{15}{6}\cdot 2\left(1 + \frac{1}{\sqrt{64} - 2}\right) = 35/6 . \blacksquare$$

This bound improves the bound $m_2 \le 6$ from [1].

Corollary 3.8. *For any* $q = p^m > 2$

$$m_q \le 3\left(1 + \frac{1}{q - 2}\right) .$$

Proof: For any $q = p^m > 2$ we have $q^2 = p^{2m} \ge 9$ and thus Theorem 3.1 gives

$$m_{q^2} \le 2\left(1 + \frac{1}{q - 2}\right) .$$

By Corollary 1.3 we have

$$m_q \le m_{q^2} \cdot \mu_q(2)/2 = 3\left(1 + \frac{1}{q-2}\right) ,$$

since it is well known that $\mu_q(2) = 3$ for any q . ∎

In particular, we have proved that $m_q < \infty$ for any q .

Now we are going to consider the more complex and interesting case of M_q . We are going to show that for M_q the same upper bound as for m_q can be proved though only in the case of q being an even power of a prime.

Theorem 3.9. *Let* $q = p^{2m} \ge 9$. *Then*

$$M_q \le 2\left(1 + \frac{1}{\sqrt{q} - 2}\right) . \blacksquare$$

Proof: Let us first deduce the theorem from the following

Claim. For $q = p^{2m} \ge 9$ there exists a family $\{X_s\}$ of absolutely irreducible smooth curves of genus $g = g_s$ with

$$\lim_{g\to\infty} |X(\mathbb{F}_q)|/g = \sqrt{q} - 1 , \tag{3.2}$$

and

$$\lim_{s\to\infty} g_s/g_{s+1} = 1 . \tag{3.3}$$

This means that we can choose a family of curves with the maximum possible ratio of the number of $\mathbb{F}_q$-rational points to the genus and such that the sequence of its genera is sufficiently dense.

Assume the claim to hold. Fix $\varepsilon > 0$ and consider an arbitrary k . Let $g(k)$ be the least genus of a curve in our family such that

$$2k + g(k)(1 + 3\varepsilon) \le n(k)$$

where $n(k)$ is the number of points on the corresponding curve. From (3.2) and (3.3) we get

$$k = \big(n(k) - g(k)(1 + 3\varepsilon)\big)/2 + o(g(k)) \qquad (3.4)$$

and

$$\lim_{k\to\infty} \quad n(k)/g(k) = \sqrt{q} - 1 > 1. \qquad (3.5)$$

It is clear that for sufficiently large k we have

$$q^k - 2g(k)q^{k/2} + 1 - \sum_{d\le k/2} (q^d + 2g(k)q^{d/2} + 1) > 0$$

since q^k grows faster than the other terms. Therefore there exists a point $Q = Q(k)$ of degree k. Set

$$a(k) = \lceil k + g(k)(1 - \varepsilon)\rceil$$

and consider divisors of degree $a(k)$. Using the same arguments as in the proof of Theorem 3.1 we get $\mu_q(k) \le n(k)$. Thus from (3.4) and (3.5) we get

$$\mu_q(k)/k \le 2n(k)/\big(n(k) - g(k)(1 + 3\varepsilon) + o(g(k))\big).$$

Using (3.2) we see that

$$M_q \le 2\,\frac{\sqrt{q} - 1}{\sqrt{q} - 2 - 3\varepsilon}.$$

Since $\varepsilon > 0$ is arbitrary, then from this inequality and from (3.2) we obtain the bound of the theorem.

Proof of the claim: First note that for $q = p^2$ the existence of the required family is clear: it is sufficient to put $X = X_0(11\ell_k)$ where ℓ_k is the k-th prime number. Then $g = \ell = \ell_k$ and it is well-known (see [3] 4.1.2) that

$$|X(\mathbb{F}_{p^2})| \ge (p - 1)(\ell + 1).$$

For the general case $q = p^{2m}$ the situation is similar but a bit more complicated since one has to use Shimura curves.

Let $q = p^{2m}$ and let F be a totally real abelian over $\mathbb{Q}$ number field of degree m in which p is inert, thus $O_F/(p) \cong \mathbb{F}_{p^m}$ where O_F is the ring of integers of F. Let $\mathfrak{p}$ be a prime of F which does not divide p and let B be a quarternion algebra for which

$$B\otimes_F\mathbb{R} = M_2(\mathbb{R})\times\mathbb{H}\times \dots \times\mathbb{H}$$

where $\mathbb{H}$ is the skew field of Hamilton quarternions. Let B be also unramified at any finite place if $(m-1)$ is even; let B be unramified outside infinity and $\mathfrak{p}$ if $(m-1)$ is odd. Then, over F one can define the Shimura curve by its complex points

$$X_\Gamma(\mathbb{C}) = \mathfrak{H}/\Gamma ,$$

where $\mathfrak{H}$ is the upper half-plane and Γ is the group of units of a maximal order $\mathcal{O}$ of B with totally positive norm modulo its center. Moreover, it is well known that its reduction $X_{\Gamma,p}$ modulo p is good and (see [2])

$$|X_{\Gamma,p}(\mathbb{F}_{p^{2m}})| \geq (p^m - 1)(g + 1) .$$

Let now ℓ be a prime which is greater than the maximum order of stabilizers Γ_z, where $z \in \mathfrak{H}$ is a fixed point of Γ and let $\mathfrak{p} \nmid \ell$.

Let $\Gamma_0(\ell)_\ell$ be the following subgroup of $GL_2(\mathbb{Z}_\ell)$

$$\Gamma_0(\ell)_\ell = \left\{\begin{pmatrix} a & b\\ c & d\end{pmatrix} \in GL_2(\mathbb{Z}_\ell) \mid c \equiv 0 \pmod{\ell}\right\} .$$

Suppose that ℓ splits completely in F. Then there exists an embedding $F \hookrightarrow \mathbb{Q}_\ell$, and since $B\otimes_F\mathbb{Q}_\ell = M_2(\mathbb{Q}_\ell)$ we have a natural map

$$\varphi_\ell : \Gamma \longrightarrow GL_2(\mathbb{Z}_\ell) .$$

Let Γ_ℓ be the inverse image of $\Gamma_0(\ell)_\ell$ in Γ under φ_ℓ. Then Γ_ℓ is a subgroup of Γ of index ℓ. We consider the Shimura curve X_ℓ with

$$X_\ell(\mathbb{C}) = \mathfrak{H}/\Gamma_\ell .$$

It can be defined over F and its reduction $X_{\ell,p}$ modulo p is good. Moreover it is not hard to show that "supersingular" $\mathbb{F}_p$-points of $X_{\ell,p}$ split completely in the natural projection

$$\pi_\ell : X_{\ell,p} \longrightarrow X_{\Gamma,p} .$$

Thus, their number is at least $\ell\cdot(\sqrt{q}-1)\cdot(g+1)$. Since ℓ is greater than the maximum order of a fixed point of Γ on $\mathfrak{H}$, the projection π_ℓ is unramified and thus by the Hurwitz formula

$$g_\ell = 1 + \ell(g - 1)$$

where g_ℓ is the genus of X_ℓ . This shows that

$$\lim_{l\to\infty} |X_{\ell,p}(\mathbb{F}_q)|/g_\ell = \sqrt{q} - 1 .$$

Moreover, since ℓ runs over primes in an arithmetical progression (F being abelian over $\mathbb{Q}$) the ratio of two consecutive genera in the sequence $\{X_{\ell,p}\}$ tends to 1 and we are done. ■

Corollary 3.10. *For any* $q = p^m > 2$

$$M_q \le 6\left(1 + \frac{1}{q-2}\right) .$$

For $q = 2$

$$M_2 \le 27 .$$

Proof: Both inequalities follow from Corollary 1.3 and Theorem 3.9. For $q > 2$ we put $k = 2$, for $q = 2$ we have to put $k = 4$. Then we use $\mu_q(2) = 3$ and $\mu_2(4) = 9$ (cf. [1]). ■

4. Remarks

One can ask what is new in this paper in comparison with [1]? The principal idea is contained in [1]. There are however some points to be mentioned.

1. We divide the idea of David and Gregory Chudnovsky into two parts, that of linear algebra and that of algebraic geometry. The first one leads us to the notion of supercode which may be useful even without algebraic geometry. All statements of our paper concerning supercodes are new.

2. They do not distinguish between m_q and M_q . Their statements concern M_q but the proofs are good only for m_q . All statements in our paper

concerning M_q are therefore also new.

3. For $q = p^{2m}$ our estimate for m_q is better. Their estimate is obtained by a slightly different method and is

$$m_q \leq 2\left(1 + \frac{1}{\sqrt{q} - 3}\right) .$$

It is valid for $q = p^{2m} \geq 25$ (and the case $q = 25$ is not completely proved in [1]).

4. For q being an odd power of a prime they consider only the case $q = 2$. Their estimate for $q = 2$ is

$$m_2 \leq 6$$

which is slightly worse than ours. Our estimates for other odd powers (and also for $q = 4, 9, 16$) are new.

5. They claim that one can construct a polynomial complexity bilinear multiplication algorithm realizing the upper bound for m_q . This last assertion seems to be unproved. Let us explain this point.

The argument of David and Gregory Chudnovsky is that the curves we need can be constructed effectively (i.e. in polynomial time). This is true (see [3] 4.3) but unfortunately unsufficient in two respects:

First, the arguments used both in [1] and in this paper involve some "random choice" over an exponentially large set, since they prove only existence of a divisor with some prescribed properties and the set of divisors under consideration is exponentially large.

At the expense of constants being worse one can overcome this difficulty (provided one can explicitly construct points Q of degree we need). This can be done as follows. Let $q = p^{2m} \geq 49$ and let $\{X_i\}$ be a family of absolutely irreducible smooth curves of genus $g = g_i$ with

$$\lim_{g \longrightarrow \infty} |X(\mathbb{F}_q)|/g = \sqrt{q} - 1 ,$$

such that one can polynomially work with points and linear systems on these curves. It is known that such families do exist (see [3] 4.3). For example one can choose $X_i = X_0(11\ell_i)$ to be the reduction of the classical modular curve, ℓ_i being the i-th prime (for $q = p^2$), or $X_i = X_0(p_i)$ where p_i is an irreducible polynomial over $\mathbb{F}_q$ of odd degree coprime with $(q - 1)$ (for $q = p^{2m}$), here $X_0(p_i)$ is the reduction of the Drinfeld modular curve. One can prove

Proposition 4.1. *Suppose that for a family of modular curves described above for any* $X \in \{X_i\}$ *there is given an explicit point* Q *of* X *of some degree* k *such that*

$$g\cdot(\sqrt{q} - 5)/2 - o(g) \le k \le g\cdot(\sqrt{q} - 5)/2 .$$

(Let Q *be defined, say, by its coordinates in some projective embedding). Then one can polynomially construct a sequence* $\mathfrak{A} = \mathfrak{A}_i$ *of bilinear multiplication algorithms in finite fields* $\mathbb{F}_{q^k}$ *for the given sequence of* $k \longrightarrow \infty$ *such that*

$$\lim_{g\longrightarrow\infty} \mu(\mathfrak{A})/k = 2\left(1 + \frac{4}{\sqrt{q} - 5}\right) .$$

Proof: Since Q is given and we can polynomially compute on X and work with linear systems on them including computation of evaluation maps, it is sufficient to show that one can find appropriate divisors D and sets $\mathcal{P}$ of $\mathbb{F}_q$-points of X. In fact, one can take $\mathcal{P}$ to be the set of "supersingular" points of X (see [3] 4.2.2) and $D = a\cdot R$, where R is any $\mathbb{F}_q$-point of X. Put $a = \lceil (N - 1)/2 \rceil$, where $N = |\mathcal{P}|$. For modular curves it is known that $N \ge g\cdot(\sqrt{q} - 1)$. Let us check that for the triple $(Q, D, \mathfrak{P})$ Conditions A and B of Section 2 hold. Indeed, we have

$$\deg\,(D - Q) = a - k \ge$$

$$\ge \big(g\cdot(\sqrt{q} - 1) - 2\big)/2 - g\cdot(\sqrt{q} - 5)/2 \ge 2g - 1$$

and

$$\deg\,(2D - \mathrm{P}) = 2\cdot\lceil (N - 1)/2\rceil - N < 0$$

where

$$\mathrm{P} = \sum_{P\in\mathcal{P}} P .$$

Therefore, by the Riemann-Roch theorem,

$$\ell(D - Q) = \deg(D - Q) - g =$$

$$= \deg D - g + 1 - k = \ell(D) - k$$

and $\ell(2D - \mathrm{P}) = 0$. Thus Conditions α and β of Section 2 (and therefore A and B) are satisfied. By Proposition 2.1 we see that the multiplication algorithm $\mathfrak{A}_i$ corresponding to the triple $(Q, D, \mathcal{P})$ has multiplicative complexity

$$\mu(\mathfrak{A}) \le N = g\cdot(\sqrt{q} - 1) + o(g) =$$

$$= 2k\cdot(\sqrt{q} - 1)/(\sqrt{q} - 5) + o(g) .$$

and involves only computation of bases of $L(Q)$ and $L(2D)$ and computation of evaluation maps which can be done polynomially. ∎

This proposition means that to get a polynomially constructable algorithm with linear complexity, one needs to construct explicitly (i.e. polynomially) points of corresponding degrees on modular curves (or on other curves with many points).

Unfortunately, it is completely unclear how to produce such points and this is the second difficulty which as yet we are unable to overcome. Moreover we do not know a single polynomial construction of bilinear multiplication algorithm with linear multiplicative complexity.

References

1. **Chudnovsky D.V., Chudnovsky G.V.** *Algebraic complexity and algebraic curves over finite fields*. J. Complexity, 4, 285-316 (1988)
2. **Ihara Y.** *Some remarks on the number of rational points of algebraic curves over finite fields*. J.Fac.Sci.Tokyo, IA, 28, 721-724 (1981)
3. **Tsfasman M.A., Vlăduţ S.G.** *Algebraic-geometric codes*. Kluwer Acad. Publ., Dordrecht/Boston/London, 1991
4. **Tsfasman M.A.** *Some remarks on the asymptotic number of points*. This volume

The Domain of Covering Codes

Philip Stokes. C.N.R.S. - U.R.A.1376, Laboratoire I.3.S.,
250 rue Albert Einstein, 06560 Valbonne, France.

Abstract

In this paper the relationship between the normalised covering radius and the rate is considered for both linear and unrestricted codes. We characterise explicitly, for both cases, the region in the unit square where this type of behaviour is possible and show that certain types of asymptotic properties are wholly dependent upon it.

1 Introduction

The *covering radius*, $r(C)$, of a code C of length n over an arbitrary alphabet F is defined by

$$r(C) = \max_{v \in F^n} \min_{c \in C} d(v, c)$$

where $d(v, c)$ is the Hamming distance between the vectors v and c.

There is an extensive literature devoted to the study of this parameter and the reader who is unacquainted with it is referred to [1].

For the time being, we shall consider exclusively the class of linear codes and shall use $[n, k]r$ to denote the parameters of a linear code of length n, dimension k and covering radius r. The class of unrestricted codes will be addressed in Section 4.

Let C be an $[n, k]r$ code. The *rate* of C, $R(C)$, is defined by $R(C) = k/n$ and its *covering fraction* by $\rho = r/n$.

In this paper we shall determine:

1. For which points of the (ρ, R)-plane it is possible to find a sequence of q-ary linear codes $(C_i)_{i=1}^{\infty}$ such that $\frac{r(C_i)}{n(C_i)} \longrightarrow \rho$, $\frac{k(C_i)}{n(C_i)} \longrightarrow R$ and $n(C_i) \longrightarrow \infty$ as $i \longrightarrow \infty$.

2. Properties of the set of points in the (ρ, R)-plane which are realised by q-ary linear codes.

2 Preliminaries

We state here some elementary results on coding theory which will be needed later. Firstly, we define the *q-ary entropy function* $H_q : [0, \frac{q-1}{q}] \longrightarrow [0, 1]$ by

$$H_q(x) = \begin{cases} 0 & \text{if } x = 0, \\ x \log_q(q-1) - x \log_q x - (1-x)\log_q(1-x) & \text{if } 0 < x \leq \frac{q-1}{q}. \end{cases}$$

[3, p. 55].

The proofs of the following are omitted as they are well-known.

Lemma 2.1 [7, p. 2615]
Let $0 \leq r \leq n(\frac{q-1}{q})$. *Then*

$$\frac{1}{n+1} q^{nH_q(\frac{r}{n})} < \sum_{j=0}^{r} \binom{n}{j} (q-1)^j \leq q^{nH_q(\frac{r}{n})}.$$

□

Lemma 2.2 [2, p. 331] [The Redundancy Bound.]
Let C *be an* $[n,k]r$ *code. Then* $r \leq n-k$. □

Lemma 2.3 [2, p. 329] [The Sphere-Covering Bound.]
Let C *be an* $[n,k]r$ *code. Then*

$$\sum_{j=0}^{r} \binom{n}{j} (q-1)^j \geq q^{n-k}.$$

□

We now turn our attention to a simple method of combining codes which will be used to prove our main results.

Let C_1, C_2 be q-ary linear codes. The *direct sum* of C_1 and C_2, $C_1 \bigoplus C_2$, is defined by $C_1 \bigoplus C_2 = \{(c_1 \mid c_2) : c_1 \in C_1, c_2 \in C_2\}$ [4, p. 76].

Proposition 2.4
Let C_1 *be an* $[n_1, k_1]r_1$ *code,* C_2 *be an* $[n_2, k_2]r_2$ *code.*
Then $C_1 \bigoplus C_2$ *is an* $[n_1 + n_2, k_1 + k_2]r_1 + r_2$ *code.*

Proof
See [4, p. 76] and [2, p. 329]. □

Corollary 2.5
Let C *be an* $[n,k]r$ *code. Then for any* $a \in \mathbf{N}$ *there exists an* $[an, ak]ar$ *code.*

Proof
The code

$$C' = \bigoplus_{i=1}^{a} C$$

is such a code, by Proposition 2.4. □

Proposition 2.6
There exist codes C_1 *and* C_2 *such that* C_1 *has parameters* $[1,0]1$ *and* C_2 *has parameters* $[1,1]0$.

Proof
Choose C_1 , C_2 to be the subspaces $\langle 0 \rangle$, $\langle 1 \rangle$ of $GF(q)$ respectively. □

Finally, we shall need the following result.

Theorem 2.7
Let $t[n,k]$ *denote the minimum possible covering radius of an* $[n,k]$ *linear code. Then, for fixed* $\frac{k}{n} = R$,

$$\lim_{n \to \infty} \frac{t[n, Rn]}{n} = H_q^{-1}(1-R).$$

Proof
The proof is given in [1, p. 127]. □

3 Linear Codes

We are now in a position to start to tackle the problems raised in the introduction. Let (X, d) be a metric space and let $A \subseteq X$. An element $x \in X$ is called a *limit point* of A if each open sphere centred on x contains at least one point of A different from x [6, p. 65]. The *closure* of A, denoted by $\overline{A}$, is the union of A and the set of all its limit points [6, p. 68]. A is said to be *dense* in X if $\overline{A} = X$ [6, p. 70]. Sets K_q, L_q are defined as follows:

$$\begin{aligned} K_q &= \{(\rho, R) : \rho(C) = \rho, R(C) = R \text{ for some q - ary linear code C}\}. \\ L_q &= \{(\rho', R') : (\rho', R') \text{ is a limit point of } K_q\}. \end{aligned}$$

When we consider a point $(\frac{r}{n}, \frac{k}{n}) \in K_q$ arising from some q-ary linear code C, we shall assume, without loss of generality, that no cancellation of common factors of $n(C)$, $k(C)$ and $r(C)$ has taken place. In this way it will be permissible to consider C to be an $[n, k]r$ code. Before proceeding further we define a function $J_q : [0,1] \longrightarrow [0,1]$ by

$$J_q(x) = \begin{cases} 1 - H_q(x) & \text{if } 0 \leq x \leq \frac{q-1}{q}; \\ 0 & \text{if } \frac{q-1}{q} < x \leq 1. \end{cases}$$

which will be of use shortly.

There is a close inter-relationship between the set L_q and certain sequences of q-ary linear codes, as the following results indicate.

Lemma 3.1
Let $(\rho, R) \in L_q$. Then there exists a sequence of q-ary linear codes $(C_i)_{i=1}^{\infty}$ such that $\frac{r(C_i)}{n(C_i)} \longrightarrow \rho$, $\frac{k(C_i)}{n(C_i)} \longrightarrow R$ and $n(C_i) \longrightarrow \infty$ as $i \longrightarrow \infty$.

Proof
As $(\rho, R) \in L_q$ it is possible , by the definition of L_q, to find a sequence of points $(\rho_i, R_i) \in K_q$ with the property that $(\rho_i, R_i) \longrightarrow (\rho, R)$ as $i \longrightarrow \infty$ and such that $(\rho_i, R_i) \neq (\rho, R)$ for all i. Now, for each i, there is a q-ary linear code, B_i say, which gives rise to the point (ρ_i, R_i), by the definition of K_q. Hence we have established the existence of a sequence of q-ary linear codes $(B_i)_{i=1}^{\infty}$ such that $\frac{r(B_i)}{n(B_i)} \longrightarrow \rho$ and $\frac{k(B_i)}{n(B_i)} \longrightarrow R$ as $i \longrightarrow \infty$. It remains to ensure that a sequence of codes $(C_i)_{i=1}^{\infty}$ can be found which has all the properties of the sequence $(B_i)_{i=1}^{\infty}$ but that has the additional property that $n(C_i) \longrightarrow \infty$ as $i \longrightarrow \infty$. To construct such a sequence we alter the sequence $(B_i)_{i=1}^{\infty}$ in the following way. Let $C_1 = B_1$ and for $i \geq 1$ let

$$C_{i+1} = \bigoplus_{j=1}^{a} B_{i+1}$$

for any $a \in \mathbf{N}$ such that $a.n(B_{i+1}) > n(C_i)$. Hence $n(C_{i+1}) = a.n(B_{i+1}) > n(C_i)$ for all $i \geq 1$ by the above construction. Thus $n(C_i) \longrightarrow \infty$ as $i \longrightarrow \infty$. The result now follows.

□

Lemma 3.2
Let $(C_i)_{i=1}^{\infty}$ be a sequence of q-ary linear codes with the property that $\frac{r(C_i)}{n(C_i)} \longrightarrow \rho$, $\frac{k(C_i)}{n(C_i)} \longrightarrow R$ and $n(C_i) \longrightarrow \infty$ as $i \longrightarrow \infty$. Then

. $R \geq J_q(\rho)$ if $0 \leq \rho \leq 1$.
2. $R \leq 1-\rho$ if $0 \leq \rho \leq 1$.

Proof

. When $0 \leq \rho \leq \frac{q-1}{q}$, by estimating the sum in Lemma 2.3 by Lemma 2.1, we have for ach code C_i in the sequence

$$q^{n-k} \leq \sum_{j=0}^{r} \binom{n}{j} (q-1)^j \leq q^{nH_q(\frac{r}{n})}$$

nd hence $1 - \frac{k}{n} \leq H_q\left(\frac{r}{n}\right)$. By taking limits as $i \longrightarrow \infty$ the result follows.

For $\frac{q-1}{q} < \rho \leq 1$ the result is trivial.

2. From Lemma 2.2 we have $\frac{r}{n} \leq 1 - \frac{k}{n}$ and the result now follows upon letting $i \longrightarrow \infty$. □

Corollary 3.3

$L_q \subseteq \{(\rho, R) \in [0,1]^2 : J_q(\rho) \leq R \leq 1-\rho\}$.

Proof

mmediate from Lemma 3.1 and Lemma 3.2. □

The next result is fundamental to the approach we employ.

Lemma 3.4

The line segment joining any two points of K_q lies in L_q.

Proof

et $\left(\frac{r_1}{n_1}, \frac{k_1}{n_1}\right), \left(\frac{r_2}{n_2}, \frac{k_2}{n_2}\right) \in K_q$. We may assume, without loss of generality, that $n_1 = n_2 = n$ ay, because if $n_1 \neq n_2$ we know from Corollary 2.5 that $\left(\frac{r_1 n_2}{n_1 n_2}, \frac{k_1 n_2}{n_1 n_2}\right)$ and $\left(\frac{r_2 n_1}{n_1 n_2}, \frac{k_2 n_1}{n_1 n_2}\right) \in K_q$. Hence we are able to assume that $[n, k_1]r_1$ and $[n, k_2]r_2$ codes exist. It again follows rom Corollary 2.5 that for any $u, v \in \mathbf{N}$ we can construct $[nu, k_1 u]r_1 u$ and $[nv, k_2 v]r_2 v$ odes. Hence by Proposition 2.4 an $[n(u+v), k_1 u + k_2 v]r_1 u + r_2 v$ code exists and so $\left(\frac{r_1 u + r_2 v}{n(u+v)}, \frac{k_1 u + k_2 v}{n(u+v)}\right) \in K_q$. Now set $\lambda = \frac{u}{u+v}$. Hence we have shown that for any points $A, B \in K_q$ and any $\lambda \in \mathbf{Q} \bigcap (0,1)$, $\lambda A + (1-\lambda)B \in K_q$. As $\mathbf{Q}$ is dense in $\mathbf{R}$, these points re dense on the line AB and hence this line lies in L_q. □

There are several consequences of this result.

Corollary 3.5

$K_q \subseteq L_q$. □

Corollary 3.6

$\overline{K_q} = L_q$. □

So, as may have been expected, K_q is dense in L_q.

Corollary 3.7

L_q is convex.

Proof

mmediate from Lemma 3.4 and the definition of L_q. □

Lemma 3.8
1. $\{(\rho, J_q(\rho)) : 0 \leq \rho \leq 1\} \subseteq L_q$.
2. $\{(\rho, 1-\rho) : 0 \leq \rho \leq 1\} \subseteq L_q$.

Proof
1.(a) When $0 \leq \rho \leq \frac{q-1}{q}$.
This follows from Theorem 2.7
(b) When $\frac{q-1}{q} < \rho \leq 1$.
From (a) above we know that $(\frac{q-1}{q}, 0) \in L_q$ and from Proposition 2.6 and Corollary 3.5 we know that $(1,0) \in L_q$. The result now follows by Corollary 3.7.
2. As $(1,0)$ and $(0,1) \in L_q$, applying Corollary 3.7 gives the result. □

Lemma 3.9
$\{(\rho, R) \in [0,1]^2 : J_q(\rho) \leq R \leq 1-\rho\} \subseteq L_q$.

Proof
Let $(a,b) \in \{(\rho, R) \in [0,1]^2 : J_q(\rho) \leq R \leq 1-\rho\}$. Trivially, the line joining $(0,1)$ to (a,b) intersects the curve $R = J_q(\rho)$ at some point $(\rho_0, J_q(\rho_0))$ for some ρ_0 such that $a \leq \rho_0 \leq 1$. Now, $(0,1)$ and $(\rho_0, J_q(\rho_0)) \in L_q$ by Lemma 3.8. Hence, by Corollary 3.7, $(a,b) \in L_q$. □

We are now able to precisely determine the elements of L_q.

Theorem 3.10
$L_q = \{(\rho, R) \in [0,1]^2 : J_q(\rho) \leq R \leq 1-\rho\}$.

Proof
Immediate from Corollary 3.3 and Lemma 3.9. □

This result enables us to answer completely the first question raised in the introduction. For given any point $(\rho, R) \in L_q$ we know from Lemma 3.1 that there exists a sequence of q - ary linear codes $(C_i)_{i=1}^{\infty}$ such that $\frac{r(C_i)}{n(C_i)} \longrightarrow \rho$, $\frac{k(C_i)}{n(C_i)} \longrightarrow R$ and $n(C_i) \longrightarrow \infty$ as $i \longrightarrow \infty$. Conversely, given a sequence of q-ary linear codes $(C_i)_{i=1}^{\infty}$ such that $\frac{r(C_i)}{n(C_i)} \longrightarrow \rho$, $\frac{k(C_i)}{n(C_i)} \longrightarrow R$ and $n(C_i) \longrightarrow \infty$ as $i \longrightarrow \infty$, we know from Lemma 3.2 and Theorem 3.10 that $(\rho, R) \in L_q$. Hence a sequence of codes with the specified properties can be found if and only if $(\rho, R) \in L_q$.

4 Unrestricted Codes

The results that have been presented here can be extended to cover the class of unrestricted codes, as will now be shown. For the remainder of this section F will denote an alphabet of size q. Let C be an (n, M) code over F, i.e. a block code of length n and with M codewords. The *generalised dimension* of C, $k(C)$, is defined to be $\log_q M$. The notation $(n,k)r$ will be used to denote a code of this type which has length n, generalised dimension k and covering radius r.

In a manner analogous to the method for linear codes, we define the sets M_q and N_q as follows :

$$\begin{aligned} M_q &= \{(\rho, R) : \rho(C) = \rho, R(C) = R \text{ for some code C over F}\}. \\ N_q &= \{(\rho', R') : (\rho', R') \text{ is a limit point of } M_q\}. \end{aligned}$$

t can be easily verified that, with the exception of Lemma 2.2, and the results consequen-ial upon it, all the proofs exhibited so far hold not only for the class of linear codes but lso for the much larger class of unrestricted codes. This should not be at all surprising s the direct sum construction, the main weapon used in the attack on the problems resented, applies equally well to both classes of codes. Hence it remains to find a tight pper bound for the domain of unrestricted codes. In order to do this we construct a ode which has a maximal number of codewords for a given length and covering radius.

Let r and n be integers such that $0 \le r \le n$ and define a code H_r^n by $H_r^n = F^n \backslash B_{r-1}(w)$ here $B_t(y)$ denotes the sphere with centre y and radius t with respect to the Hamming etric and w is some arbitrary fixed vector in F^n. Trivially, this code has length n, overing radius r and contains

$$q^n - \sum_{i=0}^{r-1} \binom{n}{i}(q-1)^i = \sum_{i=r}^{n} \binom{n}{i}(q-1)^i$$

odewords.

roposition 4.1

et C be a code over F with length n and covering radius r. Then $|C| \le |H_r^n|$.

roof

s H_r^n has covering radius r there exists an $a \in F^n$ and a $c \in H_r^n$ such that c is the earest codeword neighbour of a and such that $d(a,c) = r$. Hence $B_{r-1}(a) \subseteq F^n \backslash C$. Thus $\vert C| \le |F^n| - |B_{r-1}(a)| = |H_r^n|$. □

n order to facilitate the presentation of future results we define a function $_q : [0,1] \longrightarrow [0,1]$ by

$$U_q(x) = \begin{cases} 1 & \text{if } 0 \le x < \frac{q-1}{q}; \\ \log_q(q-1) + \{1 - \log_q(q-1)\} H_{\frac{q}{q-1}}(1-x) & \text{if } \frac{q-1}{q} \le x \le 1. \end{cases}$$

emma 4.2

$(\rho, U_q(\rho)) : \frac{q-1}{q} \le \rho \le 1\} \subseteq N_q$.

roof

et $(r_i)_{i=1}^{\infty}, (n_i)_{i=1}^{\infty}$ be sequences of integers such that $0 \le r_i \le n_i$ for all i and such that $_i \longrightarrow \infty$, $\frac{r_i}{n_i} \longrightarrow \rho$ as $i \longrightarrow \infty$. Now consider the sequence of codes $\left(H_{r_i}^{n_i}\right)_{i=1}^{\infty}$.

e have

$$\begin{aligned} |H_{r_i}^{n_i}| &= \sum_{a=r_i}^{n_i} \binom{n_i}{a}(q-1)^a \\ &= \sum_{b=0}^{n_i-r_i} \binom{n_i}{b}(q-1)^{n_i-b} \\ &= (q-1)^{n_i} \sum_{b=0}^{n_i-r_i} \binom{n_i}{b}(\gamma-1)^b, \text{ where } \gamma - 1 = (q-1)^{-1}. \end{aligned}$$

ence, by Lemma 2.1

$$\frac{1}{n_i+1}(q-1)^{n_i}\gamma^{n_i H_\gamma\left(\frac{n_i-r_i}{n_i}\right)} < |H_{r_i}^{n_i}| \le (q-1)^{n_i}\gamma^{n_i H_\gamma\left(\frac{n_i-r_i}{n_i}\right)}$$

when $\frac{q-1}{q} \le \frac{r_i}{n_i} \le 1$.
It follows from this that in the same interval

$$\frac{-\log_q(n_i+1)}{n_i} + \log_q(q-1) + H_\gamma(1-\frac{r_i}{n_i}).\log_q\gamma < \frac{\log_q|H^{n_i}_{r_i}|}{n_i}$$

and

$$\frac{\log_q|H^{n_i}_{r_i}|}{n_i} \le \log_q(q-1) + H_\gamma(1-\frac{r_i}{n_i}).\log_q\gamma.$$

Hence

$$\lim_{i\to\infty}\frac{\log_q|H^{n_i}_{r_i}|}{n_i} = \log_q(q-1) + \{1-\log_q(q-1)\}H_{\frac{q}{q-1}}(1-\rho)$$

when $\frac{q-1}{q} \le \rho \le 1$.
The result now follows. □

Corollary 4.3
Let $(\rho, R) \in N_q$. *Then* $R \le U_q(\rho)$.
Proof
1. When $0 \le \rho \le \frac{q-1}{q}$.
In this case the result is trivial.
2. When $\frac{q-1}{q} \le \rho \le 1$.
The result follows by Proposition 4.1 and Lemma 4.2. □

We are now in a position to obtain an analogue of Theorem 3.10.

Theorem 4.4
$N_q = \{(\rho, R) \in [0,1]^2 : J_q(\rho) \le R \le U_q(\rho)\}$.
Proof
Immediate from the unrestricted analogues of Corollary 3.7, Lemma 3.8(i) and from Lemma 4.2 and Corollary 4.3. □

5 Conclusion

In this article we have shown that the domains of linear and unrestricted covering codes are as depicted in Figure 1. It is interesting to observe from this that it is possible to find non-linear codes which are a lot worse than any of their linear counterparts, in the sense that they have large covering fractions for a given value of the rate.

The analogous problem to the above, in which minimum distance replaces covering radius, has been studied by Vladut and Manin [5, 7]. Whilst the problems are ostensibly similar, there turn out to be significant differences between the two situations. Indeed, in [7] it was shown that it is possible to find points in the (δ, R)-plane which are realised by codes but which are not limits of sequences of such points. This cannot happen in the (ρ, R)-plane, as Corollary 3.5 and its unrestricted analogue show.

A curious occurrence is that while, in general, it is much easier to find the minimum distance of a code than its covering radius, the domain problem for minimum distance is much harder than the corresponding one we have studied. The method we have employed hinges crucially on Theorem 2.7 and the difference is largely explained by this. The exact nature of the packing domain is still not completely determined, although it is known that the region in question is bounded by a continuous decreasing function of δ [7, p.2614].

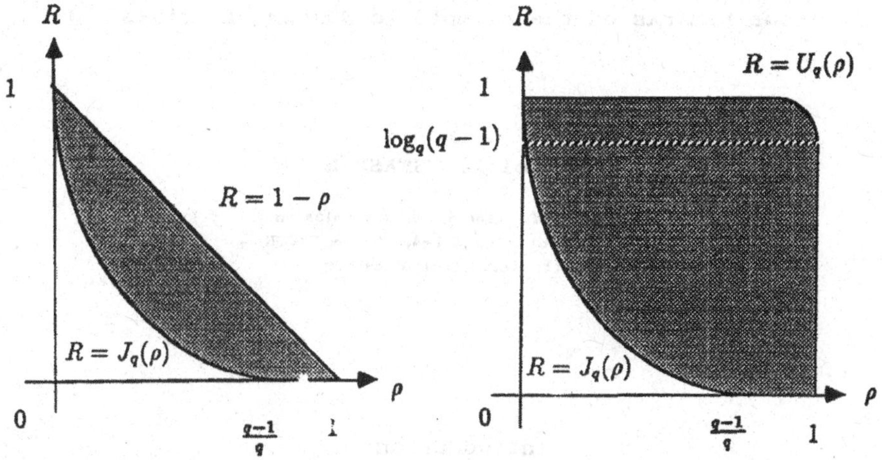

Figure 1: The domain of linear and unrestricted covering codes (left and right respectively).

Acknowledgements

The author wishes to thank Professor Peter Cameron and [illegible] Patrick Solé for their guidance and The Science and Engineering Research Council for financial assistance.

References

[1] G.Cohen and P.Frankl; Good coverings of Hamming spaces with spheres; Discrete Math. **56** (1985), 125–131.

[2] G.D.Cohen, M.R.Karpovsky, H.F.Mattson, Jr., and J.R.Schatz, Covering radius: Survey and recent results; IEEE Trans. Inform. Theory **31** (1985), 328–343.

[3] J.H.van Lint; Introduction to coding theory, New York: Springer-Verlag, 1982.

[4] F.J.MacWilliams and N.J.A.Sloane; The theory of error-correcting codes, Amsterdam: North-Holland, 1977.

[5] Yu.I.Manin, What is the maximum number of points on a curve over F_2?, J. Fac. Sci. Univ. Tokyo **28** (1981), 715–720.

[6] G.F.Simmons; Introduction to topology and modern analysis, New York: McGraw-Hill, 1963.

[7] S.G.Vladut and Yu.I.Manin; Linear codes and modular curves, Itogi Nauki i Tekhniki, Seriya Sovremennye Problemy Matematiki, Noveishie Dostizheniya **25** (1984), 209–257.

Some Remarks on the Asymptotic Number of Points

Michael A. TSFASMAN

Institute of Information Transmission
19, Ermolovoi st., GSP-4, Moscow, USSR
e-mail: tsfasman@ippi.msk.su

Introduction

The geometry of varieties defined over a finite field is somehow reflected in their numerical properties. The numerical invariants of a curve over $\mathbb{F}_q$ - such as the number of points over a given ground field extension, the number of points on its Jacobian, the number of positive divisors of a given degree, and so on - can be expressed in terms of the Galois action on cohomologies, i.e. in terms of the Frobenius roots. The behaviour of these roots is rather restricted when the genus of the curve is much larger than the cardinality of the field. When, for a given finite ground field, the genus tends to infinity, there happen to exist nice asymptotic formulae for different numerical invariants.

In this paper we generalize some results on the number of points on the curve (due to Drinfeld-Vladuţ [Dr/Vl] and Serre [Se]), and on its Jacobian (due to Vladuţ [Vl], Rosenbloom and the author [Ro/Ts]), give a formula for the asymptotic number of divisors, and some estimates for the number of points in the "Poincare filtration".

We introduce the notion of an asymptotically exact family of curves and try to use it whenever possible.

I would like to thank S.Vladuţ and the referee for many valuable remarks.

Zeta-function

It is well known that the number N_r of points over $\mathbb{F}_{q^r}$ on a curve X over $\mathbb{F}_q$ is given by the formula

$$N_r = |X(\mathbb{F}_{q^r})| = q^r + 1 - \sum_{i=1}^{2g} \omega_i^r ,$$

and the number of points on its Jacobian is given by the formula

$$h = |J_X(\mathbb{F}_q)| = \prod_{i=1}^{2g} (1 - \omega_i) ,$$

ω_i being the Frobenius roots, $\omega_{g+i} = \overline{\omega}_i$, $|\omega_i| = \sqrt{q}$.

Let B_m be the number of points of degree m on X. It is clear that $N_r = \sum_{m|r} m \cdot B_m$.

We consider the behaviour of these values in a sequence of curves X over $\mathbb{F}_q$ whose genus $g = g(X)$ tends to infinity.

To study numerical invariants it is extremely fruitful to consider the zeta-function of X .

Zeta function is defined in one of the four equivalent ways

$$Z(t) = exp\left(\sum_{m=1}^{\infty} N_m \cdot \frac{t^m}{m}\right) =$$

$$= \frac{P(t)}{(1 - t)(1 - qt)} =$$

$$= \prod_{m=1}^{\infty} (1 - t^m)^{-B_m} =$$

$$= \sum_{m=0}^{\infty} D_m \cdot t^m ,$$

here $P(t) = \prod_{i=1}^{2g}(1 - \omega_i \cdot t)$, and D_m denotes the number of effective divisors of degree m .

The zeta-function satisfies the functional equation

$$Z(q^{-1}t^{-1}) = (\sqrt{q} \cdot t)^{2-2g} \cdot Z(t) .$$

Divergent series

Let us for a while play some dubious game with the zeta-function of a curve X over $\mathbb{F}_q$.

We have the following two equalities (of formal series in t).

Proposition 1. *We have the equality*

$$\frac{1}{t^{-1}-1}+\frac{1}{q^{-1}t^{-1}-1}-\sum_{i=1}^{2g}\frac{1}{\omega_i^{-1}t^{-1}-1}=\sum_{m=1}^{\infty}m\cdot B_m\cdot\frac{1}{t^{-m}-1}\ .$$

Proof: We use the second and the third definitions of the zeta-function:

$$log\ Z(t) = -\ log\ (1-t) - log\ (1-qt) + \sum_{i=1}^{2g} log\ (1-\omega_i t) =$$

$$= -\sum_{m=1}^{\infty} B_m\cdot log\ (1-t^m)\ .$$

Taking derivatives and multiplying by t we obtain the proposition. ■

Proposition 2. *We have the equality*

$$P(t) = t^{2g-2}\cdot(1-t)\cdot(1-qt)\cdot q^{g-1}\cdot\prod_{m=1}^{\infty}(1-q^{-m}t^{-m})^{-B_m}\ .$$

Proof: This time we again use the second and the third definitions of the zeta-function, but we also make use of the functional equation:

$$P(t) = (1-t)\cdot(1-qt)\cdot Z(t) =$$

$$= (1-t)\cdot(1-qt)\cdot Z(q^{-1}t^{-1})\cdot(\sqrt{q}\cdot t)^{2g-2} =$$

$$= (1-t)\cdot(1-qt)\cdot q^{g-1}\cdot t^{2g-2}\cdot\prod_{m=1}^{\infty}(1-q^{-m}t^{-m})^{-B_m}\ .$$ ■

Now let us put $t = q^{-1/2}$ in the formula of Proposition 1. We get

$$\sum_{m=1}^{\infty} m\cdot B_m\cdot\frac{1}{q^{m/2}-1} =$$

$$= \left(\frac{1}{q^{1/2}-1}+\frac{1}{q^{-1/2}-1}\right)-\sum_{i=1}^{g}\left(\frac{1}{q^{-1/2}\omega_i-1}+\frac{1}{q^{1/2}\omega_i^{-1}-1}\right) =$$

$$= g-1\ .$$

Putting $t = 1$ in the formula of Proposition 2 we get

$$h = |J_X(\mathbb{F}_q)| = \prod_{i=1}^{2g} (1 - \omega_i) = P(1) =$$

$$= (1 - 1)\cdot(1 - q)\cdot q^{g-1}\cdot \prod_{m=1}^{\infty} (1 - q^{-m})^{-B_m} .$$

Both formulae look very suspicious. The first one, besides the fact that it can happen that $q^{-1/2}\omega_i = 1$, has the disadvantage of the series on the left being divergent. The second one needs some way to multiply $(1 - 1)$ by the divergent product on the right hand side.

The first goal of this paper is to show that nevertheless these formulae are not so far from being reasonable.

Results

Let us start with some formulae for the number of points on the curve (in terms of B_m).

Theorem 1. *Let* $g \longrightarrow \infty$ *and let* $f : \mathbb{N} \longrightarrow \mathbb{N}$ *be any function such that*

$$\lim \frac{f(g)}{\log g} = 0 .$$

Then

$$\limsup \frac{1}{g}\cdot \sum_{m=1}^{f(g)} \frac{m\cdot B_m}{q^{m/2} - 1} \le 1 . \blacksquare$$

All the proofs will be given in the next section.

Remark. For $f(g) = 1$ one gets the Drinfeld-Vladuţ inequality [Dr/Vl]

$$\limsup \frac{N_1}{g} \le \sqrt{q} - 1 .$$

For $f(g)$ equal to a constant the theorem was stated without proof in [Se].

The condition on $f(g)$ in Theorem 1 is really important. We have

Theorem 2. *Let* $g \longrightarrow \infty$ *and let* $f : \mathbb{N} \longrightarrow \mathbb{N}$ *be any function such that*

$$\lim \frac{f(g)}{\log g} = \infty .$$

Then

$$\lim \frac{1}{g} \cdot \sum_{m=1}^{f(g)} \frac{m \cdot B_m}{q^{m/2} - 1} = \infty . \blacksquare$$

A simple diagonal argument proves that each sequence of curves of growing genus has a subsequence for which for all m there exist limits

$$\beta_m = \lim \frac{B_m}{g} .$$

We shall call such sequences of curves *asymptotically exact families.*

For such families we obtain

Corollary 1. *For an asymptotically exact family of curves*

$$\sum_{m=1}^{\infty} \frac{m \cdot \beta_m}{q^{m/2} - 1} \le 1 . \blacksquare$$

The "defect" $\left(1 - \sum_{m=1}^{\infty} \frac{m \cdot \beta_m}{q^{m/2} - 1}\right)$ is in fact "the limit number of Frobenius roots equal to $\sqrt{q}$ ". We hope to return to this remark in another paper.

Now let us consider the number of points on the Jacobian. Here we shall get some equalities.

Theorem 3. *Let* $g \longrightarrow \infty$ *and let* $f : \mathbb{N} \longrightarrow \mathbb{N}$ *be any function such that*

$$\lim f(g) = \infty$$

and

$$\lim \frac{\log f(g)}{g} = 0 .$$

Then

$$\lim \left(\frac{1}{g} \cdot \log h - \frac{1}{g} \cdot \sum_{m=1}^{f(g)} B_m \cdot \log \frac{q^m}{q^m - 1}\right) = \log q . \blacksquare$$

For a family of curves let us consider the quantities

$$H_{inf} := \lim\inf \left(\frac{1}{g}\cdot\log h\right) ,$$

$$H_{sup} := \lim\sup \left(\frac{1}{g}\cdot\log h\right) .$$

If they coincide we just write

$$H := \lim \left(\frac{1}{g}\cdot\log h\right) = H_{inf} = H_{sup} .$$

Corollary 2. *For an asymptotically exact family of curves the limit H does exist and*

$$H = \log q + \sum_{m=1}^{\infty} \beta_m\cdot\log \frac{q^m}{q^m - 1} .$$

■

Remark. Even if we know only partial information, we can get something. If $\lim (B_1/g) = \beta_1$ then Theorem 3 gives us Vladuţ's inequality [Vl]

$$H_{inf} \ge \log q + \beta_1\cdot\log \frac{q}{q-1} .$$

Moreover, if $\lim\inf \frac{B_m}{g} = \beta_{m,inf}$ for $m = 1,\dots,k$ then

$$H_{inf} \ge \log q + \sum_{m=1}^{k} \beta_{m,inf}\cdot\log \frac{q^m}{q^m - 1} .$$

It can happen that "almost all points of a curve are of small degree", in this case the same is true for its Jacobian:

Theorem 4. *Let* $g \longrightarrow \infty$. *If*

$$\lim \frac{1}{g}\cdot\sum_{m=1}^{k} \frac{m\cdot B_m}{q^{m/2} - 1} = 1$$

then

$$\lim \left(\frac{1}{g}\cdot\log h - \frac{1}{g}\cdot\sum_{m=1}^{k} B_m\cdot\log \frac{q^m}{q^m - 1}\right) = \log q .$$

■

Here is a simple corollary:

Corollary 3. *If for a sequence of curves for* $m \le k$ *there exist limits*

$$\beta_m = \lim \frac{B_m}{g} ,$$

and moreover

$$\sum_{m=1}^{k} \frac{m\cdot\beta_m}{q^{m/2} - 1} = 1$$

then the limit for H *also does exist and*

$$H = \log q + \sum_{m=1}^{k} \beta_m \cdot \log \frac{q^m}{q^m - 1} . \qquad \blacksquare$$

Remark. For $k = 1$ we get the result from [Ro/Ts]: If

$$\lim\sup \frac{N_1}{g} = \sqrt{q} - 1$$

then

$$\lim \frac{1}{g} \cdot \log h = \log q + (\sqrt{q} - 1) \cdot \log \frac{q}{q-1} .$$

The following result is an easy consequence of Corollary 2.

Theorem 5. *Consider all curves over* $\mathbb{F}_q$. *Then*

$$\log q \le H_{inf} \le H_{sup} \le \log q + (\sqrt{q} - 1) \cdot \log \frac{q}{q-1} . \qquad \blacksquare$$

Remark. From $h = \prod(1 - \omega_i)$ it is obvious that H_{inf} and H_{sup} lie between $\log(q - 2\sqrt{q} + 1)$ and $\log(q + 2\sqrt{q} + 1)$. The Quebbemann's estimate [Qu] gives $\log(q + \sqrt{q})$ as an upper bound. Our lower bound follows also from Theorem 2 of Lachaud and Martin-Deschamp [La/MD]. The estimates of Theorem 5 are tight. Indeed, for an exact family with $\beta_m = 0$ for all m (such are, for example, families of abelian coverings [Fr/Pe/St]) the lower bound is attained, and if q is a square, the upper bound is attained for exact families of modular curves with $\beta_1 = \sqrt{q} - 1$ and $\beta_m = 0$ for $m \ge 2$.

Now we turn to the last definition of the zeta-function. If $m \ge 2g - 1$ then

$$D_m = h \cdot \frac{q^{m+1-g} - 1}{q - 1}$$

since any divisor class L of degree m contains $(q^{l(L)} - 1)/(q - 1)$ positive divisors and $l(L) = \deg L - g + 1$

if $\deg L \geq 2g-1$. The Riemann-Roch theorem also shows that for any m

$$D_m = q^{m+1-g} \cdot D_{2g-2-m} + h \cdot \frac{q^{m+1-g}-1}{q-1} .$$

Therefore it is enough to study $D_0, D_1, \ldots, D_{g-1}$.

In the asymptotic setting we shall look for the values

$$\Delta(\mu) := \lim \frac{1}{g} \cdot \log D_{\mu g} .$$

Here of course by μg we mean the nearest integer and for the limit to exist one needs some hypothesis on the family of curves we consider. When $\Delta(\mu)$ does not exist, we can consider

$$\Delta(\mu)_{inf} := \lim\inf \frac{1}{g} \cdot \log D_{\mu g} ,$$

$$\Delta(\mu)_{sup} := \lim\sup \frac{1}{g} \cdot \log D_{\mu g} .$$

The following result was proved by S.G.Vladuţ and the author (the proof is to appear elsewhere).

Theorem 6. *For an asymptotically exact family of curves the limit in the definition of* $\Delta(\mu)$ *does exist and if we set*

$$\mu_0 := \sum_{m=1}^{\infty} \frac{m \cdot \beta_m}{q^m - 1}$$

then for $\mu \geq \mu_0$

$$\Delta(\mu) = \mu \cdot \log q + \sum_{m=1}^{\infty} \beta_m \cdot \log \frac{q^m}{q^m - 1} = H - (1-\mu) \cdot \log q$$

and for $0 \leq \mu \leq \mu_0$

$$\Delta(\mu) = \mu \cdot \log \Lambda + \sum_{m=1}^{\infty} \beta_m \cdot \log \frac{\Lambda^m}{\Lambda^m - 1} ,$$

where $\Lambda = \Lambda(\mu)$ *is uniquely defined by the condition*

$$\sum_{m=1}^{\infty} \frac{m \cdot \beta_m}{\Lambda^m - 1} = \mu .$$

■

Remark. Let us remark that $\mu_0 \leq 1$. In fact,

$$\mu_0 := \sum_{m=1}^{\infty} \frac{m \cdot \beta_m}{q^m - 1} < \sum_{m=1}^{\infty} \frac{m \cdot \beta_m}{q^{m/2} - 1} \leq 1$$

because of Corollary 1. It is also worthwhile to remark that $\Delta(\mu)$ can be written as a sum of the linear function $\mu \cdot \log q$, which corresponds to the "no points" situation (when all β_m equal zero) and something depending on the numbers of points. The latter quantity grows for $0 \le \mu \le \mu_0$ and becomes constant for larger μ.

The next question concerns the "Poincare filtration" on the Jacobian. Let j_m be the number of classes of positive divisors of degree m. We are now interested in the values

$$\Upsilon(\mu)_{inf} := \lim\inf\left(\frac{1}{g}\cdot\log j_{\mu g}\right),$$

$$\Upsilon(\mu)_{sup} := \lim\sup\left(\frac{1}{g}\cdot\log j_{\mu g}\right)$$

here again we abuse notation writing μg for the nearest integer. When these values coincide, we write

$$\Upsilon(\mu) := \lim\left(\frac{1}{g}\cdot\log j_{\mu g}\right) = \Upsilon(\mu)_{inf} = \Upsilon(\mu)_{sup}.$$

In the most interesting case when $X(\mathbb{F}_q)$ is not empty, let us fix an $\mathbb{F}_q$-point P_0. Then any positive divisor D is mapped to the element $D - (\deg D)\cdot P_0 \in J_X(\mathbb{F}_q)$. By J_m denote the image of the set of positive divisors of degree m in J_X. Then $j_m = |J_m|$. It is well known that $J_g = J_X(\mathbb{F}_q)$. Therefore, $\Upsilon(\mu) = H$ for $\mu \ge 1$. We call

$$J_g = J_X(\mathbb{F}_q) \supseteq J_{g-1} \supseteq J_{g-2} \supseteq \ldots \supseteq J_0 = \{0\}$$

the *Poincare filtration* (the name is due to the fact that J_{g-1} is known under the name of the Poincare divisor).

Theorem 7. *For an asymptotically exact family of curves and* $\mu < 2$

$$\Delta(\mu) - \frac{\mu}{2}\cdot\log q \le \Upsilon(\mu)_{inf} \le \Upsilon(\mu)_{sup} \le \Delta(\mu).$$

Moreover, if $\mu < \beta_1/(q+1)$ *then the limit does exist and*

$$\Upsilon(\mu) = \Delta(\mu). \blacksquare$$

Remark. One can conjecture that in fact for an asymptotically exact family of curves $\Upsilon(\mu)$ does exist and

$$\Upsilon(\mu) = \Delta(\mu)$$

for all values of $\mu \le 1$. As yet I am however unable to prove this.

Proofs

We shall start with two known statements (cf. [Dr/Vl], [Vl], or section 2.3.3 of [Ts/Vl]). The proofs being very short we give them here.

Lemma 1 [Dr/Vl]. *For any integer* $b \geq 1$

$$1 \geq \sum_{j=1}^{b}\left(1 - \frac{j}{b+1}\right)\cdot\frac{N_j}{g}\cdot q^{-j/2} - \frac{1}{g}\cdot\sum_{j=1}^{b}\left(1 - \frac{j}{b+1}\right)\cdot\left(q^{j/2} + q^{-j/2}\right).$$

Proof: Let $\alpha_i = \omega_i/\sqrt{q}$. Since for any α_i

$$0 \leq |\alpha_i^b + \alpha_i^{b-1} + \ldots + 1|^2 = (b+1) + \sum_{j=1}^{b}(b+1-j)\cdot(\alpha_i^j+\alpha_i^{-j}) ,$$

we get $b+1 \geq -\sum_{j=1}^{b}(b+1-j)\cdot(\alpha_i^j+\alpha_i^{-j})$, summing it over $i = 1,\ldots,2g$ and using $\sum \alpha_i^j = \sum \alpha_i^{-j}$ and $\sum_{i=1}^{2g}\alpha_i^j = N_j\cdot q^{-j/2} - q^{j/2} - q^{-j/2}$ we get

$$g\cdot(b+1) \geq \sum_{j=1}^{b}(b+1-j)\cdot\left(N_j\cdot q^{-j/2} - q^{j/2} - q^{-j/2}\right) .\ \blacksquare$$

Lemma 2 [Vl]. *For any integer* $b \geq 1$

$$\frac{1}{g}\cdot\log h = \log q + \frac{1}{g}\cdot\sum_{j=1}^{b}\frac{q^{-j}}{j}\cdot N_j - \frac{1}{g}\cdot\sum_{j=1}^{b}\frac{1+q^{-j}}{j} - \frac{1}{g}\cdot\sum_{j=b+1}^{\infty}\frac{q^{-j/2}}{j}\cdot\sum_{i=1}^{2g}\alpha_i^j .$$

Proof: We have

$$\frac{1}{g}\cdot\log h = \frac{1}{g}\cdot\log\prod_{i=1}^{2g}(1-\omega_i) = \frac{1}{g}\cdot\log\left(q^g\cdot\prod_{i=1}^{2g}(1-\alpha_i\cdot q^{-1/2})\right) =$$

$$= \log q + \frac{1}{g}\cdot\sum_{i=1}^{2g}\log(1-\alpha_i\cdot q^{-1/2}) = \log q - \frac{1}{g}\cdot\sum_{j=1}^{\infty}\frac{q^{-j/2}}{j}\cdot\sum_{i=1}^{2g}\alpha_i^j =$$

$$= \log q + \frac{1}{g}\cdot\sum_{j=1}^{b}\frac{q^{-j/2}}{j}\cdot\left(N_j\cdot q^{-j/2} - q^{j/2} - q^{-j/2}\right) - \frac{1}{g}\cdot\sum_{j=b+1}^{\infty}\frac{q^{-j/2}}{j}\cdot\sum_{i=1}^{2g}\alpha_i^j .\ \blacksquare$$

Proof of Theorem 1: The second term in Lemma 1 is at most

$$\frac{2}{g}\cdot \sum_{j=1}^{b} q^{j/2} = \frac{2(q^{b/2} - 1)q^{1/2}}{g\cdot (q^{1/2} - 1)},$$

and for $b = f(g)$ it tends to zero because of the condition of the theorem. In fact, there exists a function $l(g)$ such that it also tends to zero for $b = l(g)$ and such that $f(g)/l(g)$ tends to zero.

Lemma 1 now shows that

$$1 + o(1) \geq \sum_{j=1}^{l(g)} \left(1 - \frac{j}{l(g) + 1}\right)\cdot \frac{N_j}{g}\cdot q^{-j/2} \geq$$

$$\geq \sum_{j=1}^{f(g)} \left(1 - \frac{j}{l(g) + 1}\right)\cdot \frac{N_j}{g}\cdot q^{-j/2} \geq \left(1 - \frac{f(g)}{l(g) + 1}\right)\cdot \sum_{j=1}^{f(g)} \frac{N_j}{g}\cdot q^{-j/2}.$$

The last expression tends to

$$\sum_{j=1}^{f(g)} \frac{N_j}{g}\cdot q^{-j/2} = \sum_{j=1}^{f(g)} \frac{1}{g}\cdot q^{-j/2}\cdot \sum_{m|j} m\cdot B_m =$$

$$= \sum_{m=1}^{f(g)} \frac{m\cdot B_m}{g}\cdot \sum_{i=1}^{[f(g)/m]} q^{-im/2} =$$

$$= \sum_{m=1}^{f(g)} \frac{m\cdot B_m}{g}\cdot q^{-m/2}\cdot \frac{1 - q^{m\cdot [f(g)/m]/2}}{1 - q^{-m/2}},$$

which tends to what we want. ∎

Proof of Theorem 2: Suppose that

$$\lim \frac{f(g)}{\log g} = \infty .$$

Then

$$\lim \frac{q^{f(g)/2}}{g} = \infty$$

(in fact this weaker condition is sufficient for Theorem 2). Let $b = f(g)$. Then

$$b\cdot B_b = N_b - \sum_{d|b} d\cdot B_d \geq N_b - \sum_{d|b} N_d \geq$$

$$\geq q^b - 2g\cdot q^{b/2} - \sum_{d|b} (q^d + 2g\cdot q^{d/2} + 1) ,$$

and since $d \le b/2$ and $\lim \frac{q^{f(g)/2}}{g} = \infty$, we see that

$$\lim \frac{1}{g}\cdot \sum_{m=1}^{f(g)} \frac{m\cdot B_m}{q^{m/2}-1} \ge \lim \frac{1}{g}\cdot\frac{b\cdot B_b}{q^{b/2}-1} \ge \lim \frac{q^{b/2}}{g} = \infty .\blacksquare$$

Proof of Theorem 3: The last term of Lemma 2 tends to zero for any b tending to infinity. Note that

$$\frac{1}{g}\cdot \sum_{j=1}^{f(g)} \frac{1+q^{-j}}{j} \le \frac{2}{g}\cdot(1+\log f(g))$$

also tends to zero because of the condition on $f(g)$. Therefore Lemma 2 gives

$$\frac{1}{g}\cdot \log h - \log q = \frac{1}{g}\cdot \sum_{j=1}^{f(g)} \frac{q^{-j}}{j}\cdot N_j = \frac{1}{g}\cdot \sum_{j=1}^{f(g)} \frac{q^{-j}}{j}\cdot \sum_{m|j} m\cdot B_m =$$

$$= \frac{1}{g}\cdot \sum_{m=1}^{f(g)} B_m\cdot \sum_{i=1}^{[f(g)/m]} \frac{q^{-im}}{i} =$$

$$= \frac{1}{g}\cdot \sum_{m=1}^{f(g)} B_m \cdot \log \frac{q^m}{q^m-1} - \frac{1}{g}\cdot \sum_{m=1}^{f(g)} B_m\cdot \sum_{[f(g)/m]+1}^{\infty} \frac{q^{-im}}{i} .$$

The last term can be estimated as follows:

$$\frac{1}{g}\cdot \sum_{m=1}^{f(g)} B_m\cdot \sum_{[f(g)/m]+1}^{\infty} \frac{q^{-im}}{i} \le \frac{1}{g}\cdot \sum_{m=1}^{f(g)} \frac{N_m}{m}\cdot\frac{q^{-m\cdot([f(g)/m]+1)}}{([f(g)/m]+1)\cdot(1-q^{-m})} .$$

It tends to

$$\frac{1}{g}\cdot \sum_{m=1}^{f(g)} N_m\cdot\frac{q^{-f(g)}}{f(g)\cdot(1-q^{-m})} \le \frac{1}{g}\cdot \sum_{m=1}^{f(g)} (q^m+1+2g\cdot q^{m/2})\cdot\frac{q^{-f(g)}}{f(g)\cdot(1-q^{-m})}$$

which is at most

$$\frac{1}{g}\cdot(q^{f(g)}+1+2g\cdot q^{f(g)/2})\cdot\frac{q^{-f(g)}}{(1-q^{-1})} .$$

The last expression tends to zero since both g and $f(g)$ tend to infinity. $\blacksquare$

Proof of Theorem 4: Let $f(g) = (\log g)^{1/2}$. Then $\lim f(g) = \infty$, $\lim f(g)/(\log g) = 0$ and $\lim (\log f(g))/g = 0$. Thus we can apply both Theorem 1 and Theorem 3. If

$$\lim \frac{1}{g}\cdot \sum_{m=1}^{k} \frac{m\cdot B_m}{q^{m/2}-1} = 1$$

then, because of Theorem 1,

$$\lim \frac{1}{g}\cdot \sum_{m=k+1}^{f(g)} \frac{m\cdot B_m}{q^{m/2}-1} = 0$$

It is clear that

$$\frac{m}{q^{m/2}-1} > \frac{m}{q^m-1} = m\cdot \sum_{i=1}^{\infty} q^{-im} > \sum_{i=1}^{\infty} q^{-im}/i = \log (1-q^{-m})^{-1} .$$

Therefore

$$\lim \frac{1}{g}\cdot \sum_{m=k+1}^{f(g)} B_m\cdot \log (1-q^{-m})^{-1} = 0 ,$$

and Theorem 3 gives the answer. ∎

Proof of Theorem 5: The left-hand inequality immediately follows from Corollary 2 since any family contains an exact one and $\beta_m \geq 0$. To estimate H_{sup} from above we should find the maximum of

$$F(\beta) = \sum_{m=1}^{\infty} \beta_m\cdot \log \frac{q^m}{q^m-1}$$

for all $\beta = (\beta_1,\beta_2,\ldots)$ subject to the condition

$$\sum_{m=1}^{\infty} \frac{m\cdot \beta_m}{q^{m/2}-1} \leq 1 .$$

If the last inequality is strict, change β_1 by a larger β_1' such that the sum equals 1, then $F(\beta)$ will also become larger. Thus we can suppose that

$$\beta_1 = (\sqrt{q}-1)\left(1 - \sum_{m=2}^{\infty} \frac{m\cdot \beta_m}{q^{m/2}-1}\right)$$

and $F(\beta)$ equals $(\sqrt{q}-1)\cdot \log \frac{q}{q-1}$ plus $G(\beta) = \sum_{m=2}^{\infty} c_m\beta_m$, where $c_m = \log \frac{q^m}{q^m - 1} - (\sqrt{q} - 1)\cdot\frac{m}{q^{m/2} - 1}\cdot\log\frac{q}{q-1}$. We shall show that $c_m \leq 0$ for any $m \geq 2$. Indeed

$$c_m = \sum_{n=1}^{\infty}\left(\frac{q^{-mn}}{n} - (\sqrt{q} - 1)\cdot\frac{m}{q^{m/2} - 1}\cdot\frac{q^{-n}}{n}\right) =$$

$$= \sum_{n=1}^{\infty}\frac{q^{-mn}}{n}\cdot\left(1 - (\sqrt{q} - 1)\cdot\frac{m}{q^{m/2} - 1}\cdot q^{(m-1)n}\right),$$

and

$$\frac{q^{m/2} - 1}{q^{1/2} - 1} = \sqrt{q}^{\,m-1} + \ldots + 1 \leq m\cdot\sqrt{q}^{\,m-1} < m\cdot q^{m-1} \leq m\cdot q^{(m-1)n}$$

Therefore, the maximum of $F(\beta)$ is attained when $\beta_1 = \sqrt{q} - 1$, and $\beta_m = 0$ for $m \geq 2$. ■

Proof of Theorem 7: We have a map

$$\varphi : Div_m^+ \longrightarrow J_m$$

from the set of positive divisors of degree m to J_m, mapping D to the class of $D - (\deg D)\cdot P_0$. If $L \in J_X(\mathbb{F}_q)$ then

$$n_L := |\varphi^{-1}(L)| = \frac{q^{l(L)} - 1}{q - 1},$$

and J_m is defined by the condition

$$J_m = \{L \in J_X(\mathbb{F}_q) \mid n_L \geq 1\}.$$

Therefore $D_m = |Div_m^+| \geq |J_m|$, and the right-hand inequality follows. On the other hand, for $m \leq 2g - 2$, we have either $l(K - L) = 0$ and $l(L) = m - g + 1 < \frac{m}{2} + 1$, or L is special and the Clifford theorem gives $l(L) \leq \frac{m}{2} + 1$. Thus $n_L \leq \frac{q}{q-1}\cdot q^{m/2}$, and $D_m \leq \frac{q}{q-1}\cdot q^{m/2}\cdot|J_m|$ which yields the left-hand inequality.

To prove the last assertion, note that if $m < B_1/(q + 1)$ then $l(L) = 1$ for any $L \in J_m$. Indeed, if $D_1 = D_2 + (f)$ and both D_1 and D_2 are positive, then $\deg(f) = \deg D_1 = m$, on the other hand $\deg(f)\cdot(q + 1) \geq |X(\mathbb{F}_q)|$ since $f : X \longrightarrow \mathbb{P}^1$ maps $\mathbb{F}_q$-points into $\mathbb{F}_q$-points.

In fact, we have just counted the number of classes of positive divisors of degree m. Therefore the assumption that there exists an $\mathbb{F}_q$-point is not important for the proof. ■

References

[Dr/Vl] Vladuţ, S.G., Drinfeld, V.G.: *Number of points of an algebraic curve.* Funct. Anal. 17, 53-54 (1983)

[Fr/Pe/St] Frey, G., Perret, M., Stichtenoth, H.: *On the different of tamely ramified abelian extensions of global fields.* This volume

[La/MD] Lachaud, G., Martin-Deschamp, M.: *Nombre de points des jacobiennes sur un corps fini.* Acta Arithm. 56, 329-340 (1990)

[Qu] Quebbemann, H.-G.: *Lattices from curves over finite fields.* Preprint (1989)

[Ro/Ts] Rosenbloom, M.Yu., Tsfasman, M.A.: *Multiplicative lattices in global fields.* Invent.Math. 101, 687-696 (1990)

[Se] Serre, J.-P. *The number of rational points on curves over finite fields.* Notes by E.Bayer, Princeton lectures (1983)

[Ts/Vl] Tsfasman, M.A., Vladuţ, S.G.: *Algebraic-Geometric Codes.* Kluwer Acad. Publ., Dordrecht/Boston/London, 1991

[Vl] Vladuţ, S.G.: *An exhaustion bound for algebraic-geometric modular codes.* Probl.Info.Trans. 23, 22-34 (1987)

ON THE WEIGHTS OF TRACE CODES

Conny Voss

1. Introduction

Let C be a linear code of length n over the field $GF(q^m)$ (where q is a power of some prime p) and $tr : GF(q^m) \to GF(q)$ denote the trace map, i.e.

$$tr(\gamma) = \gamma + \gamma^q + \ldots + \gamma^{q^{m-1}} \quad \text{for} \quad \gamma \in GF(q^m).$$

We define the trace code of C to be

$$tr(C) = \{(tr(\gamma_1), \ldots, tr(\gamma_n)) \in GF(q)^n \quad | \quad (\gamma_1, \ldots, \gamma_n) \in C\}.$$

Clearly, $tr(C)$ is a linear code of length n over $GF(q)$. Many interesting codes over $GF(q)$ can be represented as trace codes. For example, the dual code of a subfield subcode is always a trace code (Delsarte's theorem [3, p.208]). In particular, the duals of BCH codes can be considered as trace codes.
By using the Hasse-Weil-Serre bound for the number of rational places of algebraic function fields over $GF(q^m)$, we obtain an estimate for the weights of codewords in $tr(C)$ for a large class of trace codes. Our bound generalizes some recent results of Wolfmann [6], Lachaud [2] and C. Moreno and O. Moreno [4]. Trace codes also were considered by van der Geer and van der Vlugt [5].

2. Notations

Throughout this paper, we consider an algebraic function field $F/GF(q^m)$ of one variable. We assume that $GF(q^m)$ is the full constant field of F, and we denote by $g(F)$ the genus of $F/GF(q^m)$. We fix a set $S = \{P_1, \ldots, P_n\}$ of n distinct places of F of degree one and a finite-dimensional $GF(q^m)$-vector space $V \subseteq F$ satisfying the following condition:

The pole set of any $f \in V$ is disjoint from S.

Due to these assumptions, we can evaluate $f \in V$ at any place $P_i \in S$, and we define the code

$$C = \{(f(P_1), \ldots, f(P_n)) \mid f \in V\} \subseteq GF(q^m)^n$$

and its trace code

$$tr(C) = \{(tr\ f(P_1), \ldots, tr\ f(P_n)) \mid f \in V\} \subseteq GF(q)^n$$

Observe that our definition of the code C is slightly more general than Goppa's construction of algebraic geometric codes. This generalization is motivated by the fact that any cyclic code over $GF(q)$ can be represented as $tr(C)$ for an appropriate choice of F, V and S (see section 4 below).

An element $f \in V$ is said to be degenerate if there exist some $h \in F$, $\alpha \in GF(q)^*$ and $\beta \in GF(q^m)$ such that

$$f = \alpha \cdot (h^p - h) + \beta,$$

otherwise f is non-degenerate (note that $p = char\ (GF(q))$). The vector space V is called non-degenerate if V consists not only of constant elements and if any non-constant element $f \in V$ is non-degenerate. For any non-degenerate $f \in V$ we consider the function field $E_f = F(y)$, defined by the equation

$$y^q - y = f$$

Since f is non-degenerate, the extension E_f/F is an elementary abelian p-extension of degree q (cf.[1], Lemma 1.3), and $GF(q^m)$ is algebraically closed in E_f. Finally let $g = \max \{g(E_f) \mid f \in V \text{ is non-degenerate }\}$ and $N(E_f) = \sharp \{ \text{ places of } E_f/GF(q^m) \text{ of degree one } \}$.

2. Main result

2.1 Theorem. Let V be non-degenerate. Then for the weight w of any codeword of $tr(C)$ we have $w = 0$, $w = n$ or

$$\mid w - \frac{q-1}{q} \cdot n \mid \ \leq \frac{1}{q}(g - g(F))[2q^{m/2}] + \frac{q-1}{q} \cdot k,$$

where k denotes the number of places of F of degree one which are not in S.

Proof. For any non-constant element $f \in V$ we have

$$(*) \mid N(E_f) - N(F) \mid \ \leq (g(E_f) - g(F))[2q^{m/2}]$$

(cf. [2, Rem.4.4]).
Consider a codeword $c_f = (tr\ f(P_1), \ldots, tr\ f(P_n))$ of weight w.
Observe that for every $1 \leq i \leq n$

$$tr\ f(P_i) = 0 \ \text{ iff } \ f(P_i) = \gamma^q - \gamma \ \text{ for some } \ \gamma \in GF(q^m)$$

(Hilbert's theorem 90).
This is equivalent to the fact that there exist q places of degree one of E_f lying over P_i since $y^q - y = f$ is the defining equation of E_f. The number of zero components of c_f is $n - w$, and thus we have

$$N(E_f) = q(n - w) + r,$$

where r denotes the number of places of E_f of degree one which are not lying over a place of S. Setting

$$\epsilon_f = r - k,$$

nd observing that

$$0 \le r \le qk \quad \text{and} \quad N(F) = n + k,$$

ve can write

$$N(E_f) - N(F) = q(n-w) - n + \epsilon_f \quad \text{with} \quad -k \le \epsilon_f \le (q-1)k.$$

Therefore the assertion follows by (*) if we replace $g(E_f)$ by g.
For any constant $f \in V$ we have $w = 0$ or $w = n$ depending on the value of $tr(f)$. □

.2 Corollary. Let V be non-degenerate. Then the mininum distance d of $tr(C)$ satisfies

$$d \ge \frac{q-1}{q} \cdot n - \frac{1}{q}(g - g(F))[2q^{m/2}] - \frac{k}{q}.$$

.3 Remark. If $q = p$ is a prime and if $f \in V$ is degenerate then the weight w of he corresponding codeword c_f is 0 or n.

Proof. If $f = \alpha \cdot (h^p - h) + \beta$, $h \in F$, $\alpha \in GF(q)^*$ and $\beta \in GF(q^m)$, then $tr\ f(P_i) = tr(\beta)$ or every $1 \le i \le n$. □

Now we want to apply theorem 2.1 to the class of trace codes of geometric Goppa codes.

. Trace codes of geometric Goppa codes

et $D = \sum_{i=1}^n P_i$ and $G = \sum_{j=1}^r m_j Q_j$ be divisors of F with $G \ge 0$, $supp\ G \cap supp\ D = \emptyset$ nd let $t = \sum_{j=1}^r \deg\ Q_j$. Consider the geometric Goppa code

$$C = \{(f(P_1), \ldots, f(P_n)) \mid f \in L(G)\}.$$

We assume that $V = L(G)$ is non-degenerate.

.1 Proposition. For every non-constant $f \in L(G)$ we have

$$g(E_f) \le q \cdot g(F) + \frac{q-1}{q}(\deg\ G + t - 2).$$

roof. The proposition follows from the genus formula for elementary abelian p-extensions [1, Th.2.1 and Prop.2.2]. □

s an immediate consequence of theorem 2.1 and proposition 3.1 we obtain the following result (cf. Lachaud [2, Th. 6.3], Moreno's [4, Th.9]):

3.2 Theorem. Under the above assumptions the weight w of any codeword in $tr(C)$ satisfies $w = 0$, $w = n$ or

$$| w - \frac{q-1}{q} \cdot n | \leq \frac{q-1}{2q}(2g(F) + \deg\ G + t - 2)[2q^{m/2}] + \frac{q-1}{q} \cdot k$$

4. Cyclic codes

Let $n \in \mathbb{N}, (n,q) = 1$, such that $GF(q^m)$ is the splitting field of $x^n - 1$ over $GF(q)$, and let $\beta \in GF(q^m)$ be a primitive n-th root of unity. For any integer i, the cyclotomic class of i is defined to be

$$\Gamma(i) = \{0 \leq t \leq n-1 | \ t \equiv iq^j \ (mod\ n) \ \text{ for some } \ j \ \}$$

and the minimal polynomial of β^i over $GF(q)$ is

$$m_{\beta^i}(x) = \prod_{t \in \Gamma(i)} (x - \beta^t).$$

Now we consider an arbitrary cyclic code of length n over $GF(q)$. In order to apply the main theorem we first have to represent C as a trace code.
Let $x^n - 1 = g(x)h(x)$ where $g(x)$ is the generator polynomial of C. We denote the reciprocal polynomial of $h(x)$ by $h^{\perp}(x)$.
A subset $J \subseteq \{0, \ldots, n-1\}$ such that

$$h^{\perp}(x) = \prod_{j \in J} m_{\beta^j}(x),$$

is called a β-check set of C.

Let $F = GF(q^m)(x)$ be the rational function field over $GF(q^m)$ and $S = \{P_0, \ldots, P_{n-1}\}$ where P_i is the zero of $x - \beta^i$, $0 \leq i \leq n-1$. We define

$$V = \{f_a(x) = \sum_{j \in J} a_j x^j \mid a = (a_j)_{j \in J} \in GF(q^m)^J\}.$$

Then we have

$$C = tr(V) = \{(tr\ f_a(\beta^0), \ldots, tr\ f_a(\beta^{n-1})) \mid f_a \in V\}$$

(cf. [6, Prop.2.1]).
Let $r \in \mathbb{N}$ such that $nr = q^m - 1$. Then the set of n-th roots of unity in $GF(q^m)$ is also

the set of r-th powers in $GF(q^m)^*$. Therefore the elementary abelian p-extension E_{f_a} is defined by the equation

$$y^q - y = f_a(x^r).$$

Observe that

$$\sharp \{z \in GF(q^m)^* \mid tr\ f_a(z^r) = 0\} = r(n - w_a),$$

where w_a is the weight of the codeword in $tr(V)$ corresponding to f_a.
Thus we have

$$N(E_{f_a}) \geq qr(n - w_a).$$

4.1 Theorem. Let C be a cyclic code of length n over $GF(q)$ with β-check set J such that $(j, q) = 1$ for every $j \in J$, and let $\varrho = \max\ \{j \in J\}$. Then the weight of any non-zero codeword in C satisfies

$$\mid w - \frac{q^{m-1}(q-1)}{r} \mid \leq \frac{(q-1)(\varrho r - 1)}{2rq} [2q^{m/2}].$$

Proof. Since every element of J is prime to q it follows that V is non-degenerate and that for every $f_a \in V$

$$g(E_{f_a}) = \frac{(q-1)(r \cdot \deg\ f_a - 1)}{2}.$$

Thus we have

$$\mid N(E_{f_a}) - (q^m + 1) \mid \leq \frac{(q-1)(\varrho r - 1)}{2} [2q^{m/2}].$$

Moreover, since $f_a(0) = 0$, there are q places of degree one in E_{f_a} lying over the zero of x, and there is only one place of degree one lying over the pole of x. Therefore

$$N(E_{f_a}) = qr(n - w) + q + 1$$

and, as $rn = q^m - 1$, the theorem follows. □

References

[1] Garcia, A., Stichtenoth, H., Elementary Abelian p-Extensions of Algebraic Function Fields. Manuscripta Math. 72 (1991) 67-79.

[2] Lachaud, G., Artin-Schreier Curves, Exponential Sums and the Carlitz-Uchiyama Bound for Geometric Codes. J. of Number Theory 39 (1991), p. 18-40.

[3] Mac Williams, F.J., Sloane, N.J.A., The Theory of Error-correcting Codes. North Holland (1977).

[4] Moreno, C.J., Moreno, O., An Improved Bombieri-Weil Bound And Applications To Coding Theory. Preprint.

[5] van der Geer, G., van der Vlugt, M., Trace Codes and Families of Algebraic Curves. To appear in Math. Z.

[6] Wolfmann, J., New Bounds on Cyclic Codes from Algebraic Curves. Lecture Notes in Computer Science 388 (1989) 47-62.

Author's address:
Conny Voss,
Universitaet GHS Essen,
Fachbereich 6 - Mathematik,
D - 4300 Essen 1,
Germany.

MINORATION DE CERTAINES SOMMES EXPONENTIELLES BINAIRES

François RODIER
Equipe C.N.R.S. Arithmétique et Théorie de l'Information
U.F.R. de Mathématiques
Université Paris 7
75231 PARIS CEDEX 05

1. Introduction

Soit C_m un code BCH binaire, de longueur $2^m - 1 = q - 1$, de distance prescrite $= 2t+1$. Alors, d'après la borne de Carlitz-Uchiyama ([7] p. 280) le poids w d'un mot le code non nul du dual de C_m est tel que

$$|w - 2^{m-1}| \leq (t-1)2^{m/2}.$$

Dans [7] (Research Problem (9.5)), MacWilliams et Sloane suggèrent un résultat un peu plus fort:

$$|w - 2^{m-1}| \leq (t-1)2^{[m/2]},$$

où par $[a]$ on dénote la partie entière de a. Pour un nombre m impair cela veut dire

$$|w - 2^{m-1}| \leq (t-1)2^{(m-1)/2}.$$

On peut montrer que cette inégalité est vraie pour les codes de distance prescrite 3, 5 ou 7. En effet le dual du code étendu $\widehat{C}_m$ se plonge dans un code de Reed et Muller $R(2,m)$ ([7], p. 385) et les poids de ce dernier sont multiples de $2^{[(m-1)/2]}$ ([7], p. 447) c'est-à-dire de $2^{(m-1)/2}$ si le nombre m est impair. De ce fait et de l'inégalité de Carlitz-Uchiyama, on déduit, pour le code $C_m^{\perp}$, dual du code BCH de longueur $2^m - 1$ et de distance prescrite δ, que

$$\begin{array}{lcll} |w - 2^{m-1}| & \in & \{0\} & \text{si } \delta = 3, \\ |w - 2^{m-1}| & \in & \{0, 2^{(m-1)/2}\} & \text{si } \delta = 5, \\ |w - 2^{m-1}| & \in & \{0, 2^{(m-1)/2}, 2 \cdot 2^{(m-1)/2}\} & \text{si } \delta = 7, \end{array}$$

où w est le poids d'un mot non nul du code $C_m^\perp$ et où m est impair. Donc l'inégalité

$$|w - 2^{m-1}| \leq (t-1)2^{(m-1)/2}$$

est vraie dans ces trois cas-là.

Cette inégalité n'est plus vraie pour $\delta = 9$ comme on l'a montré récemment dans [8]. On va montrer ici qu'il en va de même pour d'autres valeurs de δ (par exemple $\delta = 9, 25, 49, 73, 81 \ldots$ On conjecture qu'il y en a une infinité, cf. par. 4). En effet pour ces δ-là il existe une infinité de nombres m impairs tels que

$$|w - 2^{m-1}| > (t-1)2^{(m-1)/2}$$

pour au moins un mot du dual de C_m.

Plus précisément, on a, pour de tels δ

$$\limsup_{m \text{ impair}} \frac{|w - 2^{m-1}|}{2^{(m-1)/2}} = \sqrt{2}(t-1).$$

Pour cela on montre d'abord que les mots de code c du dual du code étendu $\widehat{C}_m$ peuvent se représenter par des polynômes f dans l'espace $\mathbf{F}_q[x]$ de degré $k = \delta - 2$, puis que le poids w du mot de code c est lié à la somme

$$S_m(f) = \sum_{\mathbf{F}_q} (-1)^{\operatorname{Tr} f(x)}$$

par

$$S_m(f) = 2^m - 2w.$$

Ensuite le théorème qui sera prouvé ici donne une minoration des sommes $S_m(f)$, pour une infinité de valeurs de m:

$$\limsup_{m \text{ impair}} \frac{|S_m(f)|}{2^{m/2}} = 2(t-1),$$

où f est un polynôme particulier. C'est la meilleure minoration possible, puisqu'on a

$$\frac{|S_m(f)|}{2^{m/2}} \leq 2(t-1)$$

d'après la borne due à Weil [10] (voir aussi Serre [9]).

Rappelons que Bassalygo, Zinov'ev et Litsyn ont obtenu une autre minoration de telles sommes exponentielles [1]. En caractéristique 2 on peut déduire de leurs résultats qu'il existe un polynôme f de degré $2^r + 1$ à coefficients dans $\mathbf{F}_{2^{2r+1}}$, de trace non constante, tel que

$$|S_{2r+1}(f)| \geq 2^{2r}.$$

Ce n'est pas suffisant pour ce que l'on veut démontrer ici, puisque ça ne montre que l'inégalité

$$|w - 2^{m-1}| \geq (t-1)2^{(m-1)/2},$$

avec $m = 2r + 1$, pour un code BCH de distance prescrite $\delta = 2^r + 3 = 2t + 1$.

Par contre, en caractéristique p différente de 2, ils considèrent les sommes

$$S_m(f) = \sum_{\mathbf{F}_{q^m}} \exp\left(\frac{2\pi i \operatorname{Tr} f(x)}{p}\right),$$

où $q = p^l$. De leurs résultats, on déduit que pour tout r il existe un polynôme f de degré $d = q^r + 1$ à coefficients dans $\mathbf{F}_{q^{2r+1}}$, de trace non constante, tel que

$$| S_m(f) | = (d-1)q^{m/2},$$

pour $m = 2r + 1$.

Autrement dit les expressions $| S_m(f) |$ atteignent leur borne de Weil, pour au moins un f de degré $q^r + 1$ dans $\mathbf{F}_{q^{2r+1}}[x]$ et $m = 2r + 1$.

2. Les codes BCH binaires étendus

Soit $\widehat{C}_m$ le code BCH binaire étendu de longueur $2^m = q$ et de distance prescrite δ. On définit la trace Tr de $\mathbf{F}_q$ sur $\mathbf{F}_2$ par

$$\operatorname{Tr}(x) = \sum_{i=0}^{m-1} x^{2^i}.$$

Proposition 1.

1) Le dual de $\widehat{C}_m$ est le code associé à l'espace $\mathbf{F}_q[x]_k$ des polynômes $f \in \mathbf{F}_q[x]$ tels que $\deg f \leq k$, avec $k = \delta - 2$.

2) Les mots de code c_f peuvent s'écrire

$$c_f = (\operatorname{Tr} f(a_1), \operatorname{Tr} f(a_2), \ldots, \operatorname{Tr} f(a_q))$$

où les a_i sont les éléments de $\mathbf{F}_q$.

Démonstration.— Soit $\mathbf{C}$ le code donné par l'image de l'application

$$\mathbf{c} : \mathbf{F}_q[x]_k \mapsto (\mathbf{F}_q)^q$$

définie par $\mathbf{c}(f) = (f(a_1), \ldots, f(a_q))$. Une base de $\mathbf{F}_q[x]_k$ est donnée par la famille des monômes $1, x, \ldots, x^k$. Le code dual $\mathbf{C}^\perp$ a par conséquent une matrice de contrôle donnée par

$$H = \begin{pmatrix} 1 & \ldots & 1 \\ a_1 & \ldots & a_q \\ \ldots & \ldots & \ldots \\ a_1^k & \ldots & a_q^k \end{pmatrix}.$$

On a d'après le théorème de Delsarte ([7] chap. 7, théorème 11, p. 208)

$$(\operatorname{Tr} \mathbf{C})^\perp = (\mathbf{C}^\perp)|\mathbf{F}_2,$$

et donc $(\mathrm{Tr}\,\mathbf{C})^\perp$ est le code BCH binaire étendu, de longueur 2^m et de distance prescrite $k+2$ (cf. [5] exemple (7.2) ou [7] exemple 5, p. 345).

Soit f un polynôme dans $\mathbf{F}_q[x]_k$. On appellera c_f le mot de code associé dans le dual de $\widehat{C}_m$ et w_f le poids du mot de code c_f. Soit

$$S_m(f) = \sum_{\mathbf{F}_q} (-1)^{\mathrm{Tr}\, f(x)}.$$

Proposition 2. *Soit f un polynôme dans $\mathbf{F}_q[x]_k$. On a*

$$S_m(f) = q - 2w_f.$$

Démonstration.— D'après la proposition précédente, on a

$$\begin{aligned} S_m(f) &= \#\{x \mid \mathrm{Tr}\, f(x) = 0\} - \#\{x \mid \mathrm{Tr}\, f(x) = 1\} \\ &= (q - w_f) - w_f \\ &= q - 2w_f. \end{aligned}$$

3. Le résultat

Soit k un nombre premier. On suppose que

$$(*) \qquad \begin{cases} 1) & k \text{ est congru à } -1 \text{ (mod. 8)}, \\ 2) & k \text{ ne divise aucun des nombres } 2^i - 1 \text{ pour } 1 \le i < (k-1)/2. \end{cases}$$

Théorème. *Soit k un nombre premier. Posons $f(x) = x^k$, et soit*

$$S_m = \sum (-1)^{\mathrm{Tr}\, f(x)}$$

où la somme est sur le corps $\mathbf{F}_{2^m}$. Si la condition (*) est vérifiée alors on a:

$$\limsup_{\substack{m \to \infty \\ m \text{ impair}}} \frac{|S_m|}{2^{m/2}} = k - 1.$$

On en déduit immédiatement un corollaire sur les codes.

Corollaire 1. *Soit k un nombre premier vérifiant la condition (*). Soit C_m le code BCH binaire de longueur $2^m - 1$ et de distance prescrite $\delta = k+2$. Appelons w_c le poids du mot c du code $C_m^\perp$. On a*

$$\limsup_{m \text{ impair}} \sup_{c \neq 0} \frac{|w_c - 2^{m-1}|}{2^{(m-1)/2}} = \sqrt{2}(t-1) > (t-1),$$

où $\delta = 2t + 1$.

Par conséquent, l'inégalité espérée par MacWilliams et Sloane est fausse pour une infinité de valeurs de m.

Corollaire 2. *Soit k un nombre premier. On considère la courbe projective sur $\mathbf{F}_2$ d'équation*

$$y^2 + y = x^k$$

et on note N_m son nombre de points sur $\mathbf{F}_{2^m}$. Si la condition (*) est vérifiée on a

$$\limsup_{\substack{m\to\infty \\ m \text{ impair}}} \frac{|\, N_m - 2^m - 1 \,|}{2^{m/2}} = k - 1.$$

Démonstration.— En effet, on a

$$N_m - 2^m - 1 = S_m.$$

La démonstration du théorème se fera après une série de lemmes.

Posons $g = (k-1)/2$.

Lemme 1. *Si m n'est pas divisible par g, alors $S_m = 0$.*

Démonstration.— En effet l'application $x \mapsto x^k$ est une permutation de $\mathbf{F}_q$ sauf si $\mathbf{F}_q$ contient les racines $k^{\text{ième}}$ de l'unité, c'est-à-dire si k divise $q-1$. Or d'après la condition (*), on a $k \equiv -1 \pmod{8}$, donc 2 est un résidu quadratique modulo k, ce qui veut dire que

$$2^g = 2^{(k-1)/2} \equiv 1 \pmod{k},$$

donc k divise $2^g - 1$. Toujours d'après la condition (*) on déduit que g est le plus petit des entiers m tel que k divise $2^m - 1$, donc $\mathbf{F}_{2^g}$ est le plus petit des corps $\mathbf{F}_q = \mathbf{F}_{2^m}$ qui contiennent les racines $k^{\text{ième}}$ de l'unité. Par conséquent, $x \mapsto x^k$ est une permutation de $\mathbf{F}_q$ et donc $S_m = 0$ si et seulement si

$$\mathbf{F}_{2^m} = \mathbf{F}_q \not\supset \mathbf{F}_{2^g},$$

c'est-à-dire si et seulement si g ne divise pas m.

On va noter $\Re e$ la partie réelle d'un nombre complexe.

Lemme 2. *La somme S_{gm} est égale à $-2g(\sqrt{2})^{gm}\,\Re e\, \omega^m$, où ω est un nombre complexe de valeur absolue 1 et dont la partie réelle est donnée par*

$$\Re e\, \omega = -\frac{S_g}{2g(\sqrt{2})^g}.$$

Démonstration.— D'après [10] (voir aussi [6] p. 220, et [9]) on sait qu'il existe des nombres complexes π_i pour $1 \le i \le k-1$ tels que

$$-S_m = \sum_{1 \le i \le k-1} \pi_i^m.$$

De plus ils sont conjugués 2 à 2 et de valeur absolue égale à $\sqrt{2}$.

Les π_i sont les racines d'un polynôme unitaire $A(X)$ à coefficients entiers de degré $k-1$. On a

$$A(X) = \sum_{0\le i\le k-1} a_i X^i.$$

Les nombres S_m sont tous nuls d'après le lemme 1, sauf peut-être si g divise m. Donc d'après les relations de Newton ([2] chap. IV, par. 6, n° 4, lemme 4, p. 65), les coefficients a_i sont tous nuls, sauf peut-être a_0, a_{k-1} et a_g. On a $a_{k-1}=1$; on a donc

$$a_0 = \prod_{i=1}^{30} \pi_i = 2^g \quad \text{et} \quad a_g = \frac{S_g}{g}.$$

Par conséquent, le polynôme A s'écrit

$$A(X) = X^{2g} + \frac{S_g}{g} X^g + 2^g.$$

Ses racines sont $\pi_1 \exp(2\pi j\sqrt{-1}/g)$ et $\overline{\pi}_1 \exp(2\pi j\sqrt{-1}/g)$ pour $1 \le j \le g$.

Posons $\phi = \pi_1^g$. Alors ϕ et $\overline{\phi}$ sont racines de l'équation

$$X^2 + \frac{S_g}{g} X + 2^g = 0.$$

Soit $\omega = \phi/|\phi| = \phi/(\sqrt{2})^g$. Alors ω est racine de

$$2^g X^2 + (\sqrt{2})^g \frac{S_g}{g} X + 2^g = 0.$$

On a

$$-S_{gm} = \sum_{1\le i\le k-1} \pi_i^{gm}$$

c'est-à-dire, puisque $\phi = \pi_1^g$

$$-S_{gm} = g(\phi^m + \overline{\phi}^m) = g(\sqrt{2})^{gm}(\omega^m + \overline{\omega}^m) = 2g(\sqrt{2})^{gm}\,\Re e\,\omega^m.$$

Lemme 3. *Le nombre S_g est congru à 1 (mod. k).*

Démonstration.— Comme k divise 2^g-1 par hypothèse, l'application f restreinte au groupe multiplicatif de $\mathbf{F}_{2^g}$ qui est cyclique induit une suite exacte

$$0 \to \mathbf{Z}/k\mathbf{Z} \to \mathbf{F}_{2^g}^* \to \mathbf{F}_{2^g}^* \to \mathbf{Z}/k\mathbf{Z} \to 0.$$

Par conséquent ceci définit une partition du groupe $\mathbf{F}_{2^g}^*$ en classes de k éléments. Il y a $(2^g-1)/k$ classes et chacun des éléments x d'une même classe donne la même valeur à $(-1)^{\operatorname{Tr} f(x)}$. On en déduit que

$$S_g \equiv (-1)^{\operatorname{Tr} 0} \equiv 1 \ (\text{mod. } k).$$

Lemme 4. *Le nombre ω n'est pas une racine de l'unité.*

Démonstration.— Le nombre ω vérifie l'équation

$$2^g X^2 + (\sqrt{2})^g\, a_g X + 2^g = 0,$$

avec $a_g = S_g/g$. Cette équation peut s'écrire

$$(\sqrt{2})^g\, a_g X = -(2^g X^2 + 2^g)$$

donc, en élevant au carré, et en regroupant les termes

$$2^g X^4 + (2^{g+1} - a_g^2) X^2 + 2^g = 0$$

ou encore

$$X^4 + \left(2 - \frac{a_g^2}{2^g}\right) X^2 + 1 = 0.$$

Pour que ω soit racine de l'unité, il faut et il suffit que ω^2 le soit. Or ω^2 est racine de l'équation du second degré

$$X^2 + \left(2 - \frac{a_g^2}{2^g}\right) X + 1 = 0.$$

Cette équation admet des racines de valeur absolue égale à 1 si et seulement si son déterminant Δ, qui est égal à

$$\Delta = \left(2 - \frac{a_g^2}{2^g}\right)^2 - 4 = \left(\frac{a_g^2}{2^g} - 4\right) \frac{a_g^2}{2^g}$$

est négatif ou nul c'est-à-dire si l'on a

$$\left(\frac{a_g^2}{2^g} - 4\right) \frac{a_g^2}{2^g} \leq 0.$$

Montrons que Δ ne peut pas être nul. Comme on a $S_g \equiv 1$ (mod. k) d'après le lemme 3, on a $S_g \neq 0$ donc $a_g = S_g/g \neq 0$. De même si l'on avait $(a_g^2/2^g) - 4 = 0$, on aurait $S_g^2/g^2 = 2^{g+2}$, d'où $S_g^2 = 2^{g+2} g^2$, ce qui est impossible puisque S_g est un entier et que, g étant impair, 2^{g+2} n'est pas un carré parfait.

On a donc

$$\Delta = \left(\frac{a_g^2}{2^g} - 4\right) \frac{a_g^2}{2^g} < 0.$$

Dans ce cas l'équation est irréductible et c'est donc l'équation minimale de ω^2 sur le corps $\mathbf{Q}$. Si ω^2 était une racine de l'unité, ce serait en particulier un entier algébrique. Donc son équation minimale serait à coefficients entiers. Par conséquent $a_g^2/2^g$ serait un entier inférieur à 4. Comme de plus a_g^2 est un carré et que g est impair, on ne pourrait avoir que $a_g^2/2^g = 2$.

Supposons que ce soit le cas. Cela implique $a_g = \pm\sqrt{2^{g+1}}$, donc

$$S_g = \pm g\sqrt{2^{g+1}} = \pm(k-1)\sqrt{2^{g-1}}.$$

Par conséquent on a

$$\begin{aligned} 1 &\equiv S_g && (\text{mod. } k),\\ &\equiv \pm(k-1)\sqrt{2^{g-1}} && (\text{mod. } k),\\ &\equiv \mp\sqrt{2^{g-1}} && (\text{mod. } k),\\ &\equiv \mp 2^{(g-1)/2} && (\text{mod. } k). \end{aligned}$$

Donc on a, en élevant au carré

$$1 \equiv 2^{g-1} \ (\text{mod. } k),$$

ce qui veut dire que k divise $2^{g-1}-1$, ce qui est contraire à l'hypothèse. Par conséquent, on ne peut pas avoir $a_g^2/2^g = 2$.

Donc ω^2 n'est pas une racine de l'unité et ω non plus.

Lemme 5. *On a*

$$\limsup_{\substack{m\to\infty \\ m \text{ impair}}} |\Re e\, \omega^m| = 1.$$

Démonstration.— Comme ω n'est pas racine de l'unité, alors, d'après un théorème de Kronecker (cf. [3] théorème 439, p. 376), les ω^m pour m entier positif, sont partout denses sur le cercle $\mathbf{T}$ des nombres complexes de valeur absolue égale à 1. On va montrer qu'il en va de même des ω^m pour m entier impair et positif.

En effet les ensembles $\{\omega^m \mid m \text{ impair et positif}\}$ et $\{\omega^m \mid m \text{ impair}\}$ ont même adhérence dans $\mathbf{T}$. Cela est dû au fait qu'il existe une suite d'entiers positifs m_r tels que

$$\lim_{r\to\infty} \omega^{2m_r+m} = \omega^m$$

pour tout entier impair m (positif ou négatif).

Or, dans le groupe $\mathbf{T}$, l'adhérence de l'ensemble $\{\omega^m \mid m \text{ impair}\}$ des ω^m pour m impair est une classe latérale du sous-groupe des ω^{2m} donc aussi de son adhérence $\overline{\{\omega^{2m}\}}$. Mais ce dernier est justement le groupe $\mathbf{T}$ tout entier. On a donc

$$\overline{\{\omega^m \mid m \text{ impair et positif}\}} = \overline{\{\omega^m \mid m \text{ impair}\}} = \mathbf{T}.$$

Par conséquent on a

$$\limsup_{\substack{m\to\infty \\ m \text{ impair}}} |\Re e\, \omega^m| = 1.$$

Démonstration du théorème.—Pour montrer

$$\limsup_{\substack{m\to\infty \\ m \text{ impair}}} \frac{|S_m|}{2^{m/2}} = k-1,$$

l suffit de montrer

$$\limsup_{\substack{m\to\infty \\ m \text{ impair}}} \frac{|S_{gm}|}{2^{gm/2}} = k-1,$$

uisque $S_m = 0$ si m n'est pas divisible par g. D'après la décomposition de S_m donnée u début de la démonstration du lemme 2, cela revient à montrer

$$\limsup_{\substack{m\to\infty \\ m \text{ impair}}} 2^{-gm/2} \left| \sum_{1\le i\le k-1} \pi_i^{gm} \right| = k-1,$$

'est-à-dire, puisque $\phi = \pi_1^g$

$$\limsup_{\substack{m\to\infty \\ m \text{ impair}}} 2^{-gm/2} \left| \phi^m + \overline{\phi}^m \right| = 2.$$

omme $\omega = \phi/(\sqrt{2})^g$, l'équation précédente peut s'écrire

$$\limsup_{\substack{m\to\infty \\ m \text{ impair}}} |\omega^m + \overline{\omega}^m| = 2,$$

u

$$\limsup_{\substack{m\to\infty \\ m \text{ impair}}} |\Re e\, \omega^m| = 1,$$

e qui est la conclusion du lemme 5.

4. Discussion sur la condition (*)

La condition (*) est vérifiée pour les nombres premiers

$$k = 7, 23, 47, 71, 79, 103, 167, 191, 199\ldots$$

n peut conjecturer qu'il y en a une infinité, et que de plus leur densité est égale à

$$\frac{1}{2}\prod_{p \text{ premier}} \left(1 - \frac{1}{p(p-1)}\right) = 0,1869\ldots$$

Cette conjecture est analogue à la conjecture d'Artin suivante.

onjecture (Artin). *Pour tout nombre entier a distinct* de 1, -1 *et qui ne soit pas n carré parfait, il existe une infinité de nombres premiers k tels que a soit une racine* *primitive* (mod. k) *. De plus ces nombres ont une densité positive. Si $a = 2$ elle est égale à*

$$\prod_{p \text{ premier}} \left(1 - \frac{1}{p(p-1)}\right).$$

En effet pour que a soit une racine primitive modulo k (c'est-à-dire que a engendre le groupe $\mathbf{Z}_k^*$), il faut et il suffit que a ne soit pas multiple de k et que k ne divise aucun des nombres $a^m - 1$ pour $1 \leq m < k-2$.

C. Hooley a montré que l'hypothèse de Riemann généralisée impliquait la conjecture d'Artin [4]. On peut montrer également qu'elle implique la conjecture que nous avons faite ci-dessus.

Schéma de démonstration.—

La condition (*) comporte 2 sous-conditions. La première ($k \equiv -1$ (mod. 8)) peut se décomposer en

$$2^{(k-1)/2} \equiv 1 \text{ (mod. } k) \quad \text{et} \quad (-1)^{(k-1)/2} \not\equiv 1 \text{ (mod. } k),$$

d'après la théorie des résidus quadratiques. Elle implique en particulier que le nombre 2 est dans le sous-groupe d'indice 2 du groupe cyclique $(\mathbf{Z}/k\mathbf{Z})^*$.

La deuxième sous-condition est que

$$2^m \not\equiv 1 \text{ (mod. } k) \quad \text{pour} \quad 1 \leq m < (k-1)/2$$

c'est-à-dire que 2 engendre le sous-groupe d'indice 2 du groupe $(\mathbf{Z}/k\mathbf{Z})^*$. Cela équivaut à dire que

$$2^{(k-1)/p} \not\equiv 1 \text{ (mod. } k) \text{ si } p \text{ divise } k-1 \text{ et } p \neq 2.$$

D'après la théorie algébrique des nombres, k se décompose complètement dans $K_p = \mathbf{Q}(\sqrt[p]{1}, \sqrt[p]{2})$ si et seulement si $2^{(k-1)/p} \equiv 1$ (mod. k). De même k se décompose complètement dans $\mathbf{Q}(\sqrt{-1})$ si et seulement si $(-1)^{(k-1)/2} \equiv 1$ (mod. k).

On montre alors que la densité des nombres premiers qui se décomposent complètement dans K_p est égale à $1/[K_p : \mathbf{Q}] = 1/p(p-1)$.

En suivant les raisonnements de Hooley utilisant l'hypothèse de Riemann généralisée [4] on montre que la densité des k qui vérifient (*) est égale à

$$\left(1 - \frac{1}{[\mathbf{Q}(\sqrt{-1}) : \mathbf{Q}]}\right) \frac{1}{[K_2 : \mathbf{Q}]} \prod_{\substack{p \text{ premier} \\ p \neq 2}} \left(1 - \frac{1}{[K_p : \mathbf{Q}]}\right)$$

$$= \frac{1}{2}\frac{1}{2} \prod_{\substack{p \text{ premier} \\ p \neq 2}} \left(1 - \frac{1}{p(p-1)}\right).$$

5. Références

[1] L.A. Bassalygo, V.A. Zinov'ev et S.N. Litsyn: "A lower estimate of complete trigonometric sums in terms of multiple sums", *Dokl. Acad. Nauk SSSR*, vol. 33, n° 5 (1988); traduction anglaise, *Soviet Math. Dokl.*, vol. 37, n° 3 (1988) p. 756-759.

[2] N. Bourbaki: *Algèbre Chap. 4 à 7*, Masson, Paris, 1981.

[3] G.H. Hardy et E.M. Wright: *An introduction to the theory of numbers*, Oxford University Press, Londres, 1971.

[4] C. Hooley: "On Artin's conjecture", *J. reine angew. Math.*, vol. 225 (1967), 209-20.

[5] G. Lachaud: "Exponential sums, algebraic curves and linear codes", preprint (1989).

[6] R. Lidl et H. Niederreiter: *Finite Fields*, Encyclopedia of mathematics and its applications, vol. 20, Cambridge University Press, Cambridge, 1983.

[7] F.J. MacWilliams et N.J.A. Sloane: *The Theory of Error-Correcting Codes*, North-Holland, Amsterdam, 1977.

[8] F. Rodier: "On the spectra of the duals of binary BCH codes of designed distance $\delta = 9$", à paraître aux *IEEE transactions on Information Theory* (1991).

[9] J.P. Serre: "Majoration de sommes exponentielles", *Astérisque 41-42* (1977), p. 111-126.

[10] A. Weil: *Variétés abéliennes et courbes algébriques*, Hermann, Paris, 1948.

LINEAR CODES, STRATA OF GRASSMANNIANS, AND THE PROBLEMS OF SEGRE

Alexei N. Skorobogatov

Institute for Problems of Information Transmission
USSR Academy of Sciences
19 Ermolovoy, Moscow 101447 USSR
skoro@ippi.msk.su

0. Introduction

In this talk we explore an approach to the problem of classification of linear codes, which in some sense combines combinatorics and algebraic geometry. We also discuss a connection between the three classical problems of B.Segre and the étale cohomology of certain open subsets of Grassmannians.

Let F be a field. It is possible to define linear $[n,k]$-codes over F as k-dimensional vector subspaces C in the coordinatized n-dimensional vector space $V=F^n$, that is, in the space equipped with a basis. (In this definition we do not specify F). There are many natural connections of linear codes with classical geometric objects. For example, intersecting our code with the coordinate hyperplanes we obtain an arrangement of hyperplanes in C. Another possibility is to consider the corresponding point set in the projective space of hyperplanes in C.

Each of these viewpoints presumes its own perspective of study. The problem of existence and construction of codes with certain good properties over finite fields belongs to the realm of coding theory [12]. The combinatorics and the topology of arrangements are typically studied over the field of real or complex numbers. The configurations of points in the projective spaces are interesting as "finite projective geometries" [10], as well as for their applications to algebraic geometry (cf. [3]).

In each case, there exists a certain combinatorial structure, say, a matroid or a geometric lattice, which "controls" the geometry of the object of study. If we fix this structure, we are led to consider "the moduli scheme" parametrizing all objects of given combinatorial type defined over all posssible fields F, or, in other terms, the representations of a given matroid or a geometric lattice over all fields F. (See [17] for an extensive introduction to this philosophy.)

The "moduli scheme" of all linear $[n,k]$-codes is the Grassmannian $G(k,n)$, so that $[n,k]$-codes over F are in 1-to-1 correspondence with the set of F-points of $G(k,n)$. e_i be the vector whose i-th coordinate is 1, and the other coordinates are 0. For $I\subseteq\{1,\dots,n\}$ define V_I to be the subspace of F^n generated by e_i, $i\in I$. Let f be a function from the subsets of $\{1,\dots,n\}$ to the non-negative integers. Define a constructible algebraic set $U_f(k,n)$ as the set of vector subspaces $C\subseteq F^n$ such that $\dim(C\cap V_I)=f(I)$ for all subsets I. We have a decomposition

$$G(k,n) = \bigcup_f U_f(k,n).$$

The sets $U_f(k,n)$ were considered by Gelfand [7], Gelfand, Goreski, MacPherson, and Serganova [8], [9], who called them "strata", and by Mnëv [11], [12] and Vershik [17], under the name of "determinantal varieties". Each $U_f(k,n)$ consists of subspaces represented by points of Grassmannian belonging to an intersection of Schubert cells for all possible orderings of the basis $\{e_1,\dots,e_n\}$ (see [9] for other equivalent definitions). However, our classification problem is "wild" in the following sense [17]. According to a powerful theorem of Mnëv [11], [12], for $k\geq 3$, and for any closed projective algebraic set X defined over the prime subfield of F, there exists n and f such that X is isomorphic to the Zarisky closure of the factor of $U_f(k,n)$ under the action of the group of diagonal matrices. This gives very little hope to study the problem in such a generality.

Let us now focus on the unique Zarisky dense open set

$$U(k,n):=U_s(k,n),\ s(I)=\max\{0,k+\#I-n\}.$$

It parametrizes subspaces in the general position with respect to the coordinate simplex, which over finite fields are the same as MDS codes. Let p be a prime, $q=p^s$, and define $f_{n,k,p}(s)$ to be the number of $\mathbb{F}_q$-points on $U(k,n)$, which is equal to the mysterious number of MDS $[n,k]$-codes over $\mathbb{F}_q$. This is a function from positive integers to nonnegative integers, and we know from the general theory [2] that $f_{n,k,p}(s)$ is a sum of exponents. We can restate now the first of the three fundamental problems of B.Segre [15] in the following equivalent form: for given n, k, and p, determine the maximal zero of $f_{n,k,p}(s)$. By the Grothendieck trace formula, a computation of the Frobenius eigenvalues on the ℓ-adic ($\ell \neq p$) cohomology groups of $U(k,n)$ over an algebraic closure of $\mathbb{F}_p$ gives a full desription of $f_{n,k,p}(s)$. However this seems to be far beyond our present knowledge. We refer the reader to the papers of Segre, Thas, Casse, Glynn, and their colleagues (see [15], [16], [1], and the references in that paper) for a fruitful combinatorial treatment of the Segre problems.

In Section 1 we define the combinatorial equivalence on the set of linear codes, and discuss its elementary properties in the framework of coding theory. The relation between n-arcs and MDS codes is recalled in Section 2 in connection with the covering radius of MDS codes. In Section 3 we reformulate the problems of Segre in terms of the functions $f_{n,k,p}(s)$, and discuss the known cases. Our References are far from being complete. Let us just note that the connection between $U(k,n)$ and MDS codes was discussed in [16] (in terms of n-arcs), [14], [4], and was reformulated in the language of matroid theory in [5].

1. Combinatorial equivalence of linear codes

Let F be a field. A linear $[n,k]$-code over F is a k-dimensional linear subspace of F^n. Denote $I_n=\{1,\dots,n\}$. Let e_i be the vector whose i-th coordinate is 1, and the other coordinates are 0. For $I \subseteq I_n$ define V_I to be the subspace of F^n generated by e_i, $i \in I$. We call these V_I the coordinate subspaces.

Definition 1.1. Let C_1 and C_2 be linear $[n,k]$-codes over F. The codes C_1 and C_2 are called combinatorially equivalent iff

$$\dim(C_1 \cap V_I)=\dim(C_2 \cap V_I) \text{ for any } I \subseteq I_n.$$

Let us denote by $[C]$ the combinatorial equivalence class of C. The set of linear $[n,k]$-codes over F is the union of combinatorial equivalence classes. As follows from Definition 1.1, each class **C** is defined by its dimension function, that is a function from the subsets of I_n to non-negative integers given by

$$d_{\mathbf{C}}(I)=d_C(I)=\dim(C \cap V_I),\ I \subseteq I_n, \text{ for any } C \in \mathbf{C}.$$

A dimension function of a non-empty class **C** satisfies the following obvious properties:

(1) $\#I \geq d_{\mathbf{C}}(I) \geq 0,\ d_{\mathbf{C}}(I \cup J) \geq d_{\mathbf{C}}(I)+d_{\mathbf{C}}(J)-d_{\mathbf{C}}(I \cap J),\ d_{\mathbf{C}}(I) \leq d_{\mathbf{C}}(J)$ for $I \subseteq J$.

We also have

(2) $$d_{\mathbf{C}}(I_n)=k.$$

If we define $r_{\mathbf{C}}(I)=k-d_{\mathbf{C}}(I_n \setminus I)$, then one easily sees that (1) and (2) are equivalent to the set of axioms of a matroid of rank k with n elements in terms of its rank function $r_{\mathbf{C}}(I)$ ([18], 1.2.3). We shall call it the associated matroid $\mathbf{M}(C)=\mathbf{M}(\mathbf{C})$ of C. It will be convenient to use the dimension function instead of the rank function, but this is clearly an equivalent language. A matroid **M** is called representable over a field F if there is a linear code C over F such that $\mathbf{M}(C)=\mathbf{M}$. Clearly the set of linear codes representing a given matroid over a given field is no other than a combinatorial equivalence class.

Remark 1.2. The dimension function of a MDS code is $s(I)=\max\{0,k+\#I-n\}$. In particular, all MDS $[n,k]$-codes have the same associated matroid (the uniform matroid $U_{k,n}$, [18] 1.3).□

Let $a=(a_1,\dots,a_n)$, $a_i \neq 0$. For $v=(v_1,\dots,v_n) \in F^{*n}$ we define $av=(a_1v_1,\dots,a_nv_n)$. Let $a^{-1}=(a_1^{-1},\dots,a_n^{-1})$.

Lemma 1.3. Let C_1 and C_2 be linear codes over F. Assume that for some $a \in F^{*n}$ we have $aC_1=C_2$, then $\mathbf{M}(C_1)=\mathbf{M}(C_2)$.

Proof. Clearly the coordinate subspaces are invariant under the action of F^{*n}. Thus $\dim((aC_I)\cap V_I)=\dim(C_I\cap(a^{-1}V_I))=\dim(C_I\cap V_I)$.□

The converse of Lemma 1.3 is true for MDS [4,2]-codes over $\mathbb{F}_3$, and is not true in general, e.g. for MDS [4,2]-codes over $\mathbb{F}_q$ for $q\geq4$. (In fact the orbits of the action of F^{*n} on the elements of the combinatorial equivalence class of MDS [4,2]-codes over F bijectively correspond to $F\setminus\{0,1\}$.)

Recall that the dual code $C^{\perp}$ is defined as the set of vectors $v\in F^n$ such that the scalar product $(v.c)=v_1c_1+\ldots+v_nc_n$ is zero for all $c\in C$.

Proposition 1.4. The associated matroid of C uniquely determines the associated matroid of its dual code $C^{\perp}$.

Proof. It is easy to see that the dimension function of the dual code $C^{\perp}$ is given by $d_{C^{\perp}}(I)=\#I-k+d_C(I_n\setminus I)$.□

Recall that the support of $v\in F^n$ is defined as

$$\mathrm{supp}(v):=\{i\in I_n \mid v_i\neq0\}.$$

Let S_C be the set of supports of non-zero code words of C. It has a natural structure of a partially ordered set: $I\leq J$ if $I\subseteq J$. Let Min_C be the set of minimal elements of S_C.

Theorem 1.5. Let C_1, respectively C_2, be a linear $[n,k]$-code over a field F_1, respectively F_2. Then $Min_{C_1} = Min_{C_2}$ if and only if $M(C_1)=M(C_2)$.

Proof. Firstly, we show by induction in $\#I$ that the dimension function $d_C(I)$ of a code C is uniquely determined by Min_C. We know that $d_C(\emptyset)=0$ for any C. Assume that $d_C(J)$ is determined for all J, $\#J<\#I$. We have to find $d_C(I)=\dim(C\cap V_I)$. There are three possibilities:

(a) There is no $J\in Min_C$ such that $J\subset I$.

(b) $I\in Min_C$.

(c) There exists $J\in Min_C$ such that $J\subset I$.

Clearly in the case (a) we have $d_C(I)=0$. Let us prove that (b) implies $d_C(I)=1$. Indeed, by (b) any vector of $C\cap V_I$ must have its support equal to I. If there are two linearly independent code vectors with the same support, then their appropriate non-zero linear combination has a strictly smaller support, which is a contradiction. Now let us assume (c). To simplify our notation, we denote $V=C\cap V_I$, $H_i=V_J$ for $J=I\setminus\{i\}$, $i\in I$. Now H_i are hyperplanes in V_I with common intersection equal to 0. By the inductive hypothesis we know $\dim(V\cap H_i)=d_C(I\setminus\{i\})$. Since V is either contained in H_i or not, we obtain that $\dim(V)$ equals $\dim(V\cap H_i)$ or $\dim(V\cap H_i)+1$. Therefore if among $\dim(V\cap H_i)$ there are two different numbers we have to conclude that $\dim(V)$ is the maximum of $\{\dim(V\cap H_i)\}$. Now we explore the possibility when all $\dim(V\cap H_i)$ coincide. We claim that in this case $\dim(V)=\dim(V\cap H_i)+1$. Indeed, $\dim(V)=\dim(V\cap H_i)$ would imply that V is contained in every H_i, thus $V=0$. But this contradicts (c).

Secondly, we note that $I\in Min_C$ if and only if $d_C(I)=1$ and $d_C(J)=0$ for all $J\subset I$. Thus Min_C is uniquely determined by $M(C)$.□

Remark 1.6. If F is infinite, then S_C is also determined by $M(C)$. In fact, we have

(3) $$S_C=\{I\subseteq I_n \text{ such that } d_C(I)>d_C(J) \text{ for any } J\subset I\}.$$

Indeed, over an infinite field the compliment to a finite arrangement of hyperplanes always contains a non-zero vector. This is no more true over finite fields unless we fix the field F (see Corollary 1.8). Let L/F be an extension field. Given a code $C\subseteq F^n$ we can consider the tensor product $C\otimes L\subseteq L^n$, which by definition is the linear code over L with the same generator (or the parity check) matrix as $C\subseteq F^n$. An obvious comment is that $M(C\otimes L)=M(C)$. However, $S_{C\otimes L}$ is not necessarily the same as S_C. If L is large enough, $S_{C\otimes L}$ is given by (3).□

Recall that the Hamming weight of v is the cardinality of $\mathrm{supp}(v)$. For $I\subseteq I_n$ define

$$A_C(I):=\#\{c\in C \mid \mathrm{supp}(c)=I\}.$$

Let us denote by $A_C(i)$, $i=0,1,\dots,n$, the spectrum of C, that is the number of code words of C of Hamming weight i. We have

$$A_C(i) = \sum_{\#I=i} A_C(I).$$

Proposition 1.7. Let C be a linear code over a finite field $F=\mathbb{F}_q$. Then

(a) the function $A_C(I)$, $I\subseteq I_n$, is uniquely determined by $M(C)$ and q;

(b) the spectrum $A_C(i)$, $i=0,1,\ldots,n$, is uniquely determined by $M(C)$ and q, and is given by the formula

$$A_C(i) = \sum_{j=0}^{i} (-1)^{i-j} \begin{bmatrix} n-j \\ i-j \end{bmatrix} \sum_{\#J=j} q^{d_C(J)}.$$

Proof. Note that $q^{d_C(I)} = \sum_{J\subseteq I} A_C(J)$. Now the statements (a) and (b) follow from the inclusion-exclusion principle.□

Note that is possible to recover the MacWilliams identity relating $A_C(i)$ and $A_{C^\perp}(i)$ from the general results relating the Tutte polynomials of $M(C)$ and its dual matroid (see [18], 15.7).

Corollary 1.8. Let C_1 and C_2 be linear codes over a field F (not necessarily finite). Then the following conditions are equivalent:

(i) $M(C_1)=M(C_2)$, that is C_1 and C_2 are combinatorially equivalent;

(ii) $S_{C_1}=S_{C_2}$, that is C_1 and C_2 have the same set of supports of code words;

(iii) $Min_{C_1}=Min_{C_2}$.

Proof. The equivalence of (i) and (iii) was proved in Theorem 1.5, and the implication (ii)⇒(iii) is trivial, so it is enough to show that (i)⇒(ii). For finite F this follows from Proposition 1.7 (a), and for infinite F from Remark 1.6.□

Corollary 1.9. Let C be a non-empty combinatorial equivalence class over $\mathbb{F}_2$, then C consists of a single element.

Proof. To determine a binary code C is equivalent to determining the set S_C of supports of its nonzero code words. Thus the statement follows from Corollary 1.8.□

We have seen that the four fundamental parameters of a linear code over a given finite field $F=\mathbb{F}_q$ (the minimum distance and the number of non-zero weights of the code and its dual) are uniquely determined by its combinatorial equivalence class. In this connection one may ask the same question about other parameters, say the covering radius. In the next section we shall negatively answer this question by pointing at two MDS $[n,k]$-codes with different covering radii.

2. n-arcs and the covering radius of MDS codes

In this section we assume that F is a finite field, $F=\mathbb{F}_q$. Let W be a vector space of dimension m. Consider the corresponding projective space $\mathbb{P}^{m-1}=\mathbb{P}(W)$. Recall that a (ordered) n-arc in $\mathbb{P}^{m-1}$ is an ordered set $\mathbf{P}=\{P_1,\dots,P_n\}$ of $\mathbb{F}_q$-points of $\mathbb{P}^{m-1}$ such that for any $I\subset I_n$, $\#I=m$, the points P_i, $i\in I$, are linearly independent. A n-arc is called complete if it can not be extended to a $n+1$-arc in $\mathbb{P}^{m-1}$. Let us define $S(I)$ to be the projective subspace spanned by P_i, $i\in I$. Then $\mathbf{P}$ is a n-arc if and only if $\dim(S(I))=\min(m,\#I)-1$, and $\mathbf{P}$ is complete if and only if the union of $S(I)$, $\#I=m-1$, is the set of all $\mathbb{F}_q$-points of $\mathbb{P}^{m-1}$.

Now consider a parity check matrix of a $[n,k]$-code C, and think of its columns as points $P_1,\dots,P_n$ in the $n-k-1$-dimensional projective space $\mathbb{P}^{n-k-1}$ over $\mathbb{F}_q$. Almost by definition, the covering radius R_C of C is the least integer R such that the union of $S(I)$ for $I\subseteq I_n$, $\#I=R$, is the set of all $\mathbb{F}_q$-points of $\mathbb{P}^{n-k-1}$.

It is well known and easy to see that $\mathbf{P}=\{P_1,\dots,P_n\}$ is a n-arc if and only if C is a MDS $[n,k]$-code (cf. [13], 11.2, Cor.3, [1]). We shall denote this arc by $\mathbf{P}(C)$. It is clear from the previous discussion that if $\mathbf{P}(C)$ is not complete, then R_C equals $n-k$, whereas if $\mathbf{P}(C)$ is complete then $R_C\le n-k-1$.

A classical example of a complete $q+1$-arc in $\mathbb{P}^{m-1}$, $m\le q-1$, for odd q is given by the point set of a rational normal curve. This arc corresponds to the extended Reed-Solomon code. It can be given in coordinates as follows: $P_0=(0,\dots,0,1)$, and $P_t=(1,t,t^2,\dots,t^{m-1})$ for $t\in\mathbb{F}_q$. If q is power of 2, such an arc is complete except for $m=3$ or $m=q-2$, when it can be extended to a $q+2$-arc (cf. [13], [1], [16]).

On the other hand, there is a plenty of complete n-arcs, $n\leq q$, which *a fortiori* can not be embedded into a rational normal curve. Thus any such n-arc and a subset of the point set of a rational normal curve of cardinality n provide two MDS $[n,k]$-codes with different covering radii, since the latter arc (corresponding to a Reed-Solomon code) is obviously not complete. For instance, let q be a square, $n=q-\sqrt{q}+1$, $k=q-\sqrt{q}-2$, then a complete $q-\sqrt{q}+1$-arc in $\mathbb{P}^2$ was constructed in [6].

We now formulate the three fundamental problems of B.Segre.

1. For which n there exists a n-arc in $\mathbb{P}^{k-1}$ over $\mathbb{F}_q$?

2. For which k, $k<q$, every $q+1$-arc in $\mathbb{P}^{k-1}$ is the point set of a rational normal curve?

3. For which n and k, $k<q$, every n-arc in $\mathbb{P}^{k-1}$ is a subset of the point set of a rational normal curve?

3. The generic part of Grassmannian

Consider the Grassmannian variety $G(k,n)$, that is the variety of k-dimensional subspaces of the n-dimensional space V over a field F. We fix a decomposition of V into the direct sum of one-dimensional subspaces, $V=F^n=F\oplus\ldots\oplus F$. This defines the decomposition:

$$G(k,n) = \bigcup_f U_f(k,n)$$

(see the Introduction). The set of F-points in $U_f(k,n)$ is a combinatorial equivalence class, or, in other terms, the set of representations of the matroid with the dimension function f over F. Motivated by Mnëv's theorem [11] we shall only consider from now on the set $U(k,n)$ corresponding to the function $s(I)=\max\{0,k+\#I-n\}$. This set parametrizes MDS codes over finite fields, or, equivalently, the representations of the uniform matroid of rank k with n elements. The condition $f(I)=s(I)$ describes the subspaces which are in the general position with respect to all the coordinate subspaces V_I. Thus $U(k,n)$ is Zarisky open and dense in $G(k,n)$. We shall call it the generic part of $G(k,n)$. (It is also often called the open stratum of $G(k,n)$.) Let us denote by D_n the group of non-degenerate diagonal matrices in $GL(n)$ modulo the scalar matrices. (This is a split maximal torus in PGL(n).)

Next consider the open subvariety of $(\mathbb{P}^{k-1})^n$ which parametrizes n-tuples of points in $\mathbb{P}^{k-1}$ such that every k out of them are linearly independent. (Over a finite field these are the n-arcs in $\mathbb{P}^{k-1}$.) Define $\mathrm{P}_{k,n}$ as its factor with respect to the diagonal action of PGL(k).

Lemma 3.1. (a) $U(k,n)$ is isomorphic to $\mathrm{P}_{k,n}\times D_n$.
(b) $U(k,n)$ and $U(n-k,n)$ are isomorphic.

Proof. (a) D_n freely acts on $U(k,n)$, and the factor is isomorphic to $\mathrm{P}_{k,n}$. (A basis in $C\subset F^n$ is given by a $k\times n$-matrix. Consider the set of its columns as an element of $\mathrm{P}_{k,n}$.) This action makes $U(k,n)$ into a D_n-torsor over $\mathrm{P}_{k,n}$. It is easy to see that $H^1(U(k,n),\mathbb{G}_m)=\mathrm{Pic}(U(k,n))=0$, hence any torsor under $D_n\cong\mathbb{G}_m^{n-1}$ must be trivial by Hilbert's theorem 90.

(b) The map which sends $V\subset F^n$ to its orthogonal $V^{\perp}=\{v'\in F^n, (v.v')=0 \text{ for any } v\in V\}$ is an isomorphism.□

Now let $F=\mathbb{F}_p$. Of coarse, the relation between $U(k,n)$ and $\mathrm{P}_{k,n}$ is a generalization of the relation between MDS codes and arcs on the level of finite fields. Define $f_{n,k,p}(s)$ as the number of $\mathbb{F}_{p^s}$-points of $U(k,n)$. Then Lemma 3.1 (a) implies that this function can be written as the product of $(p^s-1)^{n-1}$ and the number of (ordered) n-arcs in $\mathbb{P}^{k-1}$ over $\mathbb{F}_{p^s}$. Here is the standard interpretation of $f_{n,k,p}(s)$ in terms of the cohomology of $U=U(k,n)\times_{\mathbb{F}_p}k_p$, where k_p is an algebraic closure of $\mathbb{F}_p$. For a prime $\ell\neq p$ let $\mathrm{H}^i_c(U,\mathbb{Q}_\ell)$ be the ℓ-adic étale cohomology groups of U with compact supports. Since U is a smooth affine variety of dimension $d=k(n-k)$ we have $\mathrm{H}^i_c(U,\mathbb{Q}_\ell)=0$ for all i except, may be, $d\leq i\leq 2d$ (affine Lefschetz theorem and Poincaré duality). Let us denote by $h_i=\dim(\mathrm{H}^i_c(U,\mathbb{Q}_\ell))$ the Betti numbers of U. By the Grothendieck trace formula we have

$$f_{n,k,p}(s) = \sum_{i=d}^{2d}(-1)^i\sum_{j=1}^{h_i}\omega_{i,j}^{\,s},$$

where $\omega_{i,1},\dots,\omega_{i,h_i}$ are the p-Frobenius eigenvalues on $\mathrm{H}^i_c(U,\mathbb{Q}_\ell)$. By

Cor. 3.3.4 of [2] each of these is an algebraic number such that all its complex conjugates are of absolute value $p^{w/2}$ for some integer w, $2(i-d)\leq w\leq i$. Clearly, $h_{2d}=1$ and $\omega_{2d}=p^d$.

Let us now discuss what is know about $f_{n,k,p}(s)$. By Lemma 3.1 (a) this function has a trivial factor $(p^s-1)^{n-1}$. By duality (Lemma 3.1 (b)) we have

(4) $$f_{n,k,p}(s)=f_{n,n-k,p}(s).$$

Obviously

$$f_{n,1,p}(s)=f_{n,n-1,p}(s)=(p^s-1)^{n-1}.$$

To exclude the trivial cases we shall assume from now on that $2\leq k\leq n-2$. To go a little farther, one easily sees that $P_{2,n}$ is isomorphic to $(\mathbb{P}^1\setminus\{0,1,\infty\})^{n-3}$, hence

(5) $$f_{n,2,p}(s)=(p^s-1)^{n-1}(p^s-2)\dots(p^s-n+2).$$

Thus $f_{5,3,p}(s)=(p^s-1)^4(p^s-2)(p^s-3)$ by (4), and by simple calculations in the projective plane one obtains

$$f_{6,3,p}(s)=(p^s-1)^6(p^s-2)(p^s-3)(p^{2s}-9p^s+21).$$

After the conference Ruud Pellikaan and Aart Blokhuis communicated to me the following formulae reproduced here with their kind permission. For p different from 2, one has

$$f_{7,3,p}(s)=(p^s-1)^6(p^s-3)(p^s-5)(p^{4s}-20p^{3s}+148p^{2s}-468p^s+498),$$

and for $p=2$

$$f_{7,3,2}(s)=(2^s-1)^6(2^s-2)(2^s-4)(2^{4s}-22.2^{3s}+183.2^{2s}-678.2^s+930).$$

These examples may lead one to speculate whether $f_{n,k,p}(s)$ is in general a polynomial in p^s with alternating coefficients. A closely related question is whether the p-Frobenius acts on $H^i_c(U,\mathbb{Q}_\ell)$ by multiplication by p^{i-d}.

Let us now reformulate the Segre problems in terms of $f_{n,k,p}(s)$.

Proposition 3.2. (a) There exists a n-arc in $\mathbb{P}^{k-1}$ over $\mathbb{F}_{p^s}$ if and only if

$$f_{n,k,p}(s)>0.$$

(b) For $p^s+1\geq k+2$, every p^s+1-arc in $\mathbb{P}^{k-1}$ is the point set of a rational normal curve if and only if

$$f_{p^s+1,k,p}(s)=f_{p^s+1,2,p}(s)=(p^s-1)^{p^s}(p^s-2)!$$

(c) For $p^s+1\geq n\geq k+2$, every n-arc in $\mathbb{P}^{k-1}$ is a subset of the point set of a rational normal curve if and only if

$$f_{n,k,p}(s)=f_{n,2,p}(s)=(p^s-1)^{n-1}(p^s-2)\dots(p^s-n+2).$$

Proof. Part (a) is obvious, and (b) is a particular case of (c). In order to prove (c) we have to compute the number of subsets of cardinality n of the point sets of all rational normal curves in $\mathbb{P}^{k-1}$, up to the action of PGL(k). Recall a well-known fact that every $k+2$ points in $\mathbb{P}^{k-1}$ uniquely determine a rational normal curve which contains them. Thus a n-arc, $n\geq k+2$, uniquely determines the rational normal curve to which it belongs. It is well known that the rational normal curves in $\mathbb{P}^{k-1}$ form one orbit under PGL(k) (such a curve is the embedding of $\mathbb{P}^1$ by means of the sheaf $\mathcal{O}(k-1)$). Note that the subgroup of PGL(k) which leaves one such curve invariant is PGL(2). Now it remains to compute the number of subsets of cardinality n of the set of $\mathbb{F}_{p^s}$-points of $\mathbb{P}^1$ up to the action of PGL(2), which equals $f_{n,2,p}(s)$ divided by $(p^s-1)^{n-1}$ (see (5)).□

According to a well known conjecture (cf. [13], Ch.11), for $(k,p)\neq(3,2)$, $(k,p)\neq(n-3,2)$ we should have

$$f_{n,k,p}(s)=0 \text{ for } s=1,\dots,[\log_p(n-2)],$$

$$f_{n,k,p}(s)\geq 1 \text{ for } s\geq[\log_p(n-2)]+1.$$

If $(k,p)=(3,2)$ or $(k,p)=(n-3,2)$ the same is known to be true with $n-1$ instead of $n-2$. The above conjecture is proved for $k\leq 5$ or $q\leq 11$ ([15], [16], [1], and the references in [13], Ch.11, [10], [1]).

Our present understanding of the algebraic geometry of $U(k,n)$ is surprisingly insufficient to deal with these problems. To the best of my knowledge, the only general result proved so far is a theorem of Mnëv [11], which can be interpreted in the following way: the connected components of the set of real points of $U(k,n)$, for $k\geq 3$ and large n, can be topologically as complicated as any open real semi-algebraic set defined over $\mathbb{Q}$.

The author is grateful to Michael Tsfasman for enumerable

discussions. An attempt to understand his approach to coding theory was one of the motivations for this work. I would like to thank Vera Serganova, Serge Vladut, and the participants of the conference, in particular, A.A.Bruen, Ruud Pellikaan, and Jacques Wolfmann, for their help with references and nice discussions. I am grateful to Université Paris-VII for its hospitality.

References

[1] A.A.Bruen, J.A.Thas, A.Blokhuis, On M.D.S. codes, arcs in *PG(n,q)* with *q* even, and a solution of the three fundamental problems of B.Segre. Inv. Math. 92 (1988), 441-459.

[2] P.Deligne, La conjecture de Weil.II. Publ. Math. IHES 52 (1980), 137-252.

[3] I.Dolgachev and D.Ortland. Point sets in projective spaces and theta functions. Astérisque 165 (1988).

[4] A.Dür, The automorphism group of Reed-Solomon codes. J. Comb. Th., Series A 44 (1987), 69-82.

[5] N.E.Fenton, P.Vamos, Matroid interpretation of maximal *K*-arcs in projective spaces. Rend. Mat. 2 (1982) Ser.VII, 575-580.

[6] J.C.Fisher, J.W.P.Hirschfeld, and J.A.Thas, Complete arcs in planes of square order. Ann. Discrete Math. 30 (1986), 243-250.

[7] I.M.Gelfand, General theory of hypergeometric functions. Doklady AN SSSR 288:1 (1986), 14-18. (In Russian) = I.M.Gelfand. Collected Papers. Springer, 1989. Vol.3, 877-881.

[8] I.M.Gelfand, M.Goreski, R.D.MacPherson, and V.V.Serganova, Combinatorial geometries, convex polyhedra, and Schubert cells. Adv. Math. 63 (3) (1987), 301-316. = I.M.Gelfand. Collected Papers. Springer, 1989. Vol.3, 906-921.

[9] I.M.Gelfand, V.V.Serganova, Combinatorial geometries and torus strata on homogeneous compact manifolds. Uspekhi Mat. Nauk 42:2 (1987), 107-134. (In Russsian) = I.M.Gelfand. Collected Papers. Springer, 1989. Vol.3, 926-958.

[10] J.W.P.Hirschfeld. Projective geometries over finite fields. Oxford, Clarendon Press, 1979.

[11] N.E.Mnёv, On manifolds of combinatorial types of projective configurations and convex polytopes. Doklady AN SSSR, 283:6 (1985),

1312–1314. (In Russian) = Soviet Math. Doklady 32:1 (1985), 335–337.

[12] N.E.Mnёv, The universality theorems on the classification problem of configuration varieties and convex polytopes varieties. In: Lect. Notes in Math. 1346, Topology and Geometry – Rohlin Seminar, O.Ya.Viro (Ed.), p.527–543.

[13] F.J.MacWilliams, N.J.A.Sloane. The theory of error-correcting codes. Parts I,II. North-Holland, 1977.

[14] U.Oberst and A.Dür, A constructive characterization of all optimal linear codes. Sém. d'Algèbre P.Dubreil et M.-P.Malliavin (1983–1984), Lect. Notes in Math. 1146 (1985), 176–213.

[15] B.Segre, Curve razionali normali e k-archi negli spazi finiti. Ann. Math. Pura Appl. 39 (1955), 357–379.

[16] J.A.Thas, Connection between the Grassmannian $G_{k-1,n}$ and the set of the k-arcs of the Galois space $S_{n,k}$. Rend. Math. 2 (1969) Ser. VI, 121–134.

[17] A.M.Vershik, Topology of the convex polytopes' manifolds, the manifold of the projective configurations of a given combinatorial type and representations of lattices. In: Lect. Notes in Math. 1346, Topology and Geometry – Rohlin Seminar, O.Ya.Viro (Ed.), p.557–581.

[18] D.J.A.Welsh. Matroid theory. Academic Press, 1976.

Lecture Notes in Mathematics

For information about Vols. 1–1323
please contact your bookseller or Springer-Verlag

Vol. 1324: F. Cardoso, D.G. de Figueiredo, R. Iório, O. Lopes (Eds.), Partial Differential Equations. Proceedings, 1986. VIII, 433 pages. 1988.

Vol. 1325: A. Truman, I.M. Davies (Eds.), Stochastic Mechanics and Stochastic Processes. Proceedings, 1986. V, 220 pages. 1988.

Vol. 1326: P.S. Landweber (Ed.), Elliptic Curves and Modular Forms in Algebraic Topology. Proceedings, 1986. V, 224 pages. 1988.

Vol. 1327: W. Bruns, U. Vetter, Determinantal Rings. VII, 236 pages. 1988.

Vol. 1328: J.L. Bueso, P. Jara, B. Torrecillas (Eds.), Ring Theory. Proceedings, 1986. IX, 331 pages. 1988.

Vol. 1329: M. Alfaro, J.S. Dehesa, F.J. Marcellan, J.L. Rubio de Francia, J. Vinuesa (Eds.): Orthogonal Polynomials and their Applications. Proceedings, 1986. XV, 334 pages. 1988.

Vol. 1330: A. Ambrosetti, F. Gori, R. Lucchetti (Eds.), Mathematical Economics. Montecatini Terme 1986. Seminar. VII, 137 pages. 1988.

Vol. 1331: R. Bamón, R. Labarca, J. Palis Jr. (Eds.), Dynamical Systems, Valparaiso 1986. Proceedings. VI, 250 pages. 1988.

Vol. 1332: E. Odell, H. Rosenthal (Eds.), Functional Analysis. Proceedings. 1986–87. V, 202 pages. 1988.

Vol. 1333: A.S. Kechris, D.A. Martin, J.R. Steel (Eds.), Cabal Seminar 81–85. Proceedings, 1981–85. V, 224 pages. 1988.

Vol. 1334: Yu.G. Borisovich, Yu.E. Gliklikh (Eds.), Global Analysis – Studies and Applications III. V, 331 pages. 1988.

Vol. 1335: F. Guillén, V. Navarro Aznar, P. Pascual-Gainza, F. Puerta, Hyperrésolutions cubiques et descente cohomologique. XII, 192 pages. 1988.

Vol. 1336: B. Helffer, Semi-Classical Analysis for the Schrödinger Operator and Applications. V, 107 pages. 1988.

Vol. 1337: E. Sernesi (Ed.), Theory of Moduli. Seminar, 1985. VIII, 232 pages. 1988.

Vol. 1338: A.B. Mingarelli, S.G. Halvorsen. Non-Oscillation Domains of Differential Equations with Two Parameters. XI, 109 pages. 1988.

Vol. 1339: T. Sunada (Ed.), Geometry and Analysis of Manifolds. Proceedings, 1987. IX, 277 pages. 1988.

Vol. 1340: S. Hildebrandt, D.S. Kinderlehrer, M. Miranda (Eds.), Calculus of Variations and Partial Differential Equations. Proceedings, 1986. IX, 301 pages. 1988.

Vol. 1341: M. Dauge, Elliptic Boundary Value Problems on Corner Domains. VIII, 259 pages. 1988.

Vol. 1342: J.C. Alexander (Ed.), Dynamical Systems. Proceedings, 1986–87. VIII, 726 pages. 1988.

Vol. 1343: H. Ulrich, Fixed Point Theory of Parametrized Equivariant Maps. VII, 147 pages. 1988.

Vol. 1344: J. Král, J. Lukes, J. Netuka, J. Vesely´ (Eds.), Potential Theory – Surveys and Problems. Proceedings, 1987. VIII, 271 pages. 1988.

Vol. 1345: X. Gomez-Mont, J. Seade, A. Verjovski (Eds.), Holomorphic Dynamics. Proceedings, 1986. VII. 321 pages. 1988.

Vol. 1346: O.Ya. Viro (Ed.), Topology and Geometry – Rohlin Seminar. XI, 581 pages. 1988.

Vol. 1347: C. Preston, Iterates of Piecewise Monotone Mappings on an Interval. V, 166 pages. 1988.

Vol. 1348: F. Borceux (Ed.), Categorical Algebra and its Applications. Proceedings, 1987. VIII, 375 pages. 1988.

Vol. 1349: E. Novak, Deterministic and Stochastic Error Bounds in Numerical Analysis. V, 113 pages. 1988.

Vol. 1350: U. Koschorke (Ed.), Differential Topology Proceedings, 1987, VI, 269 pages. 1988.

Vol. 1351: I. Laine, S. Rickman, T. Sorvali (Eds.), Complex Analysis, Joensuu 1987. Proceedings. XV, 378 pages. 1988.

Vol. 1352: L.L. Avramov, K.B. Tchakerian (Eds.), Algebra – Some Current Trends. Proceedings. 1986. IX, 240 Seiten. 1988.

Vol. 1353: R.S. Palais, Ch.-l. Teng, Critical Point Theory and Submanifold Geometry. X, 272 pages. 1988.

Vol. 1354: A. Gómez, F. Guerra, M.A. Jiménez, G. López (Eds.), Approximation and Optimization. Proceedings, 1987. VI, 280 pages. 1988.

Vol. 1355: J. Bokowski, B. Sturmfels, Computational Synthetic Geometry. V, 168 pages. 1989.

Vol. 1356: H. Volkmer, Multiparameter Eigenvalue Problems and Expansion Theorems. VI, 157 pages. 1988.

Vol. 1357: S. Hildebrandt, R. Leis (Eds.), Partial Differential Equations and Calculus of Variations. VI, 423 pages. 1988.

Vol. 1358: D. Mumford, The Red Book of Varieties and Schemes. V, 309 pages. 1988.

Vol. 1359: P. Eymard, J.-P. Pier (Eds.) Harmonic Analysis. Proceedings, 1987. VIII, 287 pages. 1988.

Vol. 1360: G. Anderson, C. Greengard (Eds.), Vortex Methods. Proceedings, 1987. V, 141 pages. 1988.

Vol. 1361: T. tom Dieck (Ed.), Algebraic Topology and Transformation Groups. Proceedings. 1987. VI, 298 pages. 1988.

Vol. 1362: P. Diaconis, D. Elworthy, H. Föllmer, E. Nelson, G.C. Papanicolaou, S.R.S. Varadhan. École d´ Été de Probabilités de Saint-Flour XV–XVII. 1985–87 Editor: P.L. Hennequin. V, 459 pages. 1988.

Vol. 1363: P.G. Casazza, T.J. Shura, Tsirelson´s Space. VIII, 204 pages. 1988.

Vol. 1364: R.R. Phelps, Convex Functions, Monotone Operators and Differentiability. IX, 115 pages. 1989.

Vol. 1365: M. Giaquinta (Ed.), Topics in Calculus of Variations. Seminar, 1987. X, 196 pages. 1989.

Vol. 1366: N. Levitt, Grassmannians and Gauss Maps in PL-Topology. V, 203 pages. 1989.

Vol. 1367: M. Knebusch, Weakly Semialgebraic Spaces. XX, 376 pages. 1989.

Vol. 1368: R. Hübl, Traces of Differential Forms and Hochschild Homology. III, 111 pages. 1989.

Vol. 1369: B. Jiang, Ch.-K. Peng, Z. Hou (Eds.), Differential Geometry and Topology. Proceedings, 1986–87. VI, 366 pages. 1989.

Vol. 1370: G. Carlsson, R.L. Cohen, H.R. Miller, D.C. Ravenel (Eds.), Algebraic Topology. Proceedings, 1986. IX, 456 pages. 1989.

Vol. 1371: S. Glaz, Commutative Coherent Rings. XI, 347 pages. 1989.

Vol. 1372: J. Azéma, P.A. Meyer, M. Yor (Eds.), Séminaire de Probabilités XXIII. Proceedings. IV, 583 pages. 1989.

Vol. 1373: G. Benkart, J.M. Osborn (Eds.), Lie Algebras. Madison 1987. Proceedings. V, 145 pages. 1989.

Vol. 1374: R.C. Kirby, The Topology of 4-Manifolds. VI, 108 pages. 1989.

Vol. 1375: K. Kawakubo (Ed.), Transformation Groups. Proceedings, 1987. VIII, 394 pages, 1989.

Vol. 1376: J. Lindenstrauss, V.D. Milman (Eds.), Geometric Aspects of Functional Analysis. Seminar (GAFA) 1987–88. VII, 288 pages. 1989.

Vol. 1377: J.F. Pierce, Singularity Theory, Rod Theory, and Symmetry-Breaking Loads. IV, 177 pages. 1989.

Vol. 1378: R.S. Rumely, Capacity Theory on Algebraic Curves. III, 437 pages. 1989.

Vol. 1379: H. Heyer (Ed.), Probability Measures on Groups IX. Proceedings, 1988. VIII, 437 pages. 1989.

Vol. 1380: H.P. Schlickewei, E. Wirsing (Eds.), Number Theory, Ulm 1987. Proceedings. V, 266 pages. 1989.

Vol. 1381: J.-O. Strömberg, A. Torchinsky, Weighted Hardy Spaces. V, 193 pages. 1989.

Vol. 1382: H. Reiter, Metaplectic Groups and Segal Algebras. XI, 128 pages. 1989.

Vol. 1383: D.V. Chudnovsky, G.V. Chudnovsky, H. Cohn, M.B. Nathanson (Eds.), Number Theory, New York 1985–88. Seminar. V, 256 pages. 1989.

Vol. 1384: J. Garcia-Cuerva (Ed.), Harmonic Analysis and Partial Differential Equations. Proceedings, 1987. VII, 213 pages. 1989.

Vol. 1385: A.M. Anile, Y. Choquet-Bruhat (Eds.), Relativistic Fluid Dynamics. Seminar, 1987. V, 308 pages. 1989.

Vol. 1386: A. Bellen, C.W. Gear, E. Russo (Eds.), Numerical Methods for Ordinary Differential Equations. Proceedings, 1987. VII, 136 pages. 1989.

Vol. 1387: M. Petkovi´c, Iterative Methods for Simultaneous Inclusion of Polynomial Zeros. X, 263 pages. 1989.

Vol. 1388: J. Shinoda, T.A. Slaman, T. Tugué (Eds.), Mathematical Logic and Applications. Proceedings, 1987. V, 223 pages. 1989.

Vol. 1000: Second Edition. H. Hopf, Differential Geometry in the Large. VII, 184 pages. 1989.

Vol. 1389: E. Ballico, C. Ciliberto (Eds.), Algebraic Curves and Projective Geometry. Proceedings, 1988. V, 288 pages. 1989.

Vol. 1390: G. Da Prato, L. Tubaro (Eds.), Stochastic Partial Differential Equations and Applications II. Proceedings, 1988. VI, 258 pages. 1989.

Vol. 1391: S. Cambanis, A. Weron (Eds.), Probability Theory on Vector Spaces IV. Proceedings, 1987. VIII, 424 pages. 1989.

Vol. 1392: R. Silhol, Real Algebraic Surfaces. X, 215 pages. 1989.

Vol. 1393: N. Bouleau, D. Feyel, F. Hirsch, G. Mokobodzki (Eds.), Séminaire de Théorie du Potentiel Paris. No. 9. Proceedings. VI, 265 pages. 1989.

Vol. 1394: T.L. Gill, W.W. Zachary (Eds.), Nonlinear Semigroups, Partial Differential Equations and Attractors. Proceedings, 1987. IX, 233 pages. 1989.

Vol. 1395: K. Alladi (Ed.), Number Theory, Madras 1987. Proceedings. VII, 234 pages. 1989.

Vol. 1396: L. Accardi, W. von Waldenfels (Eds.), Quantum Probability and Applications IV. Proceedings, 1987. VI, 355 pages. 1989.

Vol. 1397: P.R. Turner (Ed.), Numerical Analysis and Parallel Processing. Seminar, 1987. VI, 264 pages. 1989.

Vol. 1398: A.C. Kim, B.H. Neumann (Eds.), Groups – Korea 1988. Proceedings. V, 189 pages. 1989.

Vol. 1399: W.-P. Barth, H. Lange (Eds.), Arithmetic of Complex Manifolds. Proceedings, 1988. V, 171 pages. 1989.

Vol. 1400: U. Jannsen. Mixed Motives and Algebraic K-Theory. XIII, 246 pages. 1990.

Vol. 1401: J. Steprans, S. Watson (Eds.), Set Theory and its Applications. Proceedings, 1987. V, 227 pages. 1989.

Vol. 1402: C. Carasso, P. Charrier, B. Hanouzet, J.-L. Joly (Eds.), Nonlinear Hyperbolic Problems. Proceedings, 1988. V, 249 pages. 1989.

Vol. 1403: B. Simeone (Ed.), Combinatorial Optimization. Seminar, 1986. V, 314 pages. 1989.

Vol. 1404: M.-P. Malliavin (Ed.), Séminaire d´Algèbre Paul Dubreil et Marie-Paul Malliavin. Proceedings, 1987–1988. IV, 410 pages. 1989.

Vol. 1405: S. Dolecki (Ed.), Optimization. Proceedings, 1988. V, 223 pages. 1989. Vol. 1406: L. Jacobsen (Ed.), Analytic Theory of Continued Fractions III. Proceedings, 1988. VI, 142 pages. 1989.

Vol. 1407: W. Pohlers, Proof Theory. VI, 213 pages. 1989.

Vol. 1408: W. Lück, Transformation Groups and Algebraic K-Theory. XII, 443 pages. 1989.

Vol. 1409: E. Hairer, Ch. Lubich, M. Roche. The Numerical Solution of Differential-Algebraic Systems by Runge-Kutta Methods. VII, 139 pages. 1989.

Vol. 1410: F.J. Carreras, O. Gil-Medrano, A.M. Naveira (Eds.), Differential Geometry. Proceedings, 1988. V, 308 pages. 1989.

Vol. 1411: B. Jiang (Ed.), Topological Fixed Point Theory and Applications. Proceedings. 1988. VI, 203 pages. 1989.

Vol. 1412: V.V. Kalashnikov, V.M. Zolotarev (Eds.), Stability Problems for Stochastic Models. Proceedings, 1987. X, 380 pages. 1989.

Vol. 1413: S. Wright, Uniqueness of the Injective III_1 Factor. III, 108 pages. 1989.

Vol. 1414: E. Ramirez de Arellano (Ed.), Algebraic Geometry and Complex Analysis. Proceedings, 1987. VI, 180 pages. 1989.

Vol. 1415: M. Langevin, M. Waldschmidt (Eds.), Cinquante Ans de Polynômes. Fifty Years of Polynomials. Proceedings, 1988. IX, 235 pages.1990.

Vol. 1416: C. Albert (Ed.), Géométrie Symplectique et Mécanique. Proceedings, 1988. V, 289 pages. 1990.

Vol. 1417: A.J. Sommese, A. Biancofiore, E.L. Livorni (Eds.), Algebraic Geometry. Proceedings, 1988. V, 320 pages. 1990.

Vol. 1418: M. Mimura (Ed.), Homotopy Theory and Related Topics. Proceedings, 1988. V, 241 pages. 1990.

Vol. 1419: P.S. Bullen, P.Y. Lee, J.L. Mawhin, P. Muldowney, W.F. Pfeffer (Eds.), New Integrals. Proceedings, 1988. V, 202 pages. 1990.

Vol. 1420: M. Galbiati, A. Tognoli (Eds.), Real Analytic Geometry. Proceedings, 1988. IV, 366 pages. 1990.

Vol. 1421: H.A. Biagioni, A Nonlinear Theory of Generalized Functions, XII, 214 pages. 1990.

Vol. 1422: V. Villani (Ed.), Complex Geometry and Analysis. Proceedings, 1988. V, 109 pages. 1990.

Vol. 1423: S.O. Kochman, Stable Homotopy Groups of Spheres: A Computer-Assisted Approach. VIII, 330 pages. 1990.

Vol. 1424: F.E. Burstall, J.H. Rawnsley, Twistor Theory for Riemannian Symmetric Spaces. III, 112 pages. 1990.

Vol. 1425: R.A. Piccinini (Ed.), Groups of Self-Equivalences and Related Topics. Proceedings, 1988. V, 214 pages. 1990.

Vol. 1426: J. Azéma, P.A. Meyer, M. Yor (Eds.), Séminaire de Probabilités XXIV, 1988/89. V, 490 pages. 1990.

Vol. 1427: A. Ancona, D. Geman, N. Ikeda, École d'Eté de Probabilités de Saint Flour XVIII, 1988. Ed.: P.L. Hennequin. VII, 330 pages. 1990.

Vol. 1428: K. Erdmann, Blocks of Tame Representation Type and Related Algebras. XV. 312 pages. 1990.

Vol. 1429: S. Homer, A. Nerode, R.A. Platek, G.E. Sacks, A. Scedrov, Logic and Computer Science. Seminar, 1988. Editor: P. Odifreddi. V, 162 pages. 1990.

Vol. 1430: W. Bruns, A. Simis (Eds.), Commutative Algebra. Proceedings. 1988. V, 160 pages. 1990.

Vol. 1431: J.G. Heywood, K. Masuda, R. Rautmann, V.A. Solonnikov (Eds.), The Navier-Stokes Equations – Theory and Numerical Methods. Proceedings, 1988. VII, 238 pages. 1990.

Vol. 1432: K. Ambos-Spies, G.H. Müller, G.E. Sacks (Eds.), Recursion Theory Week. Proceedings, 1989. VI, 393 pages. 1990.

Vol. 1433: S. Lang, W. Cherry, Topics in Nevanlinna Theory. II, 174 pages. 1990.

Vol. 1434: K. Nagasaka, E. Fouvry (Eds.), Analytic Number Theory. Proceedings, 1988. VI, 218 pages. 1990.

Vol. 1435: St. Ruscheweyh, E.B. Saff, L.C. Salinas, R.S. Varga (Eds.), Computational Methods and Function Theory. Proceedings, 1989. VI, 211 pages. 1990.

Vol. 1436: S. Xambó-Descamps (Ed.), Enumerative Geometry. Proceedings, 1987. V, 303 pages. 1990.

Vol. 1437: H. Inassaridze (Ed.), K-theory and Homological Algebra. Seminar, 1987–88. V, 313 pages. 1990.

Vol. 1438: P.G. Lemarié (Ed.) Les Ondelettes en 1989. Seminar. IV, 212 pages. 1990.

Vol. 1439: E. Bujalance, J.J. Etayo, J.M. Gamboa, G. Gromadzki. Automorphism Groups of Compact Bordered Klein Surfaces: A Combinatorial Approach. XIII, 201 pages. 1990.

Vol. 1440: P. Latiolais (Ed.), Topology and Combinatorial Groups Theory. Seminar. 1985–1988. VI, 207 pages. 1990.

Vol. 1441: M. Coornaert, T. Delzant, A. Papadopoulos. Géométrie et théorie des groupes. X, 165 pages. 1990.

Vol. 1442: L. Accardi, M. von Waldenfels (Eds.), Quantum Probability and Applications V. Proceedings, 1988. VI, 413 pages. 1990.

Vol. 1443: K.H. Dovermann, R. Schultz, Equivariant Surgery Theories and Their Periodicity Properties. VI, 227 pages. 1990.

Vol. 1444: H. Korezlioglu, A.S. Ustunel (Eds.), Stochastic Analysis and Related Topics VI. Proceedings, 1988. V, 268 pages. 1990.

Vol. 1445: F. Schulz, Regularity Theory for Quasilinear Elliptic Systems and – Monge Ampère Equations in Two Dimensions. XV, 123 pages. 1990.

Vol. 1446: Methods of Nonconvex Analysis. Seminar, 1989. Editor: A. Cellina. V, 206 pages. 1990.

Vol. 1447: J.-G. Labesse, J. Schwermer (Eds), Cohomology of Arithmetic Groups and Automorphic Forms. Proceedings. 1989. V, 358 pages. 1990.

Vol. 1448: S.K. Jain, S.R. López-Permouth (Eds.), Non-Commutative Ring Theory. Proceedings, 1989. V, 166 pages. 1990.

Vol. 1449: W. Odyniec, G. Lewicki, Minimal Projections in Banach Spaces. VIII, 168 pages. 1990.

Vol. 1450: H. Fujita, T. Ikebe, S.T. Kuroda (Eds.), Functional-Analytic Methods for Partial Differential Equations. Proceedings, 1989. VII, 252 pages. 1990.

Vol. 1451: L. Alvarez-Gaumé, E. Arbarello, C. De Concini, N.J. Hitchin, Global Geometry and Mathematical Physics. Montecatini Terme 1988. Seminar. Editors: M. Francaviglia, F. Gherardelli. IX, 197 pages. 1990.

Vol. 1452: E. Hlawka, R.F. Tichy (Eds.), Number-Theoretic Analysis. Seminar, 1988–89. V, 220 pages. 1990.

Vol. 1453: Yu.G. Borisovich, Yu.E. Gliklikh (Eds.), Global Analysis – Studies and Applications IV. V, 320 pages. 1990.

Vol. 1454: F. Baldassari, S. Bosch, B. Dwork (Eds.), p-adic Analysis. Proceedings, 1989. V, 382 pages. 1990.

Vol. 1455: J.-P. Françoise, R. Roussarie (Eds.), Bifurcations of Planar Vector Fields. Proceedings, 1989. VI, 396 pages. 1990.

Vol. 1456: L.G. Kovács (Ed.), Groups – Canberra 1989. Proceedings. XII, 198 pages. 1990.

Vol. 1457: O. Axelsson, L.Yu. Kolotilina (Eds.), Preconditioned Conjugate Gradient Methods. Proceedings, 1989. V, 196 pages. 1990.

Vol. 1458: R. Schaaf, Global Solution Branches of Two Point Boundary Value Problems. XIX, 141 pages. 1990.

Vol. 1459: D. Tiba, Optimal Control of Nonsmooth Distributed Parameter Systems. VII, 159 pages. 1990.

Vol. 1460: G. Toscani, V. Boffi, S. Rionero (Eds.), Mathematical Aspects of Fluid Plasma Dynamics. Proceedings, 1988. V, 221 pages. 1991.

Vol. 1461: R. Gorenflo, S. Vessella, Abel Integral Equations. VII, 215 pages. 1991.

Vol. 1462: D. Mond, J. Montaldi (Eds.), Singularity Theory and its Applications. Warwick 1989, Part I. VIII, 405 pages. 1991.

Vol. 1463: R. Roberts, I. Stewart (Eds.), Singularity Theory and its Applications. Warwick 1989, Part II. VIII, 322 pages. 1991.

Vol. 1464: D. L. Burkholder, E. Pardoux, A. Sznitman, Ecole d'Eté de Probabilités de Saint- Flour XIX-1989. Editor: P. L. Hennequin. VI, 256 pages. 1991.

Vol. 1465: G. David, Wavelets and Singular Integrals on Curves and Surfaces. X, 107 pages. 1991.

Vol. 1466: W. Banaszczyk, Additive Subgroups of Topological Vector Spaces. VII, 178 pages. 1991.

Vol. 1467: W. M. Schmidt, Diophantine Approximations and Diophantine Equations. VIII, 217 pages. 1991.

Vol. 1468: J. Noguchi, T. Ohsawa (Eds.), Prospects in Complex Geometry. Proceedings, 1989. VII, 421 pages. 1991.

Vol. 1469: J. Lindenstrauss, V. D. Milman (Eds.), Geometric Aspects of Functional Analysis. Seminar 1989-90. XI, 191 pages. 1991.

Vol. 1470: E. Odell, H. Rosenthal (Eds.), Functional Analysis. Proceedings, 1987-89. VII, 199 pages. 1991.

Vol. 1471: A. A. Panchishkin, Non-Archimedean L-Functions of Siegel and Hilbert Modular Forms. VII, 157 pages. 1991.

Vol. 1472: T. T. Nielsen, Bose Algebras: The Complex and Real Wave Representations. V, 132 pages. 1991.

Vol. 1473: Y. Hino, S. Murakami, T. Naito, Functional Differential Equations with Infinite Delay. X, 317 pages. 1991.

Vol. 1474: S. Jackowski, B. Oliver, K. Pawałowski (Eds.), Algebraic Topology, Poznań 1989. Proceedings. VIII, 397 pages. 1991.

Vol. 1475: S. Busenberg, M. Martelli (Eds.), Delay Differential Equations and Dynamical Systems. Proceedings, 1990. VIII, 249 pages. 1991.

Vol. 1476: M. Bekkali, Topics in Set Theory. VII, 120 pages. 1991.

Vol. 1477: R. Jajte, Strong Limit Theorems in Noncommutative L_2-Spaces. X, 113 pages. 1991.

Vol. 1478: M.-P. Malliavin (Ed.), Topics in Invariant Theory. Seminar 1989-1990. VI, 272 pages. 1991.

Vol. 1479: S. Bloch, I. Dolgachev, W. Fulton (Eds.), Algebraic Geometry. Proceedings, 1989. VII, 300 pages. 1991.

Vol. 1480: F. Dumortier, R. Roussarie, J. Sotomayor, H. Żoładek, Bifurcations of Planar Vector Fields: Nilpotent Singularities and Abelian Integrals. VIII, 226 pages. 1991.

Vol. 1481: D. Ferus, U. Pinkall, U. Simon, B. Wegner (Eds.), Global Differential Geometry and Global Analysis. Proceedings, 1991. VIII, 283 pages. 1991.

Vol. 1482: J. Chabrowski, The Dirichlet Problem with L^2-Boundary Data for Elliptic Linear Equations. VI, 173 pages. 1991.

Vol. 1483: E. Reithmeier, Periodic Solutions of Nonlinear Dynamical Systems. VI, 171 pages. 1991.

Vol. 1484: H. Delfs, Homology of Locally Semialgebraic Spaces. IX, 136 pages. 1991.

Vol. 1485: J. Azéma, P. A. Meyer, M. Yor (Eds.), Séminaire de Probabilités XXV. VIII, 440 pages. 1991.

Vol. 1486: L. Arnold, H. Crauel, J.-P. Eckmann (Eds.), Lyapunov Exponents. Proceedings, 1990. VIII, 365 pages. 1991.

Vol. 1487: E. Freitag, Singular Modular Forms and Theta Relations. VI, 172 pages. 1991.

Vol. 1488: A. Carboni, M. C. Pedicchio, G. Rosolini (Eds.), Category Theory. Proceedings, 1990. VII, 494 pages. 1991.

Vol. 1489: A. Mielke, Hamiltonian and Lagrangian Flows on Center Manifolds. X, 140 pages. 1991.

Vol. 1490: K. Metsch, Linear Spaces with Few Lines. XIII, 196 pages. 1991.

Vol. 1491: E. Lluis-Puebla, J.-L. Loday, H. Gillet, C. Soulé, V. Snaith, Higher Algebraic K-Theory: an overview. IX, 164 pages. 1992.

Vol. 1492: K. R. Wicks, Fractals and Hyperspaces. VIII, 168 pages. 1991.

Vol. 1493: E. Benoît (Ed.), Dynamic Bifurcations. Proceedings, Luminy 1990. VII, 219 pages. 1991.

Vol. 1494: M.-T. Cheng, X.-W. Zhou, D.-G. Deng (Eds.), Harmonic Analysis. Proceedings, 1988. IX, 226 pages. 1991.

Vol. 1495: J. M. Bony, G. Grubb, L. Hörmander, H. Komatsu, J. Sjöstrand, Microlocal Analysis and Applications. Montecatini Terme, 1989. Editors: L. Cattabriga, L. Rodino. VII, 349 pages. 1991.

Vol. 1496: C. Foias, B. Francis, J. W. Helton, H. Kwakernaak, J. B. Pearson, H_∞-Control Theory. Como, 1990. Editors: E. Mosca, L. Pandolfi. VII, 336 pages. 1991.

Vol. 1497: G. T. Herman, A. K. Louis, F. Natterer (Eds.), Mathematical Methods in Tomography. Proceedings 1990. X, 268 pages. 1991.

Vol. 1498: R. Lang, Spectral Theory of Random Schrödinger Operators. X, 125 pages. 1991.

Vol. 1499: K. Taira, Boundary Value Problems and Markov Processes. IX, 132 pages. 1991.

Vol. 1500: J.-P. Serre, Lie Algebras and Lie Groups. VII, 168 pages. 1992.

Vol. 1501: A. De Masi, E. Presutti, Mathematical Methods for Hydrodynamic Limits. IX, 196 pages. 1991.

Vol. 1502: C. Simpson, Asymptotic Behavior of Monodromy. V, 139 pages. 1991.

Vol. 1503: S. Shokranian, The Selberg-Arthur Trace Formula (Lectures by J. Arthur). VII, 97 pages. 1991.

Vol. 1504: J. Cheeger, M. Gromov, C. Okonek, P. Pansu, Geometric Topology: Recent Developments. Editors: P. de Bartolomeis, F. Tricerri. VII, 197 pages. 1991.

Vol. 1505: K. Kajitani, T. Nishitani, The Hyperbolic Cauchy Problem. VII, 168 pages. 1991.

Vol. 1506: A. Buium, Differential Algebraic Groups of Finite Dimension. XV, 145 pages. 1992.

Vol. 1507: K. Hulek, T. Peternell, M. Schneider, F.-O. Schreyer (Eds.), Complex Algebraic Varieties. Proceedings, 1990. VII, 179 pages. 1992.

Vol. 1508: M. Vuorinen (Ed.), Quasiconformal Space Mappings. A Collection of Surveys 1960-1990. IX, 148 pages. 1992.

Vol. 1509: J. Aguadé, M. Castellet, F. R. Cohen (Eds.), Algebraic Topology - Homotopy and Group Cohomology. Proceedings, 1990. X, 330 pages. 1992.

Vol. 1510: P. P. Kulish (Ed.), Quantum Groups. Proceedings, 1990. XII, 398 pages. 1992.

Vol. 1511: B. S. Yadav, D. Singh (Eds.), Functional Analysis and Operator Theory. Proceedings, 1990. VIII, 223 pages. 1992.

Vol. 1512: L. M. Adleman, M.-D. A. Huang, Primality Testing and Abelian Varieties Over Finite Fields. VII, 142 pages. 1992.

Vol. 1513: L. S. Block, W. A. Coppel, Dynamics in One Dimension. VIII, 249 pages. 1992.

Vol. 1514: U. Krengel, K. Richter, V. Warstat (Eds.), Ergodic Theory and Related Topics III, Proceedings, 1990. VIII, 236 pages. 1992.

Vol. 1515: E. Ballico, F. Catanese, C. Ciliberto (Eds.), Classification of Irregular Varieties. Proceedings, 1990. VII, 149 pages. 1992.

Vol. 1517: K. Keimel, W. Roth, Ordered Cones and Approximation. VI, 134 pages. 1992.

Vol. 1518: H. Stichtenoth, M. A. Tsfasman (Eds.), Coding Theory and Algebraic Geometry. Proceedings, 1991. VIII, 223 pages. 1992.